Bioethics and Biolaw through Literature

Law & Literature

Edited by
Daniela Carpi · Klaus Stierstorfer

Volume 2

De Gruyter

Bioethics and Biolaw through Literature

Edited by

Daniela Carpi

De Gruyter

ISBN 978-3-11-025284-2
e-ISBN 978-3-11-025285-9
ISSN 2191-8457

Library of Congress Cataloging-in-Publication Data

Bioethics and biolaw through literature / edited by Daniela Carpi.
   p. cm. – (Law & literature ; 2)
ISBN 978-3-11-025284-2 (acid-free paper)
1. Law and literature.   2. Bioethics in literature.   3. Self in literature.
I. Carpi, Daniela.
PN56.L33B56   2011
809'.933561 – dc22
                                                                  2011007823

Bibliographic information published by the Deutsche Nationalbibliothek

The Deutsche Nationalbibliothek lists this publication in the Deutsche Nationalbibliografie;
detailed bibliographic data are available in the Internet at http://dnb.d-nb.de.

Printing: Hubert & Co. GmbH & Co. KG, Göttingen
∞ Printed on acid-free paper

Printed in Germany

www.degruyter.com

Table of Contents

Daniela Carpi

Introduction

In recent years the well established field of human anthropology has been put under scrutiny by the new data offered by science and technology. Scientific intervention into human life through organ transplants, euthanasia, genetic engineering, experiments connected to the genetic codes and the genome, and varied other biotechnologies have placed ethical beliefs into question and created ethical dilemmas.

Science arises out of desire, the desire to overcome natural limits. Ethics is about social life; it is the regulation of human behaviour within a social context. Since ancient times, science has always been intertwined with ethics. Ethics gives meaning and purpose to scientific enterprise.[1] The term bioethics involves a double function: it analyses the particular conditions connected to life and life experiments and it formulates specific norms regulating scientific behaviour. These scientific interventions have influenced our views on birth and death, on the construction of the body and its technical reproducibility, and have problematized the concept of the human persona. An exemplary topic is cloning, which has come to epitomize the impossibility of keeping the well-established meaning of persona, as the uniqueness of the concept is annihilated by the idea of serialization. The purpose of bioethics, the science of life, is to find new values and norms which will be valid for a multicultural society. Law arises out of the need for the regulation and rationalization of human coexistence; biolaw implies the ways in which the law must react when facing new unprecedented situations caused by scientific experimentations.

Such transformations stem from a new postmodern period, where society has passed from monolithic values to a plurality and ambiguity of values: the old world order collapses and a new problematic one arises that requires new laws to be absorbed into the polity. Facing the broadening of the known limits to human actions, man has recourse to law: law may make acceptable what seems unacceptable, may create new norms for what seems abnormal. We have recourse to law to reorganise a troubled order.

The debate is between law and conscience, law and ethics, legal law and moral law. The juridical norm is set therefore between moral law and political

---

[1] M. Bianca, *Scienza, etica, bioetica* (Firenze: Angelo Pontecorboli, 1999).

law. Law comes to be considered a defence not only against private or collective violence, but also against institutional violence.

The main conceptual transformations concern the body itself: the fact that banks have been instituted where one can deposit products of one's own body, such as gametes, blood, cells, tissues, DNA, transforms the uniqueness of human, which cannot only be reproduced and multiplied, but is also scattered across space and time. This deals with one's right to self-determination. The history of medicine and surgery is the history of interventions on the body incessantly intertwined with the development of cultural models. The interventions into the body become part of the self-determination of the individual; the protection of that persona entails both its physical and psychic integrity. This cultural evolution forces us to remap the limits and borders of physicality and of the licit and illicit uses of the human body. Therefore we must face new frontiers that concern not only the juridical status of the human body but also its own material structure.

Such innovations must go together with new juridical norms that widen the concept of human rights in this new phase toward the post-human or trans-human. The new concept of persona entails taking into consideration an evolution of the body, the expanse of its potentialities. We have to face a transformation of cultural models, which implies a transformation of the juridical and social status of persona.

The term bioethics was introduced by V. R. Potter[2] with the meaning of medical ethics or ethics of scientific research. Potter insists on the defensive nature of bioethics, a defense of persona against the excesses of science, against the violation of human dignity. Goethe's view of the Faust myth triggered fear for the dangers of medical progress, later best theorised by Hans Jonas.[3] In his *Principle of Responsibility* Jonas denounces science as a Prometheus set free and warns man against a dangerous use of science.

Bioethics and biolaw are two philosophical approaches that address social tensions and conflicts caused by emerging bioscientific and biomedical research and their application. Bioethics can be defined as "the research and practice, generally interdisciplinary in nature, which aims to clarify or resolve ethical questions raised by the advances and application of biomedical and biological sciences."[4] Biolaw, on the other hand, is a philosophical concept in

---

[2] V. R. Potter, *Bioethics. Bridge to the Future* (Englewood Cliffs N.J.: Prentice-Hall, 1971).

[3] H. Jonas, *Il Principio di responsabilità. Un'etica per la civiltà tecnologica* [1979] (Torino: Einaudi, 2002).

[4] J. Miller, "Is Legislation in Bioethics Desirable? An Explanation of Aspects of the Intersection of Bioethics and Biolaw" in *Bioethics and Biolaw*, vol. I, *Judgement of Life*,

law that can be defined as "the taking of agreed upon principles and practices of bioethics into law with the sanctions that law engenders."[5]

"What is a person?" is a question that involves ethical connotations. The new biojuridical person that emerges from these questions has an ontological density.[6] The term "persona" during the last thirty years has triggered epistemological (what can I know?), moral (what is it right for me to do?) and religious (what can I hope?) speculations. Bioethics is by now a well-respected topic of research that has brought together philosophers and experts to discuss the limits of science and medicine. At the present time bioethics is changing into biolaw, a neologism that stresses the fact that the defense of life must go hand in hand with the defense and promotion of law, and that law must confront the actual results of science. All this is attuned to the general characteristics of postmodernity, which challenges the very possibility of reaching universal truths; it is rooted in the contingent, with a general loss of objective knowledge. Postmodernity is rooted in theoretical non-cognition. As a consequence, even the concepts that have always been taken for granted are put into question. In fact, what is now at stake in bioethics/biolaw is the concept of persona.

The concept of persona is linked to a problematic attitude towards human life derived from theology. Linguistically, the concept of persona comes from *prosopon*, mask, the one used by actors to hide their faces: the very semantic meaning of the term suggests ambiguity, because if the mask is the space of revelation, it is also the place of concealment. On the one hand, persona suggests something that appears outwardly through gesture and behaviour; on the other hand it suggests a secret noumenic reality of the I, hardly perceivable from the outside. If, according to Boetius, persona is *"rationalis naturae individua substantia"*, it suggests something that as substance it is individual and as rationality it is relational. Therefore persona is characterised by two contrasting modes: it tends to be individual, but it also needs to be put in relation with the "other". The fact that persona as *prosopon* is something that clarifies and conveys a meaning explains why for the Romans theatrical action and judicial action could be conflated. Also during trials persons "act": the defendant, the prosecutor, the lawyer who, in turn, can

---

eds. P. Kemp, J. Rendtorff and N. Mattsson Johanssen (Copenhagen: Rhodos International Science and Arts Publishers, and Centre for Ethics and Law, 2000) 313.

5 Kemp, Rendtorff, Johanssen, 246. Cfr. Susan Cartier Poland, *Bioethics, Biolaw and Western Legal Heritage*, Scope Note 45, National reference Center for Bioethics Literature, in *Kennedy Institute of Ethics Journal*, Sept. Part I and Dec. Part II, 2002.

6 S. Bauzon, *La personne biojuridique* (Paris: PUF, 2006).

embody different *personae*, that is, different roles. The various conflicts represented by the different *personae* must in the end find their resolution in the final judgment.

What each one of us is and can become revolves around four main elements that are essential for the definition of persona: 1) what is shared by all men through the use of reason; 2) the subjective way in which the universal category of rationality is applied; 3) the external elements (fate, chance, circumstances, etc.) that influence the course of our lives; 4) what is rooted in freedom and in the personal responsibility for our choices. These four elements derive from Cicero's positive conception of the term "persona". According to Cicero and to the Roman view, man is formed by a plurality of *personae*. It is only with Christianity that "persona" starts to acquire the characteristic of uniqueness.

In addition, across centuries we may discern two main attitudes towards the value of human life: the "separationist" position, which purports to see a separation between "persona" and human being, and the "personalist" position, where person and human being coincide.[7] The theories that separate persona from human being consider life not to be a value in itself, but only in the presence of certain qualities. These theories involving a separation between body and persona date far back in time: to Plato's conception of the body as the prison-house of the soul; to Descartes' distinction between *res cogitans* and *res extensa*. The body is thus depersonalised. The persona is reduced to its mechanistic functions. Once these functions occur (coma, pre-natal situation, etc.) we cannot speak of a persona.

As for the "personalist" theories, they claim that the very scientific description of the body makes us realize the insufficiency of a mechanistic explanation of persona and the necessity to formulate metabiological hypotheses. Being a persona is a radical ontological status. In the context of a personalist view human life must be protected at all cost, which entails therapeutical interventions and maintains the intrinsic dignity of human life.

Greek philosophers often speak of the idea of persona, especially after Socrates, but they do not consider the problem of subjectivity, which is part of the modern concept of persona. The intersubjective or relational aspect becomes the most radical characteristic of persona in modern times. Accord-

---

[7] L. Palazzani, *Introduzione alla biogiuridica* (Torino: Giappichelli, 2002), Ch. 1.4 "Il valore della vita"; "Bioetica e persona" monographic issue of the journal *Per la filosofia. Filosofia e insegnamento* IX.25 (1992); V. Melchiorre ed., *L'idea di persona, Metafisica e storia della metafisica, Vita e pensiero* 16 (Milano: Pubblicazioni dell'Università Cattolica, 1996); S. Maffettone, *Valori comuni* (Milano: Arnoldo Mondadori, Il Saggiatore, 1989).

ing to Saint Thomas Aquinas, persona is *"rationalis naturae individua substantia"*,[8] that is to say that persona as a definition entails the concepts of individuality, substance, and rationality. Individuality means an inner coherence and unity of being. Therefore self-sufficiency, self-construction, autonomy, and totality are essential for the existence of the persona. In his *Summa Contra Gentiles* Saint Thomas Aquinas asserts that God Himself deals with Man *magna cum reverentia* and in order to define persona Saint Thomas has recourse to the elements of freedom, immortality, and ability to respond to the world in its totality.

Among the many theoreticians who deal with the concept of persona, Peter Singer[9] focuses on a biological definition: a persona is one who may experience suffering or joy, one who is self-sufficient and self-conscious. This erases the distinction between man and animal and does not consider the ontological density of the biojuridical person. In contrast, Francesco D'Agostino's perspective is centred on the relational aspects of persona, on its ontological dimension.[10]

Biolaw is a very recent term that combines the field of bioethics with that of law, with the intention of giving juridical answers to bioethical dilemmas. It arises in cases of so-called juridical void, that is to say when the law is called to give practical ruling to a changed social situation. Biolaw pertains to the field of human rights and must guarantee man the possibility to realize fully his own humanity. The biojurist must question the validity of the current legal system and must strive to adapt positive law to natural law. Human rights must be reformulated according to the new techno-scientific discoveries. Bioethics and biolaw are intrinsically connected, even if they are epistemologically different. As has been asserted, "Biolaw without bioethics is blind; bioethics without biolaw is a void."[11]

The biojuridical debate also takes into consideration new forms of subjectivity, such as animal rights. Discussing the rights and dignity of animals means discussing the possibility or impossibility of considering animals not only as objects of the law, but also as juridical subjects. Up to what point can

---

[8] *Summa Theologiae*, I, q.29, a.4.

[9] P. Singer, *Practical Ethics* (Cambridge: CUP, 1993); P. Singer, *Rethinking Life and Death* (Melbourne: Text Publishing Company, 1994).

[10] See Palazzani, *Introduzione alla Biogiuridica*; L. Palazzani, *Il concetto di persona fra bioetica e diritto* (Torino: Giappichelli, 1996). See also F. D'Agostino, *Filosofia del diritto* (Torino: Giappichelli, 2000); F. D'Agostino, *Bioetica nella prospettiva della filosofia del diritto* (Torino: Giappichelli, 1983).

[11] D. Gracia, *Fondamenti di Bioetica. Sviluppo storico e metodo* (Milano: San Paolo, 1993), 683. My translation.

animals be considered mere objects of manipulation? Do they have rights? How can we balance the sacrificing of animals for scientific reasons with the necessity of protecting them as legal subjects?[12] The contemporary movement called "rightism" is the moral outspokesperson for animal rights. The movement speaks of animals not as mere machines devoid of conscience, but as new juridical subjects. In fact discussing animals means discussing man and man's role in nature.

The connection between these various fields of inquiry and research can be found in literature. Literature brings together the discussion about science, with its intrinsic dangers; law, with its attempt at systematizing the problems inherent in scientific-technological experiments; and ethics, dealing with the moral response to such innovations. The cultural result is the creation of new semantic fields, latest of all the field of biolaw and biojuridical person. Literature is helpful in creating mental experiments that alert us to problems in the real world.

In fact, literature has often anticipated such existential problems and questioned the ethical and legal limits we should set for science. Let us consider, for instance, Mary Shelley's *Frankenstein*, where there is an experiment on the creation of life through the collection of organs and body parts; in H. G. Wells's science-fiction novel *The Island of Doctor Moreau*, the mad physician, Doctor Moreau, wants to transform animals into human beings through long and painful explants and transplants; there are also similar genetic experiments in *Brave New World* by T. Huxley. As early as the Renaissance, Thomas More's *Utopia* highlighted the relationship between nature, social organization, and penal control. In *New Atlantis*, Francis Bacon revealed the link between science, nature, and control, thus adding to the socio-juridical control the perspective of the socio-scientific one, opening up new and fertile opportunities for reflection.

The strong opposition of literature to scientific experiments considered as dangerous can be ascribed to the two cultures debate that engaged T. H. Huxley, F. R. Leavis, and C. P. Snow. The discussion was centred on which of the two cultures should prevail, the scientific or the humanistic one, with the scales in turn leaning from one to the other. The debate originated in a famous lecture by C. P. Snow in Cambridge in 1959; it transmitted the idea of the two cultures struggling for predominance, with no possibility of reaching a consensus.

Later on Aldous Huxley took active part in the debate. He asserted:

---

12 R. Rodd, *Biology, Ethics and Animals* (Oxford: Clarendon Press, 1990).

> I feel strongly that the man of letters should be intensely aware of the problems which surround him, of which technological and scientific problems are most urgent. It is his business to communicate his awareness and concern.[13]

He invited scholars of both fields to advance together:

> Let's advance together, men of letters and men of science, further and further into the ever expanding regions of the unknown.[14]

Aldous Huxley's position within the two cultures debate makes him compare science to civilization, with science really meaning technology: for him technology is science without morality. Science and technology are positive elements if they do not become instruments of civilization, if they do not deprive man of his freedom.

> Technological advance is rapid. But without progress in charity, technological advance is useless. Technological progress has merely provided us with more efficient means for going backward.[15]

Aldous Huxley's position is that of fostering the use of science and technology so as to improve man's life, while being aware of the possible dangers of applied science.[16]

> He [Huxley] wants scientists to be more actively responsible for the technological improvements they help to bring into existence; in other words, he wants them to be morally responsible for their actions, or, as has been the case in the past, for their lack of active protests against producing more destructive weapons of mass annihilation.[17]

Of late, the debate over the two cultures has become much more heated, feeding off reciprocal accusations different from those expressed by Arnold and Huxley or by Snow and Leavis. They no longer accuse each other of not building a cultural heritage, but instead of being useless or even dangerous within the historical process. But nowadays the resistance to science from the humanistic field has been transformed into resistance to technology. Issues linked to technology are the latest aspect of the science/literature controversy. Technology is the main new force against which human exist-

---

13 J. Huxley ed., *Aldous Huxley: 1984–1963 Memorial Volume* (New York: Harper and Row, 1965), 100.

14 Huxley, *Memorial Volume*, 99.

15 A. Huxley, "Goals, Road and the Contemporary Starting Point" in *Ends and Means* (London: Chatto and Windus, 1937), 8.

16 For a thorough analysis of this topic see: C. Battisti, *Civiltà come manipolazione, cultura come redenzione* (Ravenna: Longo, 2004).

17 M. Birnbaum, *Aldous Huxley's Quest for Values* (Knoxville: University of Tennessee Press, 1971), 149.

ence must measure itself. Long considered subordinate to science, today technology is considered on par with it.[18]

Much philosophy of the Twentieth Century has interpreted technology in a distorted manner, demonizing it and considering it the cause of all humanity's ills (think of Benjamin, Leavis, and Spengler); however, philosophers such as Dewey have stressed how art makes use of technology, art at times successfully reaching a perfect balance between instrumental moments and moments of consumption during enjoyment of an artwork. For Dewey, continuity exists between science and nature, between man and technology.[19]

In her book *How We Became Posthuman* Katherine Hayles deals specifically with the relationship between man and technology, and with the renegotiation of the boundaries between man and machine.[20] This posthumanistic phase favours information structures that render biological incarnation a mere accidental fact. The body tends to be considered a prosthesis which we all learn to manipulate, removing the distinction between bodily existence and computer simulation. The humanistic-liberal subject comes to an end, leaving room for an amalgam, a grouping of heterogeneous components, a material-informational entity whose boundaries are continually subject to construction and reconstruction. Man leaves room for the cyborg.

While having its antecedents in the literature of past centuries, contemporary literature has started to deal extensively with the various human and ethical facets of liminal situations, where the possibility that robots, cyborgs and clones might have a soul is suggested. Let us consider, *in primis*, Philip Dick's famous short story "Do Androyds Dream of Electric Sheep?" (1968), later transformed into the equally famous movie by Ridley Scott, *Blade Runner*, which focuses on the suffering of the replicant for not being considered human. Also the computer mind in *2001: A Space Odyssey* is worth mentioning, where the computer protects its own existence, crying out its agony.

Kazuo Ishiguro's *Never Let Me Go* explicitly deals with human beings created as clones ready for explantation whenever it is necessary to create superior human beings, raising questions on the possible rights of a clone. We must also mention some women writers, such as the Canadian writer Margaret Atwood, who has dealt with bioethical issues in *The Handmaid's Tale* (1985), and, more recently in *Oryx and Crake* (2003), and with environmental

---

[18] G. Vattimo, *Tecnica ed esistenza* (Torino: Paravia, 1997); M. Nacci, *Pensare la tecnica. Un secolo di incomprensioni* (Bari: Laterza, 2000).

[19] See D. Carpi, "Literature and Science: the State of the Art in Contemporary Criticism", *Anglistik* 1 (2004): 51–61.

[20] K. Hayles, *How We Became Posthuman. Virtual Bodies in Cybernetics, Literature and Informatics* (Chicago and London: University of Chicago Press, 1999).

bioethics in *The Year of the Flood* (2009). The American novelist Jodi Picoult has frequently dealt with controversial issues such as euthanasia (*Mercy*, 1996), sterilization laws (*Second Glance*, 2003), and genetic planning (*My Sister's Keeper*, 2004). In *Correspondence* (1993), the first novel by British writer Sue Thomas, the protagonist gradually replaces her body parts with cybernetic prostheses, raising relevant questions about the status of the cyborg, a liminal being between man (or woman, as in this case) and machine. But many more examples could be added.

The aim of this book is to merge the two fields of bioethics and law (or biolaw) through the literary text, by taking into consideration the transformations of the concept of persona at which we have nowadays arrived. The new meaning of the term "persona" represents in fact the final point of a long-standing quest for man's sense of his own being and human dignity, and of his capacity to live in social interrelations. Is it legitimate to distinguish between man and *homo sapiens*? We may start with the assertion that not every person is a human being, and not every human being is a person. In fact human embryos, mentally handicapped people, people in a coma are not considered *personae*, because they are not characterised by self-consciousness and self-sufficiency. The debates concerning artificial intelligence assert that in the future robots may be endowed with self-consciousness and in this case they will become legal persons and will have civil rights.

One of the definitions to start from is that "persona" is a being aware of his/her own existence, developing across time, and endowed with desires and projects for the future, also characterized by rationality and freedom.[21] "Persona" is self-conscious, rational, capable of moral activities, self-sufficient, endowed with a body, developing through time, interacting with his/her surroundings and human beings.[22]

When we come to a more Christian definition, the concept of "persona" is strictly connected to the idea of God; God is persona *par excellence*. Here the moral question is central to the definition of persona: man is a persona because he can distinguish between good and evil. The ambiguities that arise from the term persona (it can denote also a non-human being) can be avoided by using persona to mean human being alone.

In April 1997 the Council of Europe signed a convention for the protection of human rights and the dignity of the human being with regard to the application of biology and medicine: "Convention on Human Rights and Biomedicine". One of the objectives is the "need to respect the human being

---

21 S. Maffettone, *Valori Comuni* (Milano: Mondadori, 1989), 218.
22 A. Pessina, *Bioetica. L'uomo sperimentale* (Milano: Bruno Mondadori, 1999), 86.

both as an individual and as a member of the human species and recognizing
the importance of ensuring the dignity of the human being."[23] Even in this
Convention what emerged is the ambiguity between the so-called "human
being" and the individual and it is only the individual that represents a juridical person.[24]

> The human being, according to the Convention could be described as follows: a
> being that is human possesses, to the extent that it is human, an essential dignity
> and identity, because it is "human", i.e. of the human species [...] an individual
> human being, which in turn ensures its inherent integrity, demanding respect and,
> thus, establishing subsequent fundamental rights and freedoms of the individual.[25]

"Persona" is also strictly connected with morality and politics. According to
Paul Ricoeur, ethics has to do with a good life, while morality has to do with
duty. The connection between morality and politics is in itself constitutive of
a persona. The desire for a good life can be realised only if one is part of a
political community. The necessity for the political stems from the fact that
persona is constituted by his/her capacities and fields of power, attitudes
that can develop only in a political context. The capable man can be recognised by his actions, is responsible for his actions, and bears their consequences on the level of penal law through punishment.[26] It is here that persona becomes a legal entity, responsible for his/her actions: it is here that
justice steps in to regulate the relationship of persona with other persons, in
connection with the idea of a good life (ethics), in a harmony of soul and
politics. John Rawls asserts that justice is the first virtue of social institutions;
it marks the connection between the social bond and the political bond
within moral judgement.[27] Persona therefore becomes a political entity coinciding with the notion of citizen.

This volume presents a wide range of perspectives: from more methodological approaches which try to define what bioethics and biolaw are, to applied
methodological approaches which illustrate the theoretical assumptions;
from more legal aspects, analysing real cases that problematize the appli-

---

23 Council of Europe, 1997, 3.
24 L. Reuter, "Human is What is Born of a Human: Personhood, Rationality, and an
    European Convention", *Journal of Medicine and Philosophy* 25.2 (2000): 181–194,
    185.
25 Reuter, "Human is What is Born of a Human", 186.
26 P. Ricoeur, "La persona: sviluppo morale e politico" in V. Melchiorre, *L'idea di
    persona* (Milano: Pubblicazioni dell'Università Cattolica, Collana Vita e Pensiero,
    Milano, 1996), 163–173.
27 J. Rawls, *A Theory of Justice* (Cambridge, Mass: Harvard University Press, 1971).

cation of the law, to more literary aspects, demonstrating the careful attention to these problems in literature. The volume ends with a very subtle interpretation of biolaw from the perspective of cartography.

JEANNE GAAKEER in "The Genetics of Law and Literature: What is Man?" discusses the use of forensic technology which in some way foretells the judicial invasion of the body. Law and forensics often collaborate nowadays in order to assess the identification of the defendant, but in these cases we are split between the bodily evidence the defendant gives and his own narrative of facts. We are caught between the clash of two opposing situations: the right to silence and the evidence given by scientific testing such as DNA which goes against the person's right to silence. So who are we? Are we material objects embedded in persons? "If we are enhanced by technological possibilities what does this mean for the question of our identity?" Gaakeer wonders. Therefore the questions move on to how law itself is influenced by technology. Gaakeer tries to answer the questions she poses by analysing Michel Houellebecq's novel *Atomised* where the central issue is whether the legal subject absorbs the human person. In fact new technologies seem to have resulted in their arbitrary incorporation into law. The philosophical core of Gaakeer's essay is the analysis of Heidegger's definition of technology which combines the meaning of "a means to an end" and "a human activity". *Technē* means bringing forth, *poiēsis,* but it is also linked to *epistēmē*, knowing, revealing. But this also entails Aristotle's distinction between theoretical and practical knowledge; thus modern sciences represent practical wisdom – the law is never general knowledge, but knowledge put into practice. From Heidegger's definition we derive the idea that technology is connected to knowing, to bringing forth, not to manufacturing. So the question becomes: how far can we challenge forensic technology? And this is where we must have recourse to the humanities, because through literature we can reflect on man as a moral being. Gaakeer also quotes Martha Nussbaum who asserts that literary texts are indispensable to a philosophical inquiry in the ethical sphere; they serve as a moral engagement in forensic technological issues. Through art we can develop cognitive competencies. The same is true for bioethics: the importance of literature in this field was already highlighted by Stephen Toulmin as far back as 1982. Literature "can also help us understand how distortions can occur in our perception of forensic technology. We are very much inside the project of the invention of the human."

IAN WARD in "Ghostly Presences: the Case of Bertha Mason" examines the accounts of sane people confined to lunatic asylums by revisiting the case of Bertha Mason in Charlotte Brontë's novel *Jane Eyre*. Bertha Mason epitomizes the tensions concerning the problem of madness in the Mid-Nineteenth Century and of the jurisprudence it involved. On the one hand madness was considered a moral failing, thus taking on hues of cultural condemnation; on the other hand it was viewed in terms of physiology and psychology; but most of all it entailed a patriarchal perspective on woman and its repressive consequences. Since Charlotte Brontë herself suffered from bouts of depression and *anorexia nervosa*; as an author she behaved as a literary surgeon responsible for diagnosing and treating social ills. In fact, *Jane Eyre* is saturated with surgical and psychopathic metaphors. Another fundamental issue in the text is the issue of confinement and of the control of property by the father or husband. The presentation of Bertha Mason is very much in character with the stereotypes of the period when considering the phenomenon of "the madwoman in the attic": sexually obsessed, violent, neurotic etc. The case of Bertha Mason chronicles the Victorian debates regarding insanity and the most appropriate strategies for its containment, while at the same time exemplifying the legal and medical discourses on insanity.

GARY WATT in "The Case of Conjoined Twins: Medical Dilemma in Law and Literature" speaks of the bio-ethical and legal problems of separating two conjoined twins, at the risk of killing one of the two. In the case of Mary and Jodie that Watt analyses, Mary was alive as a distinct personality but was not viable as a separate human being. That is to say that identity and personhood were at variance. To better explain the conundrum physicians and jurists had to face in this case of "conjoined twins", Watt has recourse to some literary and philosophical examples. In Plato's *Symposium* the character of Aristophanes employs the image of a conjoined Androgyny to explain the constant human quest for the ideal other; in *Twelfth Night or What You Will* Shakespeare presents the case of a reunion of twin brother and sister, with a direct allusion to Plato's myth of the Androgyny. Watt also quotes the political metaphor of the union of the body politic and body poetic, very significant in the Renaissance, intended as a union of nations that are unlike. The metaphor is made even more evident in shields divided into quarters, as in the Royal Coat of Arms of England and Scotland United, quoted in *A Midsummer Night's Dream*. Ruskin uses the metaphor of "reciprocal interference" to stress the political aspect of twin colours in some coats of arms: the metaphor of the importance of the strict link between "giving and receiving" comments upon the court's choice to separate the conjoined twins by depicting Jodie as the life-support machine for Mary. Such a motivation is questioned and the essay closes with the query: can two be bound into one noble whole?

MICHELE SESTA in "*Vida interminable*. Patients and Family Members between the Right to Live and the Obligation not to Die" has recourse to Allende's story *Vida interminable* in order to debate the problem of the ethical and juridical issues embedded in decisions to terminate life. That text narrates the story of two spouses and of their decision to die together. It is emblematic of the relationship between ethical evaluations and juridical regulations. Do physicians have to respect the patient's will to accelerate his/her death? The conflict between opposing juridical solutions mirrors the current bioethical debate.

JANE BRYAN in "Reading Beyond the Ratio: Searching for the Subtext in the 'Enforced Caesarean' Cases" takes into consideration some cases which came before the English courts in the 1990's concerning "enforced caesarean" cases. Such cases have to be treated beyond the *ratio,* that is, we must unearth the subtext of the judgment and try to read between the lines, as we would in the case of a literary text. Adopting Foucault's theory of the relationship between knowledge and power, Bryan explores some caesarean cases as power relations between the patient (the pregnant woman that refuses treatment) and the medical profession. The literary text she uses to enforce her legal views is Shakespeare's *A Midsummer Night's Dream,* where Theseus would like to force his daughter Hermia into a marriage she does not want. It is a clash between two different perspectives of what a legal persona is. According to Theseus Hermia belongs to him and he can order her against her will. According to Hermia herself, her freedom as a legal persona allows her to disobey her father. A similar situation is the one where a woman refuses caesarean treatment notwithstanding the danger facing her foetus. She does not respect the foetus' right to life. In the same way if the physician forces the woman into caesarean treatment not respecting her will he is not treating her as a legal persona.

ERIC RABKIN in "Science Fiction and Bioethical Knowledge" observes that bioethics concerns the questions of right and wrong in biotechnological matters and stresses the fact that the issue hinges on problems of knowing. He also arrives at some definitions that help us discern between ethics and bioethics: ethics, he claims, is mainly conservative: it speaks of the way we should behave according to some well-assessed rules and accepted behaviour. Therefore ethics is stabilizing, while biotechnology is inherently destabilizing because it promotes change. By taking into consideration a number of science fiction novels, such as *Victor Frankenstein,* for example, he ascertains that biotechnology deracinates the individual from his community: the loneliness of Mary Shelley's monster, the family-lessness of Dr. Moreau which brings him to conceive a very personal, anti-social sense of ethics, etc. Then Rabkin contrasts the figure of the scientist to that of the physician: if the scientist has a single-minded devotion to science, the physician must function socially. At this point the problem of law sets in: in the face of the rapid transformations brought about by modern technology is informed consent possible? Can the law be large enough to include the technological future?

JOHN DRAKAKIS in "Shaping Personhood: Problems of Subjectivity and the Self in Shakespeare's *The Taming of The Shrew* and *Much Ado About Nothing*" centres his discourse on Foucault's principles concerning how the body should be disciplined, thus voicing the scientific processes that began to emerge with the Enlightenment. Drakakis stresses the fact that Foucault links bio-power with racism and with the sovereign capacity to kill. Bio-power and bio-ethics are entangled in various forms of the regulation of subjectivity that starts in the early modern period. Drakakis exemplifies his assumptions by taking into consideration the anonymous text of *The Taming of the Shrew* and comparing it to Shakespeare's: both texts epitomize the indignity of having one's body "read" and "interpreted" so as to "translate" it into a more acceptable form for society. The same happens in *Much Ado About Nothing*, where the containment of sexual energy presents the problematization of subjectivity and selfhood.

PATRIZIA NEROZZI BELLMAN in "On the Sciences of Man in Eighteenth Century English Literature and Art: Anatomizing the Self" affirms that the idea of the Self, which marks the transition to modernity, takes shape in the Eighteenth Century, where a host of disciplines such as literature, law, art and medicine, collaborated in creating the "science of man". Moreover the construction of the first automata triggered the debate on what it is to be human that has become so central to the contemporary scene. The automaton becomes the great symbolic product setting itself at the intersection of many disciplines and anticipates the contemporary bioethical dilemmas. Nerozzi Bellman draws a brief diachronical description of the evolution of the concept of Self, focusing her attention on Locke's *Essay Concerning Human Understanding* (1694). The budding spectacularity of science causes new questions regarding the constitution of personhood. Swift shows how fascinated the century is by measures and models and how it is the very uniqueness and consistency of the body that starts to fall apart. The frequent mechanical similes that we find in Swift's work suggest the impact of mechanical improvements on culture.

DANIELA CARPI in "The Beyond: Science and Law in *The Island of Doctor Moreau* by H. G. Wells" asserts that Wells' novel anticipates many of the legal problems that the latest scientific discoveries are posing in our century. There is a close fit between Wells' specific criticisms of chimeras and the concerns that trouble medical ethicists today. What is mainly at stake is an ever-changing concept of "persona" that is extended by the new cloning experiments and by the recourse to organ transplants. Wells's novel actually speaks of vivisection, but the connection with genetic experiments is very strong. Such experiments, being extreme, undermine human beings' uniqueness by suggesting the possibility of the serialization of beings. A new law is necessary to keep these new beings within society, or a new concept of society is required so as to include them.

YVONNE BEZRUCKA in "Bio-ethics *avant la lettre*: Ninenteenth-Century Instances in Post-Darwinian Literature" highlights how the publication of Darwin's *The Origin of Species* (1859) fostered an attention to people as ontological beings. The new scientific hypotheses which developed at the end of the Nineteenth Century and the related embodiment of man's fears about the period's pre-eminence of science disjointed from ethics in science fiction works led to a reconceptualization of human beings and identity. The debate between monogenesis and polygenesis in the origin of men, pseudo-sciences such as anthropometry, phrenology, and physiognomy, all focus on taxonomical human subdivisions and point to biological differences between peoples as innate differences. In such a hierarchical classification, somatic features become racial ones and support a specific body politics supporting the supremacy of a specific group of people. Eugenics, in particular, with its emphasis on selective breeding for the creation of an eu-genic (that is, good, pure) race is seen as an example of bioethics *avant la lettre*, an example of scientific manipulation on human life. As a reaction to such an attitude, the end of the century saw also a proliferation of works linking ethics to evolution, works aiming at safeguarding the principle of life as well as human bio-essence.

SILVIA MONTI in "Rhetoric, Lexicography and Bioethics in Shelley Jackson's Hypertext *Patchwork Girl*" observes that some literary works such as *Frankenstein* seem to represent the testing ground for contemporary bioethical experiments. She notices that we cannot discern any absolute demarcations between bodily existence and computer simulation, between the human and non-human of the cyborg. Shelley Jackson's novel *Patchwork Girl*, that appeared only in hypertextual form, clearly epitomizes the problems lying behind artificial creation and the potential of science to manipulate life forms. *Patchwork Girl* becomes the emblem of man's estrangement from nature, hence the inherent sense of loss and frustration. In the text we often find a comparison between a literary composition and the human body: as the text is formed by different lexias, so the body is composed of different parts or organs. Monti analyses the bioethical implications of the novel from a linguistic perspective, noticing how rhetorical figures such as metaphors, similes, and synedoches express the hybridity of *Patchwork Girl*. For instance the generative metaphor represents a parallel between the character's body and the hypertext. The different body parts correspond to the letters of the alphabet that give form to the text. However, Monti concludes, as the biological body cannot be reduced to a mechanical organism made of dismountable parts, so we must be careful not to indulge in the excesses of artificial manipulation of man.

PAOLA CARBONE in "One Monstrous Ogre and One Patchwork Girl: Two Nameless Beings" states that a proper name affects the representation of the self, because it gives identity to a human being. A name not only identifies a child but it also gives him/her a status inside a summa of beliefs, symbols, rules so as to suggest his/her uniqueness. Therefore a name is a cognitive necessity. What happens in the case of hybrids such as Frankenstein or its later epitome Patchwork Girl? It is important to probe the extent to which we can consider a person a human being. All hybrids are crossbreeds of nature and are resistant to specific identification. Carbone compares the different sense of self that we discern in *Frankenstein* and in *Patchwork Girl*. In both cases we are in front of simulacra of human beings and of aesthetic constructions. The aesthetics of the monster demonstrates a transition towards a new awareness of the human standard. In the case of Frankenstein, Victor's manipulation of the monster's body condemns it to solitude, while in Patchwork Girl her multiple nature describes her as a harmonious compound of different pieces: she cannot feel lonely because she is the summa of many different persons that all coexist. The two characters are consequently "cultural outsiders" while remaining "juridical insiders". They represent a concept of person as a bio-cultural-legal complex. Frankenstein's monster interprets personhood as sameness, Patchwork Girl intends it as interacting patches.

PAUL CHEUNG in "A Serious Reading of Biotechnology in Japanese Graphic Novels: Weak Thoughts Regarding Ethics, Literature and Medicine" argues that in post-war Japanese manga, in particular in Tezuka's, the problem of the liminal situation between man and cyborg is analysed. Astro Boy and Mitchy, two characters, occasionally wonder about their human or non-human identity, developing a flair for the cautious assessment of science and technology. In particular biotechnology is taken into consideration. In *Astro Boy* and *Black Jack*, for instance, we find the practising of surgery without consent, and the disconcerting possibilities of the human, come to the forefront long before bioethics, becomes a distinct discourse. Cheung bases his critical analysis on Vattimo's concept of "weak thought", observing how the relationship between philosophy, literature, ethics and medicine has been a concern in medical education.

LAURA APOSTOLI in "Fulfilling Personhood at the Margins of Life: Anna Quindlen's *One True Thing*" starts with some definitions of what a person is. In particular she centres her attention on the legal recognition of who counts as a person. She is aware of the fact that we have moved into a post-human stage of history (to use an assertion by Francis Fukuyama). The attainment of a biolegal guardianship able to deal with new forms of liminal beings is felt as an urgent need. Literature shares the idea that human beings are full entities formed by the harmonious sum of body and mind in a worthy life. Grand narratives support a rethinking of what constitutes a person both in moral and legal terms. Anna Quindlen's novel is focused on the problem of euthanasia and on the question whether any person can act as a truly autonomous agent within contemporary society. Critical illness takes man to the margins of human experience, undermining the relationship between the body, the self, and the outer world. Quindlen's passionate exploration of alienation and disability, of free will and self-determination, suggests that it is no longer possible to consider the mere value of life in itself, but that its quality and dignity should also be at the core of medical science and legal jurisprudence. The call for human dignity should represent the threshold between life and the desire for death.

VALENTINA ADAMI in "So What Is a Human Being?: an Exploration of Personhood through Jeanette Winterson's *The Stone Gods*" stresses the fact that one of the central characters of the novel, Spike, is a female robot *sapiens*: should she be entitled to personhood and moral status? The central question in the novel is in fact the definition of humanness in the context of modern technology. Should we extend the concept of person beyond the realm of the human? Some of the properties that various philosophers have considered as necessary for the definition of person are: moral status, cognitive psychological properties (such as memory and thinking), sentience, and network of relations (either social or biological). Spike presents most of the listed characteristics thus offering a model for thinking about the liminal stages of human life.

SIDIA FIORATO "The Problem of Liminal Beings in Alasdair Gray's Poor Things" focuses on the novel *Poor Things* by Alasdair Gray, a re-writing of Mary Shelley's *Frankenstein* which analyzes the social and legal consequences of the creation of a human being. The protagonist, the skilled surgeon Godwin Baxter, manages to execute a whole body transplant/brain transplant and transfers into the head of a suicide pregnant woman the brain of her unborn child. The possibility for such an act of manipulation of human life is taken for granted, while the social and legal justification of the woman represents a more problematic issue. Medical science is frequently shown to outpace the development of the laws regulating the legal persona, laws which do not apply in the case of liminal beings. Baxter himself, autonomously from the law, creates a legal persona for the woman as Bella Baxter, but then bends the law to his needs for the recognition of such a created identity. In this (deceitful) way, however, he renders his creature able to realize her own identity and engage in social relations. At the end of the novel a debate takes place about Bella's identity: is she to be considered the same person as the drowned woman and therefore be restituted to the previous life of that person in virtue of the principle of the continuation of the body, or is she a new person in virtue of the principle of psychological continuity as the criterion to determine a person's identity? In her autonomy as a social and legal being, Bella will decide for herself.

MARA LOGALDO in "Murderous Creators: How Far Can Authors Go?" observes a significant coincidence between the birth of bioethics and that of postmodernism. In fact the two positions present common assumptions. Logaldo in her critical analysis starts from Ortega y Gasset's humanistic beliefs concerning art, feeling, and the dehumanization of art. Both positions in fact review positivist epistemologies and a faith in progress; both question the central position of man in the universe; in both cases we may speak of a post-human attitude that questions man's physical and psychological integrity. Bioethics and postmodernism share a certain emphasis on self-awareness and on personhood as a narrative construction. The ambiguous relationship that links the two perspectives is centred on the loss of boundless faith both on the power of science (bioethics) and on the power of literature (postmodernism). However at the same time they are symmetrically opposed: if bioethics wishes to prevent any kind of unethical manipulation of man, literature uses characters as pawns on a chessboard and completely manipulates them.

CHIARA BATTISTI in "Fay Weldon's *The Lives and Loves of a She Devil*: Cosmetic Surgery as a Social Mask of Personhood" analyses Fay Weldon's novel as a source of inspiration for an alternative point of view. The text invites us to turn our considerations to cosmetic surgery from an extremely subjective level of judgement to a consideration on the relationship between cosmetic enhancement and identity; it allows us to expand the implications of the concept of persona. According to Battisti there is an aspect of Weldon's novel which offers a wholly innovative critical perspective by suggesting an interaction between cosmetic surgery, biolaw, and the concept of personhood. The recourse to cosmetic surgery represents a woman's voluntary act to control her identity. It is precisely the significant change in Ruth's (the main character) tone of voice that induces us to reflect upon the idea of the body as an effective mask that protects and enables us to function better as a public person by endowing us with a voice and visibility. What Battisti therefore wishes to suggest is that the body itself may become a mask, which is moulded like a voluntary act or perceived as the obligation to ostentatiously represent one's self. Clothes, cosmetics, perfume, and cosmetic surgery further shape such a mask.

LEIF DAHLBERG in "Mapping the Law – Reading Old Maps of Strasbourg as Representing and Constituting Legal Spaces and Places" intends the term "bioethics" in a very particular way: for him the connection between law and cartography fosters a new understanding of the social and legal development of constitutional nation states. Maps represent the "bios", the life of a nation: the natural evolution of states is registered in maps that mark the development of legal space. Not only do maps depict legal facts (property, for instance), administrative and political borders, but they also shape and categorize public and private spaces. Maps therefore stand for the technological innovations that force the law to transform its concepts of legal persona: if technological innovations transform the classical idea of personhood, so maps form a (technological) discourse that marks an evolution of social power structure. Dahlberg's discourse is linked to Foucault's idea of biopolitics and biopower. The discussion of the relation between maps and law therefore also includes the concept of biopower-biopolitics, biolaw, and bioethics, and in particular the affinity and analogies that can be traced between how law constitutes legal space and legal conceptions of persons and the human body. Particular attention is devoted to the intimate association between the Roman household as a private space and the inviolability of the individual body. The private sphere (a territory within a territory) is constituted by a folding of the political body onto itself, creating a pocket simultaneously inside and outside. In this way the private sphere appears as a projection of the individual body onto the social space of the city.

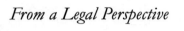

From a Legal Perspective

Jeanne Gaakeer
Erasmus University Rotterdam

The Genetics of Law and Literature: What is Man?

> From language to life is just four letters.[1]

"Let Me Hear Your Body Talk"

"Everyone charged with a criminal offence shall be presumed innocent until proved guilty according to law", says Article 6 of the European Convention for the Protection of Human Rights and Fundamental Freedoms, thus conferring the right to fair trial.[2] The article guarantees the right to silence, the

---

[1] R. Powers, *The Gold Bug Variations* [1991], (London: Abacus, 1993), 8, from the opening "Aria" of the novel, entitled "The Perpetual Calendar".

[2] For article 6 of the Convention in an easily accessible digital form, see http://conventions.coe.int/Treaty/en/Treaties/Html/005.htm: "Article 6 – Right to a fair trial: In the determination of his civil rights and obligations or of any criminal charge against him, everyone is entitled to a fair and public hearing within a reasonable time by an independent and impartial tribunal established by law. Judgment shall be pronounced publicly but the press and public may be excluded from all or part of the trial in the interests of morals, public order or national security in a democratic society, where the interests of juveniles or the protection of the private life of the parties so require, or to the extent strictly necessary in the opinion of the court in special circumstances where publicity would prejudice the interests of justice.

1. Everyone charged with a criminal offence shall be presumed innocent until proved guilty according to law.
2. Everyone charged with a criminal offence has the following minimum rights:
    a. to be informed promptly, in a language which he understands and in detail, of the nature and cause of the accusation against him;
    b. to have adequate time and facilities for the preparation of his defence;
    c. to defend himself in person or through legal assistance of his own choosing or, if he has not sufficient means to pay for legal assistance, to be given it free when the interests of justice so require;
    d. to examine or have examined witnesses against him and to obtain the attendance and examination of witnesses on his behalf under the same conditions as witnesses against him;
    e. to have the free assistance of an interpreter if he cannot understand or speak the language used in court".

right not to incriminate oneself that is also found in the Miranda rule in American law (*Miranda v. Arizona*, 384 U.S. 435, 1966) as the requirement that a person receive certain warnings with respect to his right to remain silent, one that is also standard fare in popular TV series when a suspect is being arrested. It rests on the following related classical principles found in Roman law: *nemo tenetur prodere se ipsum* (no one is obliged to incriminate himself), *nemo tenetur edere contra se* (no one is obliged to speak against himself) and *nemo tenetur se accusare* (no one is obliged to accuse himself). As a legal right it is indissolubly connected to the rule of law in a democratic society in the protection that it guarantees against unlawful intrusions in people's lives, and the respect for the presumption of innocence it voices.[3] As a prohibition against putting pressure upon a person suspected of having committed a crime, it refers to the deference in law for the defendant as a party in criminal proceedings, when it comes to respecting human dignity in the sense of both the free will and physical and mental integrity.

This heritage of Enlightenment thought on the subject of criminal law since Cesare Beccaria urged us to abolish physical torture (on the rack) as a method to obtain convictions[4] now seems to be under pressure given the development of modern technology when that is put to forensic use. For ages,

---

[3] Since *Funke v. France*, EHRC, 25 February 1993, Series A 256-A, the standard that is set is, "[…] the right to silence and the right not to incriminate oneself are generally recognised international standards which lie at the heart of the notion of a fair procedure under Article 6". It is also found in the procedure of International Criminal Tribunal for the former Yugoslavia (art. 42 sub A sub iii ICTYS), in that of the International Criminal Tribunal for Rwanda (art. 20 lid 4 sub g ICTRS), and of the International Criminal Court in The Hague (art. 55 lid 1 sub a and b, and art. 55 lid 2 sub c ICCS). The case of *Saunders v. U.K.*, EHRC, 17 October 1996, Reports 1996-VI, is the most explicit example of the link with the presumption of innocence. See also for the recent developments with respect to fair trial, EHRC 27 November 2008 36391/02, *Salduz v. Turkey* and EHCR 11 December 2008, 4268/04, *Panovits v. Cyprus*, with, in both cases, a defendant who was a minor questioned by the police without legal counsel. For the right to fair trial for those detained at Guantanamo Bay, see *Hamdi v. Rumsfeld*, 542 US (2004), *Rumsfeld v. Padilla*, 542 US (2004), Rasul v. Bush, 542 US (2004), and *Hamdan v. Rumsfeld et alia*, US Supreme Court, nr.05–184 (29 June 2006).

[4] Cesare Bonesana, marchese Beccaria [Milan, 1738–1794], *Dei Delitti e delle Pene* [1764], *Of Crimes and Punishments* first English trans. E. D. Ingraham (Philadelphia: R. Bell, 1778). It should at once be noted, though, that Beccaria was still inclined to mete out heavier penalties to those who refused to talk at all. In contemporary Dutch law, the right to remain silent cannot be used against a defendant, but his refusal to give any reasonable explanation at all in the sense of a rebuttal of the charges against him, can be taken into consideration if there is enough other, causal evidence.

in Greek and Roman antiquity, the entrails of animals were thought to speak the truth about what had happened or would happen in the near future,[5] and for those of Christian denomination, the evidence of things not seen could be based on faith.[6] Ironically, given our Enlightenment values in criminal law, we now find ourselves in the situation that it does not matter any more if a defendant does not want to talk, because as lawyers we have the technological and legal means to say, as in Olivia Newton-John's 1981 number one hit song "Physical",[7] "Let's get physical, let me hear your body talk", and that literally, while at the same time, again as lawyers, we just have to believe what forensic sciences offer us because we lack the training and experience to fathom fully what is held before us as possible evidence.

If we ignore the actual findings or contributions from earlier body-oriented sciences such as craniology, physiognomy (starting with Aristotle and founded as a discipline by Lavater, 1741–1801), and its descendants phrenology (developed by Gall, 1758–1828) – remember the scene in *Jane Eyre* where Rochester holds up his forehead to Jane for her to "read"? – and criminal anthropology developed by Cesare Lombroso (1835–1909) which in a variety of manifestations all claimed that a person's character and disposition as well as inborn criminality can be judged from the features of his face and the outward appearance of his bodily characteristics,[8] the first "embodied" forensic standard for criminal investigation is the introduction, in the late nineteenth century, of fingerprint techniques.

In modern literature, the case in point exemplifying how fingerprints can be put to forensic use is Mark Twain's novel *The Tragedy of Pudd'nhead Wilson*. In it, fingerprints lead to the discovery of the murderer of the judge. The plot is fairly simple, at least in retrospect, given the contemporary state of the art in this technique. A changing of the clothes of the slave girl Roxy's son Chambers and the master's son Tom Driscoll when the two boys were still in the cradle has been the cause of their truly being exchanged. However, their fingerprints had been taken twice, before and after the exchange, and when

---

[5] Not to mention the Roman "augur" and "auspex" who predicted the future from the flight of birds, hence the English verb "to augur", and the adverb "auspicious".

[6] *New Testament*, the "Epistle of Paul the Apostle to the Hebrews", 11:1: "Now Faith is the substance of things hoped for, the evidence of things not seen".

[7] "Physical" (1981), written by Steve Kipner and Terry Shaddick and performed by Olivia Newton-John.

[8] For an extended discussion of this topic, see my, "'The art to find the mind's construction in the face': Lombroso's criminal anthropology and literature: the example of Zola, Dostoevsky, and Tolstoy", *Cardozo Law Review* 26 (2005): 2345–2377.

years later it comes to finding the murderer by means of the "bloodstained fingerprint upon the handle of the Indian knife", the fingerprints of the accused become proof of his innocence. Not only is the real murderer found (Chambers who was raised as Tom) but the true identities of the boys are also revealed by means of the "strange discrepancies"[9] in their physiological autographs when matched against the fingerprints taken when they were babies. This usage of fingerprint techniques as a forensic method in Twain's novel forecasts the juridical invasion of the body witnessed today with the proliferation of technological inventions facilitating forensic applications. When law and forensics cooperate to achieve the ultimate goal of identification or assessment of the defendant's individual authenticity, how are we to negotiate the tension between the defendant's narrative and the bodily evidence he "gives"?[10] There remains a dichotomy, a binary opposition that I think is all too lightly passed over in legal theory and legal practice when at the same time that the right to silence is established and guaranteed by statutes and treaties, technological developments such as the possibility of DNA testing help law to have the body talk and use genetic information in forensic settings in an effort to solve crimes.

The Double Helix of DNA

Since Watson and Crick, inspired by Erwin Schrödinger's 1944 *What is Life?*, discovered the double helix of deoxyribonucleic acid, or DNA for short,[11] technological developments have drastically augmented the possibilities of, and the perspectives on law. At the same time the meaning of genomic information has taken different shapes to such an extent that the witches' lamentation in Shakespeare's *Macbeth* also comes to mind: "Double, double, toil and trouble" (Act 4, sc.1, l.10).

---

[9] M. Twain, *The Tragedy of Pudd'nhead Wilson* (Cutchogue, New York: Buccaneer Books, 1976), 155, 163.

[10] It is interesting to note, as an aside, the comparison that can be made between the relation of the outcome of DNA research and a defendant's narrative on the one hand, and the "history and physical" format of diagnosing illness in vogue for the doctor-patient consultation since the nineteenth century, and in use today in modern medicine, with a shift in the latter from reliance on the patient's narrative to more reliance on the examination of the body (not to mention the nineteenth-century novelty of dissection after death also as a means to gain positive knowledge).

[11] See also J. D. Watson, *The Double Helix, a Personal Account of the Discovery of the Structure of DNA*, ed. G. Stent (London: Weidenfeld and Nicolson, 1981).

To start with the technical side of the matter: every DNA molecule consists of two chains of building blocks connected in a double helix. In the nucleus of each cell of any human being, DNA is distributed over the 46 chromosomes found in pairs. In each pair, one chromosome or DNA molecule comes from the father, the other from the mother. There are 22 pairs of autosomes, i.e. homologous chromosomes, and one pair of sex chromosomes, number 23, XY for the male, XX for the female. The autosomes or the chromosomes 1 to 22, consist of repetitive pieces of DNA usually four building blocks in length and always a combination of four chemicals called bases, represented by the four letters of the genetic code: A (adenine), T (thymine), C (cytosine) and G (guanine). This, in short, is the way in which genomic information is inscribed in the molecule called DNA. In each and every cell of an individual human being, the information in the form of DNA is identical.

The order of the A, T, C and G – combinations that form the human genome is determined by the Human Genome Project which held its first conference in 1986. The order of genetic letters can be compared to the order of letters in words: any change in the genetic lettering has an impact on the functioning of the human body much in the same way that a different combination of letters can lead to different words and as a consequence different meanings: bard is not the same as drab. Or, as Richard Powers would have it in a line of poetry that I offer as an epigraph to this article, "From language to life is just four letters".

It should at once be noted that the human genes lie in only 2 % of DNA (and scattered over the chromosomes at that) that thus contains our genetic information and functions as our bodily code. As a method of identification, and, especially, of elimination of people as suspects, DNA fingerprinting (note the term!) was developed by Alec Jeffreys in 1985. The Pitchfork case (2 related murder cases in the U.K. in 1983 and 1985) was the first example in which DNA findings helped exonerate a suspect who had, falsely it turned out, confessed to one of the murders. Today, the FBI keeps a computerized databank of DNA samples (the Combined DNA Index System, or CODIS for short) that in 2003 already contained 1.7 million DNA profiles, to be used to convict or exonerate.[12]

---

[12] For a fine example of the latter, see the (1992) Innocence Project initiated by Peter J. Neufeldt and Barry C. Scheck at Benjamin Cardozo School of Law (New York), with 173 prisoners helped between 1992 and 2006, http://criminal.findlaw.com/ crimes/more-criminal-topics/evidence-witnesses/dna-as-exoneration-tool.html, accessed 1 March 2009. See also the specific reference to DNA evidence in the Brief for the Innocence Project, Inc., as *Amicus Curiae* supporting petitioner, in

In order to help solve a crime, the goal of the DNA scientist's research is to obtain a specific, individualized DNA profile from bodily matter found at the crime scene. The problem with that usually is that the amount of bodily matter found is very small and the DNA more often than not contaminated or degraded. What then is a DNA profile? There is no such thing as *the* DNA profile. What we call the DNA profile is an autosomal profile with 15 different DNA characteristics, internationally standardized, with an internationally standardized name per characteristic, and the way of denoting the variants for the same characteristic also internationally standardized. These places are called *loci*.[13] For forensic purposes, the problem most often to be solved is that of the mixture of profiles found in the body materials at the crime scene. Put differently, what if there is a mixture of *loci* at the *locus delicti*? How then can one be sure that there is not a mix-up in the suspects? Ironically, the only source of identification that remains unique is the human fingerprint, this much should be granted to Puddn'head Wilson, and think of the possibilities the fact that identical twins have the same DNA brings for planning the perfect crime (as long as you don't leave any fingerprints!).

What, then, are the other interesting shapes that DNA research has taken for law? For purposes of this article, I cannot exhaust the whole range of topics but will only mention a few examples in order to prepare the ground for a discussion of the problematics of (bio)technology in/for law. Think, firstly, of the impact of DNA research in the field of medicine. Today, variations in DNA that underlie specific diseases can be identified and the same goes for the effects of medicines. Gene therapy experiments are being done for diseases such as various forms of cancer, HIV, AIDS, Parkinson's disease, and cystic fibrosis. The good thing is of course that the hereditary component of diseases can now be the target so that treatment can become more

---

Paul Gregory House (petitioner) v. Ricky Bell (respondent), U.S. Supreme Court, 12 June 2006, no. 4–8990, and in Justice Alito's delivering of the opinion of the Court on the writ of certiorari in *Bobby Lee Holmes (petitioner) v. South Carolina*, 547 U.S. (2006).

[13] Only the 2% of our DNA mentioned above is responsible for our hereditary traits such as the colour of our eyes. The other 98% does not encode, or should we say "codify", any hereditary traits. As a consequence, such specific areas of this DNA vary greatly from one person to another. These so-called hyper variable areas consist of repetitive strings of DNA and since the number of repetitions per person also differs from person to person, it is admirably suitable for linking a biological "trace" to an individual person. The place of such a hyper variable part on DNA is the *locus* and each *locus* is given a number (or rather, two numbers since chromosomes come in pairs) indicating the number of repetitions. *Loci* are therefore admirably suited as well to distinguish between people.

precise, with the ultimate goal of prevention before our eyes. Here too, the metaphor of the letters of a language is important to note. The order of the letters A, T, C, and G can undergo change at a single location in a genetic profile. This phenomenon is called single nucleotide polymorphism (SNP) and while this is usually biologically not important, comparable as it is to the way in which the difference between the American and British spelling of a word does not *per se* affect meaning, it can sometimes change the gene's functioning.[14] The bad thing is that such research can also be held against a person. What if you know that you are genetically predisposed to diabetes, and you eat and drink until you weigh too much? The legal and ethical component enters the fray, when it comes to laying the blame on someone, as happens these days for example in the context of health and life insurance – and remember that the concepts underlying insurance used to be collectivity and solidarity – even though in the United States a law, the Genetic Information Nondiscrimination Act of 2008, was enacted to prevent discrimination from health insurers and employers so that a person's DNA information will be protected, and even though both the European Union and UNESCO have implemented declarations on the human genome containing similar nondiscrimination rules, prohibiting any form of discrimination against a person on grounds of his genetic heritage. The US law, for example, does not cover life insurance and disability insurance, and neither does it offer protection on the level of the relation between (unhealthy) life style and genetic predisposition. Another issue with a strong ethical component is that of intellectual property when DNA findings are being made fit for the market economy.

Secondly, another legal issue involved is the racial aspect. DNA can help ascertain to which section of the population a suspect belongs in terms of race. But doesn't this going ethnic contain the risk of racial bias?[15] And that at a time when given our focus on equality we thought we had become postracial in law? This is a question important not only for criminal law, but also for private law, as the 1993 Supreme Court of California case of *Johnson v. Calvert* (5 Cal. 4th 84, 851 P.2d 776) showed when a black surrogate mother tried

---

[14] My view is informed here by the data provided at GWAS Overview (NHGRI), accessed 1 March 2009. Beyond the scope of this article, but too important nevertheless not to mention it, is the context of (threats of) terrorism in which biotechnology can become an instrument of destruction.

[15] An interesting discussion is provided by K. F. C. Holloway, "Private Bodies/ Public Texts: Literature, Science, and States of Surveillance", *Literature and Medicine*, 26.1 (2007): 269–276, also accessible at http://muse.jhu.edu/journals/literature_and_medicine/v026/26.1ho.

to keep the white-looking child she had gestated for a white man and his Filipina wife. Fertility clinics may be sued for damages when *in vitro* fertilization results in "offspring of an unanticipated race".[16]

Related to this, thirdly, is the issue of DNA results in paternity suits where the scientific research is done for the purpose of determining the identity of the actual father in custody.[17]

In migration law, too, there is the acute question of forensic technology. DNA tests are used when it comes to reuniting a migrant's family, bone scans are done in order to find out a migrant's age when it comes to preventing illegal immigration as do carbon dioxide detectors in another way when they are used to scan trucks or sea containers for the presence of people. Here too, the risk of racial discrimination looms large: biometrical data, such as iris scans and fingerprints are used to distinguish between migrants on the basis of race, skin colour and ethnic background. The innocent use of the iris scan at international airports may be beneficial for the ordinary traveler who can get though customs much faster but this technological advanced method has its backdraw when it comes to the legal issue of privacy.

Yet another example of the influence of technology in and on law can be found on the interdisciplinary plane of law and the behavioral sciences. The findings in neuronal Darwinism in vogue in the behavioral sciences offer sound evidence that damage to the prefrontal cortex in first few years of our lives leads to a disturbance in social and moral behavior in adulthood. That is to say that our capabilities are always already restricted by the way our brain is being organized as we grow up. This would seem to imply that our thought on and specifically our usage of the concept of the free will, the very foundation of law in so many ways, should only deal with those of "healthy" brain.[18] Put differently, we should acknowledge that the concept of the free

---

[16] See A. E. Weinbaum, "Racial Aura: Walter Benjamin and the Work of Art in a Bio-technological Age", *Literature and Medicine* 26.1 (2007): 207–239, 212.

[17] Important here is DNA research of the mitochondrion. Apart from DNA in the nucleus of the cell, there is also DNA in the mitochondrion, the energy factory so to speak of the cell. Since both sons and daughters get their mitochondrional DNA from their mother, family members who are lineal descendants from the same mother have identical mitochondrional DNA. So Roman family lawyers were right when they claimed: *mater semper certa est*.

[18] J. Saramago, *All the Names*, trans. from the Portugese by Margaret Jull Costa (New York and London: Harcourt Inc., 1999), 28: "Moreover, if we persist in stating that we are the ones who make our decisions, then we would have to begin to explain, to discern, to distinguish, who it is in us who made the decision and who subsequently carried it out, impossible operations by anyone's standards. Strictly speaking, we do not make decisions, decisions make us".

will is and has always been a rational construct from the start rather than the absolute we took it to be, and that there is now empirical medical evidence to the contrary, that we have nothing whatsoever resembling the free will as we have long conceived it to be. The results of recent investigations in the relationship of the amygdala and the prefrontal cortex when it comes to determining irrational behavior offer proof in this direction (the fact that the amygdala is full of testosterone receptors and larger in men than in women, is part of the explanation why men generally demonstrate more aggressive behaviour than women). When it comes to determining responsibility in criminal cases in which the defendant is thought to suffer from a mental illness, we then not only have to deal with the fact that the diagnosis by forensic psychiatrists is based on a list of what may be called a mental illness according to the DSM IV (the Diagnostic and Statistical Manual of Mental Disorders, and that list is the result of a policy of those working in the field), but also with the fact that volitional impairment is no longer what we thought it was, and as a consequence diminished responsibility in law, or accountability generally, may need rethinking.[19] The same would go for our definition of criminal behavior that has been a social definition for the past few decades, wary as we have long been in the wake of fascist eugenics of the idea that some criminal behavior may be inborn. Today, the possibility of inborn criminal behavior is accepted as a serious possibility, i.e. some forms of deviant behavior have more genetic and neurobiological than social roots. For example, persons with strong psychopathic tendency have little emotion or empathy towards others, have trouble adapting to changing circumstances, have an instrumental way of dealing with people in which their own motives are dominant and all this is strongly connected to their brain which functions differently from that of other people. So for some people the root of criminal behavior is to be looked for in their neurobiological set-up. This would also lead to a related issue: Can the brain sciences change "man" if neurobiology offers proof that the structure of the brain can indeed contain a "criminal" part? Today, experiments with electromagnetic stimuli of the brain, the so-called transcranial magnetics stimulation or TMS, are already being done in order to influence the activity of the brain.

---

[19] Immanuel Kant in discussing guilt and punishment in the *Metaphysik der Sitten* (*Metaphysics of Morals*, 1797) already described forms of insanity in the context of what is now called diminished responsibility. See A. Mooij, "Kant on Criminal Law and Psychiatry", *International Journal of Law and Psychiatry*, 21 (1998): 335–341.

"Who Are You?"

So the question whether we have the same kind of body inevitably leads to the question whether we also have the same mind. If we return to the topic of DNA technology, these questions return with a vengeance on the subject of cloning. Elsewhere, in a discussion of Ishiguro's *Never Let Me Go*,[20] I have argued that the issue of cloning again brings to the fore the classical question of (the fiction of) legal personhood, the *persona juris* as an entity upon which the law confers rights and duties, i.e. "one who can act in law".[21]

That question, in the legal and philosophical sense as far as I am concerned, should of course also be raised in the context of the other examples of technological influences on law mentioned above. While we admit that who or what is to count in law as an independent unit with rights and responsibilities is a legal construct, based on legal rulings that define the legal entity of subjecthood and status, the quest after our identity and with it, the quest after self-knowledge does not evaporate. Are we persons, minds, or human animals? Minds with bodies, or bodies with minds? Are we material objects in which persons are embedded or are we as persons, as sociology would often have it, not inherently bodily materials, but rather collective representations embedded in bodily materials? And when it comes to organ donation or organ harvesting as in Ishiguro's novel, are we tenants or freeholders of our own bodies?[22] Is consciousness the key to personhood?[23]

---

[20] See my "Ishiguro's Legal Chimera: *Never Let Me Go* and the Legal Fiction of Personhood", *Pòlemos, Rivista semestrale di diritto, politica e cultura* 2 (2007): 119–132.

[21] For the concept of the persona as mask, see also J. T. Noonan jr., *Persons and Masks of the Law* [1976, 1st ed.] (Berkeley: University of California Press, 2002).

[22] For an early literary view on cloning, see Fay Weldon, *The Cloning of Joanna May* (Glasgow: Fontana/Collins, 1989), with a description of the cloning itself that is as hilarious as it is naïve: "While she was opened up we took away a nice ripe egg; whisked it down to the lab: shook it up and irritated it in amniotic fluid till the nucleus split, and split again, and then there were four" [45], and with the ethical-legal aspect described in doctor Holly's response to Joanna as, "I think 'my babies' is an unfortunate misnomer, Mrs May. I don't think ownership comes into it. Does a woman's egg, once fertilized, belong to her, or to the next generation?" [255–256], and, "These personal and ethical ramifications do keep emerging – one hardly thought about them at the time" [257]. The four clones, by the way, are all daughters, and Weldon has them all share their mother's love for cats (Joanna had "a little grey cat" [39], so have clone Jane [110], clone Julie [113], clone Gina [118], and clone Alice Morthampton [121]). Also of interest on the subject is P. Halewood, "On Commodification and Self-Ownership", *Yale Journal of Law & Humanities* 20 (2008): 131–162.

[23] See Evelyn Fox Keller's discussion of the subject with a reference to *Star Trek* in

More broadly and more importantly, in what way is our self-conception influenced by technology? If we are "enhanced" by technological possibilities, what does this mean for the question after our identity? Who am I? Or rather, who am "I"?[24] It would seem that modern technological developments have not helped us advance much on the subject of the duality of mind and body, the Cartesian dualism of *res cogitans* and *res extensa*.[25] Philosophically, the problem does not disappear even if law and other disciplines opt for the predominance of one over the other for practical purposes.

"Situating the Organism Between Telegraphs and Computers" in *The Point of Theory Practices of Cultural Analysis,* eds. M. Bal and I. M. Boer (Amsterdam: Amsterdam University Press, 1994), 271–283, 281: "In a particularly brilliant episode, a prototype of Sherlock Holmes' Professor Moriarty, recreated on the Star Trek's holodeck, acquires consciousness. And with the acquisition of consciousness, he demands, in fact asserts, his existence". See also Martha Nussbaum's "Reply" (to critics) in *For Love of Country, Debating the Limits of Patriotism*, M. C. Nussbaum with respondents, ed. J. Cohen (Boston: Beacon Press, 1996), 131–144, 133: "Human personhood, by which I mean the possession of practical reason and other basic moral capacities, is the source of our moral worth, and this worth is equal".

[24] See also D. DeGrazia, *Human Identity and Bioethics* (Cambridge: Cambridge University Press, 2005) for a discussion of these questions in relation to causal determinism: if our actions are not really free, can there be such a thing as self-creation? DeGrazia offers a wide range of thought on "certain technologies when they are employed for purposes of enhancement" [205], a term that he prefers to "biotechnological" and/or "genetic enhancements".

[25] Think of the eytmological evidence in for example the Old English word for body: "lic hama", the house (hama) of the soul (lic), the remnants of which can still be found in the contemporary Dutch word "lichaam". Edmund Waller (1606–1687) has a comparable simile in "Of the Last Verses of the Book", "The soul's dark cottage, battered and decayed [...]". Literature is indeed full of examples of how these questions have troubled us. From poets as different as Plato and Walt Whitman we sing the joys of the soul and of the Body Electric, even though Descartes has tried, and with quite some success if we look at the development of the natural sciences since the seventeenth century, to convince us that the *res cogitans* and the *res extensa* are different entities. Edward Young (1683–1765) in his poem "Night Thoughts" (on which he worked from 1742–1748) describes man as the "awful stranger" between God and the rest of the creatures including the natural environment; Shakespeare has Hamlet call man "the paragon of animals" (2.1.1296). Henry James in "The Madonna of the Future" (1879) says, "Cats and monkeys, monkeys and cats – all *human* life is here" (emphasis mine). For the recent thesis on the related subject of the untenableness of the dualism of mind versus (the rest of) the world, see A. Clark, *Supersizing the Mind: Embodiment, Action and Cognitive Extension* (Oxford: Oxford University Press, 2008). The use of "to have" is already problematic of course. Does man have a body? The French philosopher Jacques Lacan offers a critical view on the subject of Descartes' "error" as Cartesian dualism has now come to be called.

When it comes to gathering evidence in criminal cases, DNA testing, taking blood or urine samples, fingerprints, or checking people's breath for traces of alcohol are all laid down in legal rules and legitimized too by saying that this is a passive rather than an active form of self-incrimination that as such does not violate the right to fair trial, because it rests on the premise that the information gathered exists *independent of the free will of the individual* from whom it is harvested. On the meta level, therefore, the question remains how law *itself* is affected by modern technology.

How acute the topic is can also be seen from the practical side of things. As Evelyn Fox Keller quite rightly says,

> [t]he body of modern developmental biology, like the sequences of DNA – and like, too, the modern corporate or political body – has become just another part of an informational network, now machine, now message, always already for exchange, each for the other.[26]

If such an exchange is uninformed, and my thesis would be that it is as far as forensics is concerned, what then are the consequences? With the importation of modern technology in law, we are confronted with exponential changes, more specifically in genetic and robotic technology. With the traditional ontological constraints of religious eschatologies removed in our postmodern era, what, then, can or should be the reaction of law to the novel technological limitations on what has long been deemed the sovereignty of the human will? I started this article with the presumption of innocence as a genetic building block of criminal law in our legal tradition based on Enlightenment concepts and values. Now it is high time to ask whether the genetics of law are under pressure by the technological developments sketched above. Can, does or should law keep pace with the acceleration of biotechnical innovations, or have we by now reached the edges of the juridical as we know it? Has not law become vulnerable to the threat of instrumentalism and instrumentality in the sense of an ill-reflected implementation of technology that inevitably leads to tensions given the rule of law's Enlightenment DNA? Shouldn't we also ask in what sense the technological "decomposition" of the body forces us to rethink both the idea of the legal fiction of the subject and the humanities as instruments of self-exploration by means of a return to that old question: What is Man?

My answer would be affirmative, especially since judges (and juries in common law legal systems) seem to have an unhealthy respect for what they perceive as the influence of science in their field of expertise. This is by now

---

[26] Keller, *The Point of Theory: Practices of Cultural Analysis*, 282.

significantly called "the CSI effect", that is to say that popular TV series such as CSI, CSI Miami, CSI New York, Bones, and so on and so forth affect the (popular) perception of forensic science(s) to such an extent that the legal professional has to reckon with this perception when he has to deal with the findings of forensic technologies in daily practice.[27] Isn't it ironical then, and are we not back at square one now with a song entitled, "Who are you?" as the CSI opening tune, especially since erroneous forensic identifications have already begun to dampen the enthusiasm with respect to the value of forensic technologies?[28]

I will address the questions raised here by means of a reading of Michel Houellebecq's novel *Atomised* because it is a novel that painfully shows our inability to read well the book of nature and man as well as deals with profound scientific and technological developments of the kind that affect law. As such, it offers an explanatory literary situation to bring the problem to the fore and it challenges us to think about the growing disjunction between law and science. The Kuhnian paradigm shift in scientific thought that biotechnological advancements in a broad sense have brought about has not yet been fully understood by law, and therefore law's attempts at codification of new norms in its own relevant subfields suffer the consequences, some of which I discuss below before turning to Houellebecq.

---

[27] In his recent discussion of the position of *Law and Literature* as an interdisciplinary movement Peter Goodrich has also pointed to the contemporary influence of the screen versions of law. He does so in the context of the idea of law as text and speaks of, "the current move from text to screen" and calls series like *Boston Legal* and *CSI* "the new emblems of legality". See P. Goodrich, "Screening Law", *Law & Literature* 21.1 (2009): 1–23, 3. See also R. K. Sherwin, "A Manifesto for Visual Legal Realism", *Loyola of Los Angeles Law Review* 40 (2007): 539–791, esp. the text accompanying note 54.

[28] See also on the subject, M. J. Saks and J. J. Koehler, "The Coming Paradigm Shift in Forensic Identification Science", *Science* 309 (5 August 2005): 892–895, accessible at www.sciencemag.org, for the view that erroneous forensic identifications force us, "to question the core assumptions of numerous forensic sciences", the core assumption then being that, "the traditional forensic dividualization sciences rest on a central assumption: that two indistinguishable marks must have been produced by a single object" [892]. They advise us to move away from the traditional assumptions of an individual's uniqueness towards "a more defensible empirical and probabilistic foundation" [895].

The Genetics of the Rule of Law in Democratic Societies

If we return to the topic of human genes, and extend it, if only as a metaphor, to law, what, then, are the DNA building blocks of the rule of law in democratic societies? They, too, can be captured in four letters that stand for the minimal demands of the rule of law: L for legality, S for separation of powers, R for rights, and F for fair trial.

In its original meaning, the rule of law implies the sovereign's obedience to law, his rule not only being instrumentally *per leges*, but also *sub lege*. Immanuel Kant (1724–1804) elaborated on the concept in his legal theory in the first part of *Metaphysik der Sitten* (*Metaphysics of Morals*, 1797) and based it on a liberal concept of freedom and equality, with the role of the state obviously a restricted one. In the early nineteenth century, continental thought on the rule of law was therefore also aimed against the absolutism of the French rulers of the Ancien Régime. In Anglo-Saxon legal systems the concept of the rule of law was highly influenced by Albert Venn Dicey's 1895 *Law of the Constitution* in which he stated that the rule of law both implied the principle of legality and the legal responsibility of government officials for any action taken in their official capacity, even if there was no legal foundation for it in the form of a legally based authorization.

L for the principle of legality implies that any act of government must have a legal, statutory foundation. When a citizen's liberties are being restricted, these restrictions have to be laid down by the legislator in a law that applies to all. In criminal law, legality means *nullum crimen sine lege* and *nulla poena sine lege,* i.e. that human conduct is punishable only if there is an antecedent statutory provision making that conduct illegal. New provisions are not to be applied retroactively to earlier conduct. Here democracy comes in, because essentially this principle provides not only a bulwark against government's intrusions on people's lives but it also embodies the ideal of democracy as self-government in that the citizens regulate themselves through the democratic process of choosing the legislator in the form of the government and legislative bodies such as parliament, i.e. as their representatives. Legality restricts the scope of forensic DNA research. Contemporary critics of legality and diehard crime fighters argue the necessity to take bodily materials from *all* suspects at the moment that they are being arrested and taken into custody (rather than only from those suspected of having committed the more serious crimes). In the Netherlands, the Minister of Justice has called this a fundamental step in law that should only be taken in case of urgency, but what that means or when that situation comes about has so far not been explained. It strongly suggests an instrumental use of law, again, and as such poses a risk for law's underlying traditional set of values and/or the protection of the individual citizen.[29] The same might go for new legislation – in preparation in the Netherlands since March 2009 – that allows the police to use new forms of DNA research pertaining to searches for kinship of the kind already done in the field of migrant law for crime solving purposes. If a complete but unidentified DNA profile is found at a crime scene of an as yet unsolved crime, comparisons between that specific profile and those of an identified person whose profile resembles but is not 100% identical to that profile can be done in order to find out whether or not the unidentified profile can belong to a person next of kin to the identified person, all on the premise that kinship implies resemblances in genetic materials since a person inherits from both his father and his mother. From the point of the defense, one shudders to think what this might mean for individual privacy.

S stands for the separation of the three powers of the state, the legislature, the executive branch, and the least dangerous branch, the judiciary. What if neither the legislative branch, nor the judicial branch understand the full implications of technological developments, i.e. to such an extent that they can be adequately dealt with in law? The whole system of Montesquieu's *trias politica* as laid down in *The Spirit of the Laws* or the *checks and balances* as we now call them will become a charade, and the citizens will suffer the consequences. Abuse of authority and arbitrary rule rather than the rule of law loom large, and, worst of all, we may not even notice that it happens.

If who and what I am is not 100% certain, how does that affect the third building block of the rule of law, R for rights in the sense of civil rights and political rights, fundamental freedoms and constitutional rights as laid down in many written constitutions and a variety of texts of international organisations? Can new constructs, products of technological commodification be endowed with rights that presume the capacity to bind oneself in law and enter into obligations as well? Or, as George Annas puts it, "Can universal human rights and democracy, grounded on human dignity, survive genetic

---

[29] See also the recent EHRC Grand Chamber decision in the case of *S. and Marper v. The United Kingdom*, 4 December 2008, nos. 30562/04 and 30566/04, in which applicants complained under Articles 8 and 14 of the Convention that the authorities had continued to retain their fingerprints and cellular samples and DNA profiles after the criminal proceedings against them had ended with an acquittal or had been discontinued. The court held that there was indeed a violation of article 8.

engineering?"[30] What effect does technology have on the (human) rights discourse? A case in point as far as our ambiguity in these matters is concerned is our adamic naming of cloned animals in the media. Think of the bull Herman, years ago, or the lamb Dolly. We gave them names rather than numbers that suggested their authenticity as individuals.[31]

The fourth requirement, F for fair trial, shows the importance of the demand of consistency, between legal rules and underlying norms, between legal rules and the values and principles that form their background, and, last but not least, between law in the books and (law as applied to) the facts of an individual case. Access to an impartial tribunal may prove to be an empty shell if such consistency is sorely lacking, and once more, my point here is that the forensic technological influx we witness today increases that risk. As Joel Garreau quite rightly puts it, "The law is based on the Enlightenment principle that we hold a human nature in common. Increasingly, the question is whether this stills exists".[32] To me, (bio)technology seriously challenges the four letters of the DNA of law as we know it, and consequently, other traditional concepts such as equality, liberty, and fraternity are increasingly put under pressure.

Following Bernard Edelman, I would say that these questions of selfhood and commodification, with an implosion of the traditional subject as a result, when at the same time new categories are under construction in law, such as same-sex marriage, in vitro fertilization, euthanasia, wrongful life and wrongful birth, lead us, on the one hand, to the issue of the decomposition of the body into biological materials, then we cannot escape the philosophical question whether the legal subject still absorbs the human person. Liberties without a literally coherent body become abstrac-

---

[30] G. J. Annas, *American Bioethics, Crossing Human Rights and Health Law Boundaries*, (Oxford: Oxford University Press, 2005), 37. While Annas claims, "We will survive only so long as we uphold human rights" [17], he also has to admit that this poses a problem, for, "If human rights and human dignity depend on our human nature, can we change our "humanness" without undermining our dignity and our rights?" [27].

[31] See also Annas, *American Bioethics*, 33–34.

[32] J. Garreau, *Radical Evolution, the Promise and Peril of Enhancing our Minds, our Bodies – and What it Means to be Human* (New York and London: Doubleday, 2005), 8. Garreau also discusses cloning, and asks, "What will this mean? Will human nature itself change? Will we soon pass some point where we are so altered by our imaginations and inventions as to be unrecognizable to Shakespeare or the writers of the ancient Greek plays?" [21]. A very good question indeed for *Law and Literature*!

tions.[33] "To thine own self be true", Polonius admonishes his son Laertes (*Hamlet*, Act 1, sc. 3, l.78) but what is "mine own self"?

On the other hand, these questions lead us to the issue after the dominant perspective itself in current forensic DNA research and in law. It would seem that the focus has shifted to the hypervariable areas or *loci*, if we hold on to our metaphor, in that new technologies seem to have resulted in arbitrary or ill-reflected incorporation (again, we cannot escape the body in our language!) in law. There is literary-legal irony here, too, if we consider the importance attached in law to yet another form of *loci* or *topoi*, i.e. in the traditional rhetorical sense, and remember that as "the identification and description of argumentative premises", or "commonly used lines of argument"[34] legality, rights and so on are the *loci* of law's values in a double sense. This suggests that we will do well to reexamine the relationship or the bond, forced as it is in many ways, between law and forensic technology for the very reason that Quintilian gave when he warned us that *loci* are not infallible and can become obstacles when proven ill-matched to an argument because this is exactly what is the case with law as Enlightenment-based as far as its underlying values are concerned. If we have by now created a legal order that we cannot fully understand because we do not fully grasp the way in which forensic use of technology is or should be implemented, how is law at all to fulfill its task as the ordering order that it is supposed to be?

The divorce of law and morality – the consequence of the *Entzauberung der Welt*, the disenchantment of the old world in which the humanist ideal of universal knowledge and *Bildung* came to an end with the growing importance attached to the findings of the natural sciences in the course of the nineteenth century – led to one of the greatest dilemmas in modern legal the-

---

[33] B. Edelman, "Fabulation Juridique", *Droits* 41 (2005): 199–217, 216 and 217. See also his remark at 214, "Les fables technologiques qui irriguent le droit ont peut-être leur poésie propre mais ce n'est plus la même". Why? Because the way of doing law has changed because of them.

[34] M. H. Frost, *Introduction to Classical Legal Rhetoric: a Lost Heritage* (Aldershot: Ashgate, 2005), 27. Frost's basis is the famous passage in Aristotle's *The Art of Rhetoric*, trans. H. C. Lawson-Tancred tr. (London: Penguin, 1991): "[They] are the commonplaces in which are found the universal forms of argument used by all men, and in every science. And, again [they] are *special places* [like judicial or forensic *topoi*] where you naturally seek a particular argument, or an argument on some point in a more special branch of knowledge [...] *Topos*, then, may be regarded as a place or a region in the whole realm of science, or as a pigeon-hole in the mind of the speaker", xxiv, n. 35. The seminal text on the *topoi* in law is Th. Viehweg, *Topik und Jurisprudenz* (München: Beck, 1954).

ory.[35] Disenchantment demands from law as a human practice legitimacy, and law finds its legitimacy in its genetic structure of legality, legislation, democracy, rights, and fair trial at the level of the individual case. But the problem here is that the ideal of legitimacy can only be reached in that disenchanted world itself, for the very reason that law is man-made, it is both a social given and a normative system. On the premise that law is part of the humanities, I fully agree with Simon Critchley who sketches the problem succinctly when he points to the Kantian dilemma, i.e. that Kant "wished to insist on the authority of science and yet preserve the autonomy of morals". This is the gigantic task that still faces us: how are we to reconcile the disenchantment of the universe brought about by the Copernican and Newtonian revolution in natural science with the human experience of a world infused with moral, aesthetic, cultural and religious value? Is such a reconciliation possible, or are science and morals doomed to drift apart into a general nihilism? Such is still, I believe, our question.[36] This question is obviously important for law. We speak in legal terms such as privacy and rights and this presupposes the *persona*, while at the same time we use the human body as a commodity. In law, are we then literally and figuratively determined by technology? If we are "in it and of it", as John Dewey said about the world, how can we make it the object of our inquiry? Humans must inform science and it should not be the other way around, even if this is the easy way if we do not fully grasp the consequences of technology.

The divorce of the analytical and the narrative components of law, which is closely connected to, or even the consequence of this disenchantment, has by now met with resistance on the view that rationality and imagination both have a function in law in order to bring things back to a human proportion. Precisely in this world of technology, we would do well to emphasize the reciprocal relationship between those disciplines that address the human condition from all possible angles.[37]

As Stephen Toulmin pointed out, "the problem of the human understanding is a twofold one. *Man knows, and he is also conscious that he knows* (italics

---

[35] M. Weber, *Wissenschaft als Beruf* [1919] 10th ed. (Berlin: Duncker & Humblot, 1996). For an English translation, see P. Lassman, I. Velody, and H. Martins, eds., *Max Weber's Science as a Vocation*, (London: Unwin Hyman, 1989).

[36] S. Critchley, *The Book of Dead Philosophers* (London: Granta Books, 2008), 187 citing the Kant scholar W. H. Walsh.

[37] The idea can be traced back to Marc Tullius Cicero's speech, "Pro Archia Poeta", where he says, "Etenim omnes artes, quae ad humanitatem pertinent, habent quoddam *commune vinculum* (emphasis mine) et quasi cognatione quadam inter se continentur".

in original)" and, "this relationship between our knowledge of nature and our own self-knowledge has always been a tricky one to describe and discuss", because of the long-dominant idea that rational man faces nature as if it were a fixed and stable order and the unchanging object of knowledge.[38] To Toulmin, the field of *epistemics* is therefore an area of interdisciplinary enquiry because it is crucial to think of the foundations of our intellectual judgements and that inevitably leads to the question, "What is Man?". On this view, Toulmin points to the position of literature that is special in that it is, "halfway between fully-disciplined enquiries of physical science, on the one hand, and such non-disciplinable fields as ethics and philosophy on the other".[39] With this in mind we now turn to Houellebecq.

Elementary Particles

It isn't until the Epilogue of Michel Houellebecq's *Atomised*[40] that we understand that the year is 2079 and the story told by the narrator is a flashback. It is a, "last tribute to humanity" [A, 379], or, as the final line of the novel ironically has it, "this book is dedicated to mankind" [A, 379]. While there are still "some humans of the old species" in the nooks and crannies of, "[…] areas long dominated by religious doctrine" [A, 378], overall, the last of, "the brave and unfortunate species which created us" [A, 379] have been wiped away. We then learn that the main character Michel Djerzinski's life work in molecular biology has successfully been brought to its logical conclusion. On the premise that once the genome would be completely decoded, humanity would be in complete control of its evolution [A, 320] and sexuality would be

---

[38] S. Toulmin, *Human Understanding*, Volume I, General Introduction and Part I, (Oxford: Clarendon Press, 1972), 1–2, 7, and 23.

[39] S. Toulmin, *Human Understanding*, 396.

[40] M. Houellebecq, *Les particules élémentaires* (Paris: Flammarion, 1998), trans. Frank Wynne, *Atomised* (London: William Heinemann, 2000). All the quotations and references to the text will be taken from this edition and the page number will be indicated parenthetically preceded by the abbreviation of the title: *A*. Why the translator opted for this title rather than for something like *Elementary Particles*, I do not quite understand. Atomism is of course the classical philosophical doctrine of the formation of all things from indivisible particles characterized by gravity and motion, as the OED also explains, with "to atomize", as the verb for adhering to this doctrine. In the sense that "to atomise" means "to belittle", it goes back to Fulke Greville, Lord Brooke's, *A Treatie of Humane Learning* (1633), "Whereby their abstract forms yet atomis'd May be embodied" (cxx). Could that be the irony intended in this title?

seen for what it was according to Michel, "a useless, dangerous and re-
gressive function", because "all species dependent on sexual reproduction
are by definition mortal" [A, 357], Michel has dedicated his life improving on
existing genetic codes and creating new ones.[41] The consequence of Djerzinski's work was that:

> [...] any genetic code, however complex, could be noted in a standard, structurally
> stable form, isolated from any mutations. This meant that every cell contained
> within it the possibility of being perfectly copied. Every animal species, however
> highly evolved, could be transformed into a similar species, reproduced by cloning
> and therefore immortal [A, 370].

This of course met with fierce opposition of traditional humanists as well as
religious people of different denominations because the creation of this new
asexual species rather than humanity as "we" know it, did away with their
cherished notions of personal freedom and human dignity. In the end, however, Michel's successor and later biographer, Frédéric Hubczejak, obtained
UNESCO funding for the project he initiated after Michel's disappearance
on March 27, 2009,[42] on the principle that "The revolution will not be mental but genetic" [A, 377]. The result of his epoch-making project came in
2029 with the first of a new intelligent species.

What is all this and how has it come about? In the Prologue to the novel
we read that this development is to be looked upon as an example of the phenomenon of, "metaphysical mutations – that is to say radical, global transformations in the values to which the majority subscribe" and that these "are
rare in the history of humanity" [A, 4]. The rise of Christianity is the first
example of such a metaphysical mutation, the rise of modern science is the
second, and the novel's main topic, Michel Djerzinski's work, heralds the
third, for, as he himself explains, he is a minor mutation that necessarily precedes it [A, 214]. The eschatology of the third metaphysical mutation starts
to unfold when Michel during his physics studies learns about quantum
energy and the Copenhagen interpretation which calls into question established concepts of space, time and causality. He realizes that, "[...] once
biologists were forced to confront the reality of the atom, the very basis of

---

[41] In Ireland, where Michel spends the final part of his career and life, to give just an
   example, he sees back the descendants of the cows he had genetically (de)coded
   10 years earlier, "It was he who had created the genetic code which governed their
   cell production, or, more accurately, he had improved on it" [A, 348].
[42] Michel's disappearance was shrouded in mist: was it suicide or had he gone to live
   in Asia? Finally the answer is, "We now believe that Michel Djerzinski went into
   the sea" [A, 365], because that is the edge of the Western world, i.e. literally *finis
   terrae*.

modern biology would be blown away" [A, 19], and that it is his task to bring about this major paradigm shift.

The story of his scientific quest is also told in flashback. At the start of the novel we learn that at forty Michel leaves a successful post as a head of a department for research in molecular biology, where he had been busy decoding DNA. His farewell party is a complete failure, and when he returns home, the canary that he has bought for company has died. Michel takes three Xanax sleeping pills, and, "So ended his first night of freedom" [A, 14]. With this cynical comment the scene is set. The unfolding of the story of Michel's and his half-brother Bruno's lives itself shows the importance of the influence of man's hereditary building blocks. The boys' mother Janine Ceccaldi first married Serge Clément who as one of the first medical doctors saw the uses of bodily enhancements in the form of plastic surgery and made a career of it. In 1956 Bruno Clément was born and in 1958 he was sent to his maternal grandmother in Algeria when his mother was pregnant with Michel, fathered by filmmaker Marc Djerzinski. Michel, born, in 1958 was rescued from his mother's care by his father, when he found out that Janine was involved in a hippy scene, utterly neglecting her son. Michel was raised by his paternal grandmother in the French countryside. From an early moment on, Michel who looks upon nature as "a repulsive cesspit" [A, 39] realizes that nature's elementary particles could have been totally different from what they now are and that everything is the result of random events. Throughout the novel, this idea is subtly worked out on the level of the boys' lives. They, too, are elementary particles and small changes have made for huge differences, while at the same time their lives have remarkable parallels too. Is it because they share the same mother? Are matters indeed random and yet genetically predisposed? Both Bruno and Michel like bicycling at breakneck speed [A, 42, 37] as a form of release. At school, they are both the outsider, Bruno is the omega male[43] whose life is made miserable at boarding school, Michel keeps aloof from all other children except Annabelle. Bruno has Caroline for a girlfriend. Bruno whose only goal in life is sexual nevertheless acknowledges the need for love and tenderness. Annabelle loves Michel but he does not kiss her, deprived of maternal contact as he has been during infancy. Does this already prefigure his later scientific interest and goal? The boys attend the same school unaware of each other's existence. Theirs are indeed parallel lives and universes. They meet once at school and then in 1974 their lives go into different directions. For Michel it is the road of science, Bruno

---

[43] Hence also the school director Cohen's lament that he, "had no illusions about the depths to which the human animal could sink when not constrained by law" [A, 50].

opts for Kafka and masturbation. Michel uses what he learns about science to get a grip on life. When he studies quantum physics he comes across the useful concept of a "consistent Griffiths' history", i.e. a coherent narrative constructed from quantum information. He explains to Bruno that:

> [a]s a being you are self-aware, and this consciousness allows you to hypothesise that the story you've created from a given set of memories is a *consistent history* (emphasis in original), justified by a consistent narrative voice. As a unique individual having existed for a particular period and been subjected to an ontology of objects and properties, you can assert this with absolute certainty, and so automatically assume that it is a Griffiths' history. You make this hypothesis about real life, rather than the memories of dreams [A, 75–76].

And Bruno observes that there is no future in their relationship as brothers, because, "the past always seems, perhaps wrongly, to be predestined" [A, 78]. Note, then, the subtle irony when Houellebecq writes that the 1970s, especially after the year 1974, were characterized by events that, "further advanced the cause of moral relativism" [A, 80]. Why the repetition of the year 1974 throughout the novel? Is it because 1974 was the year in which the very idea of genetic manipulation became reality when it was found that "Agrobacterium tumefaciences" as a guest organism transmitted a small part of DNA from a plasmide to his host, and so showed that is was possible to enter a gene from the one organism into the genome of another? On the other hand, if the concept of a Griffiths' consistent history has any validity, moral relativism is of course not relativism at all, just the status quo.

At Orsay University, Michel finds a place to satisfy his intellectual curiosity in the physics research faculty dedicated to the study of elementary particles. He wonders about the anomalous situation that the universe is made up of about one hundred elements while as a matter of principle an infinite number of combinations is possible if one thinks of the planetary theory of the atom based on gravitational and electromagnetic fields. A caveat is in order at the same time:

> [w]hen we think about the present, we veer wildly between the belief in chance and the evidence in favour of determinism. When we think about the past, however, it seems obvious that everything happened in the way that it was intended [A, 215].

As to the concept of volition, Michel

> [...] realized that belief in the notions of reason and of free will, which are *the natural foundations of democracy* (emphasis mine), probably resulted from a confusion between the concepts of freedom and unpredictability. [A, 270].

Such questions trouble and motivate Michel's purely intellectual and very ordered existence, and they first culminate in a paper entitled, "Towards a Science of Perfect Reproduction" [A, 197].

Bruno, in the meantime, despite "DNA similarities" [A, 145], leads a hectic life full of chaos. During his college years, he gets involved with Annick whom he had already met in 1974. She commits suicide by jumping from the seventh floor, and, "So ended Bruno's first love" [A, 183]. Bruno's marriage to Anne in 1981 is not happy either, and neither is his role as a father to his son Victor. Like Michel, he cannot help himself,

> just as determining the apparatus for an experiment and choosing a method of observation made it possible to assessing a specific behaviour to an atomic system – now particle, now wave – so Bruno could be seen as an individual or as passively caught up in the sweep of history [A, 212].

Is it fate or genetic predisposition that he meets Christiane, a professor of natural science who is as critical as Michel when it comes to the uses of sexuality? Their happiness does not last long. Christiane is diagnosed with a severe illness, and when it is certain that she will be confined to a wheelchair for the rest of her life, she commits suicide. Bruno realizes that he has failed her, as Michel did Annabelle, and the reader learns that it is for the same reason, "He had no more been capable of love than his parents before him" [A, 299]. So much for the "rational certainty" [A, 322] that Michel's boss, Desplechin, has coined as the most important factor in the history of the West.

Furthering his life's project, Michel writes his first extensive work, in 2002, entitled *The Topology of Meiosis*, which

> [...] established, for the first time, on the basis of irrefutable thermodynamic arguments, that chromosomal separation at the moment of meiosis can create haploid gametes, in themselves a source of structural instability. In other words, all species dependent on sexual reproduction are by definition mortal [A, 357].

This is the death bell for (the fiction)[44] of individual existence. A positivist reading of recent developments in a manner August Comte would have approved of, or so the narrator has it, will save the day, because, "Only an ontology of states was capable of restoring the practical possibility of human relationships", since "in the ontology of states, the particles are indiscernible, and can be limited to an observable number. The only entities which can be identified and named in such an ontology are wave functions, and, using them, state vectors – from which arose the analogous concept of redefining

---

[44] See also the irony of the Epilogue [A, 369], "Though we know much about the lives, the physical appearances and the personalities of the characters in this book, it must nonetheless be considered a fiction. A plausible recreation based on recollections, rather than a definite, attestable truth". As a fiction, the novel is nevertheless a fine example of the Aristotelian "what might happen".

fraternity, sympathy and love". To us as readers who are born rather than constructed, this is a bleak picture as far as human perceptions and experiences of reality are concerned because, "all of this happens without any metaphysical intervention, without any ontology. We don't need ideas of God or Nature or Reality" [A, 360]. Michel's "gift of physical immortality" [A, 361] gives mankind more insight in the working of time,[45] and, in a way reminiscent of Immanuel Kant's theory of the *schemata* or categories of human thought used to get a grip on reality around us, he claims that natural forms are human forms, after seeing the illuminations in *The Book of Kells*, "in the midst of space, human space, we make our measurements and with these measurements we create space, the space between our instruments" [A, 362].

As irony would have it, the parallelism in the two brothers' lives comes to its natural conclusion when Annabelle has to undergo an abortion because she is diagnosed with uterine cancer. After having undergone an operation during which all her reproductive organs were removed, she attempts suicide, sinks into a coma and dies. Michel who has been unable to enter into a love relation with Annabelle, nevertheless emphasizes the possibility of love in his own work, and that has a cynical ring to it when we as readers understand that his aim to end sexual reproduction finds its counterpart in her suicide because the possibility of sexual reproduction has gone:

> [...] Djerzinski's great leap was not his rejection of the idea of personal freedom (a concept which had already been much devalued in his time (i.e. ours, my addition), and which everyone agreed, at least tacitly, could not form the basis for any kind of human progress (such as our DNA technology, my addition) but in the fact that he was able, through somewhat risky interpretations of the postulates of quantumechanics, to restore the possibility of love [...] though he had not known love himself, through Annabelle, Djerzinski had succeeded in forming an image of it [...] "His work, he knew, was done" [A, 363].

At the end of 2009, Michel's theory is considered proven, and all scientists agree that

> [t]he practical consequences of this were dizzying: any genetic code, however complex, could be noted in a standard, structurally stable form, isolated from any mutations. This meant that every cell contained within it the possibility of being perfectly copied. Every animal species, however highly evolved, could be transformed into a similar species, reproduced by cloning and therefore immortal [A, 370].

---

[45] For a cautionary tale on the human understanding of time, we should recall Nathaniel Hawthorne's short story "Dr Heidegger's Experiment" from the 1837 collection *Twice-Told Tales*, in which the water from the Fountain of Youth helps four aged friends to regain their prime of youth, which in the end proves to be illusory.

There is also criticism, of course, especially when Hubczejak goes further on the road paved by Djerzinski. One objection is that sexual difference is central to human identity and that suppression of such difference by means of cloning would end the possibility of human uniqueness at this level. The even more fundamental criticism flowing from this is that the existence of a species carrying the same genetic code implies that human individuality will disappear for good. The perversity as far as solving crimes by means of genetic technology is concerned is of course that the door of that possibility is slammed shut once we no longer have our unique genetic codes. To Hubczejak, the disappearance of the notion of human personality is not at all problematic. On the contrary, despite a shared genetic code, people can

> [...] develop[ed] different personalities while maintaining a mysterious fraternity – which, as Hjubczejak pointed out, was precisely the element necessary if humanity were to be reconciled [A, 375].

As readers of Bruno and Michel's lives, we know how human fraternity works out, I would say.

The narrator points to one major mistake that Hubczejak made, and that is his positivist,[46] literal reading of Djerzinski, which leads him

> constantly to underestimate the extent of the metaphysical change which would necessarily accompany such a biological evolution – an evolution which had, in effect, no analogue in the history of humanity. This gross misinterpretation of the philosophical subtleties of the project, and even his inability to recognize philosophical subtleties in general, in no way hampered or even delayed its implementation [A, 376].

For those of us readers involved in interdisciplinary studies in the humanities, this positivist attitude with its subsequent instrumentalism strikes a final blow in stating that the concomitant effect is that

> [t]he global ridicule inspired by the works of Foucault, Lacan, Derrida and Deleuze, after decades of reverence, far from leaving the field clear for new ideas, simply heaped contempt on all those who were active in "human sciences" [A, 376].

So ends my first reading of *Atomised*. The technological development described in the novel has an uncanny ring to it given the state of the art in contemporary genetic and robotic technology. It is, I think, a serious warning to

---

46 The theme of positivism is also exemplified in the way it is referred to when defining Hjubczejak's first article as "a long meditation on a quotation from Parmenides [...] 'that which is there to be spoken and thought of must be'" [A, 372]. The quotation also brings to mind the famous ending of Ludwig Wittgenstein's *Tractatus Logico-Philosophicus*, "Whereof one cannot speak, thereof one must be silent".

"us" in the humanities not to forget to make (more) explicit the human element, and to this end I now turn to Heidegger's view on technology.

"Let's Get Metaphysical"

What does a combined reading of Houellebecq and my findings so far suggest? To me, Michel's struggle with human relations is emblematic of the problem underlying his and Hubczejak's scientific quest. Not so many years ago, in 1998, the Council of Europe and, before that in 1997, the European Parliament declared that cloning is an instrumentalisation of human beings that offends human dignity. One of the major problems of forensic use of (bio)technology was thus succinctly pinpointed as well: that of using people and their various body parts as instruments to arrive at a supposedly greater goal.[47] This is the result of what I consider a conflation of science and technology. Science is dedicated to theoretical knowledge, knowing why, and technology is dedicated to its application, knowing how. For the likes of Hubczejak, this distinction has evaporated: if it can be done, it will be done. Houellebecq is therefore right in having his narrator speak of Hubczejak's lack of attention to possible philosophical issues involved. So we will do well to fill the void and turn to philosophy, to metaphysics. Or, as Heidegger put it, "in Greek, 'away over something', 'over beyond', is *meta*. Philosophical questioning about beings as such is *meta ta phusika*; it questions on beyond beings, it is metaphysics".[48] So back to the question, "what is technology?"

---

[47] See the "Explanatory Report to the Additional Protocol to the Convention on Human Rights and Biomedicine on the Prohibition of Cloning Human Beings" (1998) sub 3, which explicitly denotes the cloning of humans as a threat to human identity that endangers human dignity by instrumentalisation. See also D.Gurnham, "The Mysteries of Human Dignity and the Brave New World of Human Cloning", *Social & Legal Studies* 14.2 (2005): 197–212, and, generally, the oeuvre of the legal philosopher Ronald Dworkin who has consistently offered principled, normative arguments for human dignity in situations in which painful choices have to be made, e.g. in cases of abortion or euthanasia. In *Sovereign Virtue, the Theory and Practice of Equality* (Cambridge, Mass.: Harvard University Press, 2000), Dworkin addresses the issue of cloning. Also of interest on the related subject of the posthuman is F. Fukuyama, *Our Posthuman Future, Consequences of the Biotechnology Revolution* (New York: Farrar, Strauss & Giroux, 2002).

[48] M. Heidegger, *Introduction to Metaphysics*, trans. G. Fried and R. Polt (New Haven and London: Yale University Press, 2000), 18. All the quotations and references to this text will be taken from this edition.

From Heidegger, we have inherited, firstly, the notion that at the very moment that we use language to name "things" in the world, selection and restriction take place. We cannot but see things from a specific angle, from the perspective, the position that we take. This means that language, speech is a continuous process of deciding what can and will be said, what will literally not be spoken of, what will remain undiscussed. In short, language usage is in itself a selective interpretation with in the background always lingering the roads not taken. Heidegger compares this with a searchlight that sheds light on specific phenomena in a specific way, leaving others in the dark. For law and legal interpretation, this is an important notion in that it draws the attention to a form of hermeneutics characterized by skepticism.[49]

This is, as Christine Desan Husson convincingly argues,[50] the phase of autonomy before the phase of dialogue between people, or intersubjectivity that is the key to human understanding that Jürgen Habermas gives a prominent place. To Habermas, meaning is in dialogue; to Heidegger, the choice for a specific use of language precedes dialogue and is always already a decision concerning meaning. The social dimension of dialogue by means of which to arrive at intersubjective definitions of the world around us that to

---

[49] Or, as Heidegger elaborates upon in his *Introduction to Metaphysics*, our articulation of reality as we perceive it by means of language is the result of a process of selective interpretation on the basis of the perspective we take, and this opens up possibilities at the very same moment that it delimits when that which is not spoken of remains obscure in the background. See also the related view of James Boyd White on the subject. To White, any form of speech is a form of translation that has deficiencies and exuberances. He defines translation as the literary art "[...] of confronting unbridgeable discontinuities between texts, between languages, and between people", *Justice as Translation, an Essay in Cultural and Legal Criticism* (Chicago: University of Chicago Press, 1990), 257. These discontinuities are, firstly, the given that a reduction of meaning takes place whenever a person chooses the meaning he or she will use from the range of possibilities offered. Secondly, there is the idea of meaning as culture specific, a point forcefully brought home by White's example, "The German 'Wald' is different from the English 'forest', or the American 'woods', not only linguistically but physically: the trees are different", *Justice as Translation*, 235. Translation, then, just like "the art of all speech, all expression [...] lies in finding ways to recognize its omissions, its discontinuities, its false claims and pretensions, ways to acknowledge other modes of speaking that qualify or undercut it. The art of expression is the art of talking two ways at once, the art of many-voicedness", *Justice as Translation*, 26–27.

[50] For legal theory, more specifically for deconstruction and critical legal studies, this point is elaborated upon in Chr. A. Desan Husson, "Expanding the Legal Vocabulary: the Challenge Posed by the Deconstruction and Defense of Law", *The Yale Law Journal* 95 (1986): 969–991, especially 977. See also E. Holzleitner, "Ulysses ist nicht leicht zu lesen", *Juridikum, Zeitschrift im Rechtsstaat* 1 (2004): 33–40.

Habermas is the third dimension beyond those of the object and the subject of language use, does not, however, save us from the risk of conceptualization and reification. In my view, Heidegger's contribution to the issue is that he emphasizes that the process of reciprocity – man forms and is formed by language and reality – does not preclude us from falling into the trap of thinking that language is "just" the objective tool "we" use to describe "the world". In other words, it is man's *hubris* to think that this mediation is neutral, whether in dialogue or in autonomous speech.[51]

Here, I think is the root of the problem we also face in forensic applications of (bio)technology. While we are *homo faber* ever since we first created tools in order to fend off wild animals, we fail to understand fully and hence tend to underestimate the consequences of our uninformed application of technology in forensic settings, and therefore all too often do not realize what is deliberately left out, or just forgotten to incorporate, i.e. the residue and loss. In the sense that biotechnological possibilities are augmented beyond the traditional scope of human progress, and that at high speed, the constant risk that we run of being trapped in our conceptual frameworks and being ill-prepared for changes in perspective, either for good or for bad, is also a question of legal epistemology that we will do well to provide with another perspective, that of the literary imagination as an antidote. My claim, therefore, would be that the philosophical and/or metaphysical background that is now all too often sorely lacking not only merits but also desperately demands our discursive attention given the legal and ethical consequences.

Secondly and more specifically, we can learn important lessons from Heidegger's phenomenological distinction between *technè* and technology, both in the original and contemporary sense of the terms. There is good reason to turn to Heidegger's thought on technology because on the basis of this first distinction he subsequently distinguishes between two views on technology that he claims should be taken into consideration together by us: first, technology as a means to an end, and, second, technology as a human activity. It is exactly this distinction, or so I would claim, that is mistakenly pushed into

---

[51] For a comparable view with respect to the act of reading, James Boyd White offers the idea of educative friendship on the basis of the claim that there is a reciprocal relation between language and its users see *When Words Lose Their Meaning* (Chicago: University of Chicago Press, 1984). To White, our speaking and writing is always in its actual performance a claim of meaning against the odds in that the human imagination (or, in Heideggerian vein, our perspective) is translated into reality by means of the power of language; see J. B. White, "What Can a Lawyer Learn from Literature?", *Harvard Law Review* 102 (1989): 2014–2047, 2021–2022.

the background, focused as we have been on implementing technological innovations in a forensic setting by means of legal rules, with rights and guarantees prominently present, without, however, reflecting on the *what* of technology itself.

In an essay entitled, "The Question Concerning Technology" Heidegger returns to the Greek roots of the term technology with *technikon* stemming from *technè*. At the start of his inquiry, Heidegger offers a methodological *caveat* that we should take to heart for our research in the humanities:

> [q]uestioning builds a way. We would be advised, therefore, above all to pay heed to the way, and not to fix our attention on isolated sentences and topics. The way is a way of thinking.[52]

This is especially pertinent given the predilection in the legal discussion about technology not to distinguish between what it is and what it does, and to refrain from combining such questions with the discussion on the most appropriate legal methodology to be adhered to. Or, in Heidegger's thought, between the essence of something as the *what* of something that is, and the way in which as the translator explains in a note, it "pursues its course, the way in which it remains through time what it is", on the one hand, and its instrumentality on the other [QCT, 3]. Here, modern technology is a case in point because it has long been seen in terms of human progress only, while today we suffer the consequences of that view in the form of ethical issues that thrust themselves upon us now that we have become more keenly sensitive to technology's darker side effects. Or, as Heidegger claims,

---

[52] M. Heidegger, "The Question Concerning Technology" in M. Heidegger, *The Question Concerning Technology and Other Essays* [1954], trans. and introd. W. Lovitt (New York: Harper & Row, 1977), 3–35, 3. All the quotations and references to the essay will be taken from the mentioned edition and the page number will be indicated parenthetically preceded by the abbreviation of the title: *QCT*. For an interesting elucidation of Heidegger's view on technology with respect to contemporary visual arts and media projects, see T. Murray, "Artistic Simulacra in the Age of Recombinant Bodies", *Literature and Medicine* 26.1 (2007): 159–179, 159 and 162, with a reference to the Australian performance artist Stelarc who declared the demise of the human body because it is obsolete in form and function, "[i]t is time to question whether a bipedal, breathing body with binocular vision and a 1400cc brain is an adequate biological form. It cannot cope with the quantity, complexity and the quality of information it has accumulated; it is intimidated by the precision, speed and power of technology and it is biologically ill-equipped to cope with its new extraterrestrial environment". Stelarc is said to have complained that the human lack of modular design of the body is a disadvantage when it comes to replace malfunctioning organs.

[t]echnology is not equivalent to the essence of technology [...] the essence of technology is by no means anything technological. Thus we shall never experience our relationship to the essence of technology as long as we merely conceive and push forward the technological, put up with it, or evade it. [...] But we are delivered over to it in the worst possible way when we regard it as something neutral; for this conception of it, to which today we particularly like to do homage, makes us utterly blind to the essence of technology [QCT, 4].

Heidegger then proceeds from the combined definition of technology as "a means to an end" and "a human activity" [QCT, 4]. On the view that modern technology is a means to an end, "the instrumental conception of technology conditions every attempt to bring man into the right relation to technology" [QCT, 5]. For our topic of forensic (bio)technology, the question will therefore not only be "what end?", but also "in what way does an instrumental conception affect us?", given the side effect that Heidegger cautions us about when he says: "[t]he will to mastery becomes all the more urgent the more technology threatens to slip from human control" [QCT, 5]. On the view that, "[...] the correct instrumental definition of technology still does not show us technology's essence", our next question must be, "what is the instrumental itself? Within what do such things as means and end belong?" [QCT, 6], given the reciprocal relation between instrumentality and causality.

In order to show what causality means for the instrumentality of technology, Heidegger turns to the classical, philosophical view on causality that makes a distinction between are four types of causes: 1. the *causa materialis*, the material out of which something is made, 2. the *causa formalis*, the shape into which that material enters, 3. the *causa finalis*, the purpose for which that something is made, and, finally, 4. the *causa efficiens*, the cause that brings about the finished "something". So 1., for example, gold or platinum can be made into 2. a ring rather than, say, an Olympic medal or a diamond-studded platinum skull, which means that its purpose, 3. will differ depending on its form, and obviously that affects who will be 4. its maker (a jeweller or an artist such as Damien Hirst). For too long now the distinctions between these four causes have been collapsed, and we mistakenly tend to see the *causa finalis* as the standard for all causality. But originally, "the four causes are the ways, all belonging at once to each other, of being responsible for something else" [QCT, 7]. Without the gold, no ring, and the ring's image or *eidos* rather than the gold being made into something else has consequences for its purpose and hence use, and it is the maker's work that brings about the finished product. Here the notion of human activity comes in. The maker thanks to his work on the material and the choices he makes with respect to the "what" and the "how" of the coming into appearance of the ring of our example does all this by means of his *logos*: he considers carefully what he is going to bring about, what it is that will be revealed. Translated to the issue of (bio)technology, we might thus say that the problem or danger of the dominance of instrumentality is that we do not use our *logos* in the sense of "the ability to consider carefully" well enough.[53] If we keep thinking in terms of causality in terms of effecting, Heidegger claims, we shall fail to see what instrumentality merely based on such causality really is. The clarification of this problem is the main theme of his essay. To this end, he returns to the idea of the reciprocal relation between the four causes.

Each in their own way, "[t]he four ways of being responsible bring something into appearance", i.e. they occasion something, they call forth in the sense that they enable that which is not yet present to become present. The term for this event is *poiēsis*, a bringing-forth ("Her-vor-bringen" in Heidegger's German terminology) as defined in Plato's dialogue "Symposium".[54] For the Greeks, both handcraft as manufacture and artistic and poetical bringing into appearance is *poiēsis*, but *physis* is *poiēsis* too, "the arising of something from out of itself is a *poiēsis*, such as the bursting of a blossom into bloom" [QCT, 10], as Heidegger says to clarify the idea.[55] "Bringing-forth"

---

53 See also with respect to the root of *logos*, *legein*, or gathering which is an opening-up, a revealing of Being as coming-into-unconcealment, Heidegger, *Introduction to Metaphysics*, 180–181.

54 Heidegger refers to section 205b, and if one compares the various English translations it is obvious that much depends on the choice one makes (or should we say the perspective one takes?). Since Heidegger does not give the quotation himself, for purposes of this article I use Plato, *The Dialogues of Plato*, vol. II, "The Symposium", trans. with comment by R. E. Allen (New Haven and London: Yale University Press, 1991), which renders that section (in Diotima's definition of Eros (205a–206a)) thus, 149, "You know that making [*poiesis*] is something manifold: for surely the cause of passing from not being into being for anything whatever is all a making, so that the productions of all the arts are makings, and the practitioners of them all are makers [*poietai*]". Another edition, Plato, *Symposium*, trans. W. R. M. Lamb (Cambridge, Mass.: Harvard University Press, 1975), is more dense and, in my view, less helpful for a reading of Heidegger because it caters to our contemporary linguistic prejudice in translating *poiesis* as poetry, "You know that poetry is more than a single thing. For of anything whatever that passes from not being into being the whole cause is composing or poetry; so that the productions of all arts are kinds of poetry, and the craftsmen are all poets" [187].

55 See also Heidegger, *Introduction to Metaphysics*, for the notion that in the formative period of Western philosophy among the Greeks, "beings were called *phusis*" (note that depending on the translator the term is phusis or physis) [IM, 14] and that, "It was not in the natural processes that the Greeks first experienced what *phusis* is, but the other way around: on the basis of a fundamental experience of Being in

means that something that was hitherto concealed comes into unconcealment, and for this process Heidegger uses the term "revealing" which is the
Greek *alētheia*, the Roman *veritas*, and our "truth" in the sense of the correctness of an idea. This has everything to do with technology because the four
causes, or modes of occasioning, are gathered together within the very concept of "bringing-forth". So, "If we inquire, step by step, into what technology, represented as means, actually is, then we shall arrive at revealing",
and in this way, "[...] the essence of technology will open itself up to us"
[QCT, 12]. So the order is reversed: we must ask what technology in the instrumental sense is in order to arrive at the core of technology as *poiēsis* in the
sense of a human act that is ethical.

In a phenomenological manner, asking after the Greek stem of the word,
technikon, i.e. that which belongs to *technē*, Heidegger returns to the question
of what "technology" means. He points out that, "[...] *technē* is the name
not only for the activities and skills of the craftsman, but also for the arts of
the mind and the fine arts. *Technē* belongs to bringing-forth, to *poiēsis* [...]"
[QCT, 13]. The second point is that, "from earliest times until Plato the word
technē is linked with the word *epistēmē*. Both words are names for knowing in
the widest sense". And knowing is opening up, revealing.[56]

It is important to note here that Heidegger says "until Plato", because it is
Plato whom we may thank for the schism of what was originally thought of
as a unity, i.e. between knowing that and knowing how. Aristotle speaks in
terms of the distinction between *epistèmè* and *phronèsis*, between theoretical
knowledge (aimed at knowing) and practical knowledge or wisdom, i.e. the

---

poetry and thought, what they had to call *phusis* disclosed itself to them. Only on
the basis of this disclosure could they then take a look at nature in the narrower
sense. Thus *phusis* originally means both heaven and earth, both the stone and the
plant, both the animal and the human, and human history as the work of humans
and gods; and finally and first of all, it means the gods who themselves stand
under destiny. *Phusis* means the emerging sway, and the enduring over which it
thoroughly holds sway" [IM, 15].

[56] See also Heidegger, *Introduction to Metaphysics*, where *technē* is described as "[...] a
kind of knowledge, the knowing disposal over the free planning and arranging and
controlling of arrangements" [IM, 18] and, "we translate *technē* as 'knowing'. But
this requires explanation. Knowing here does not mean the result of mere observations about something present at hand that was formerly unfamiliar. [...]
Knowing, in the genuine sense of *technē*, means initially and constantly looking out
beyond what, in each case, is directly present at hand. [...] Knowing is the ability
to set Being into work as something that in each case *is* in such and such a way"
[IM, 169–170].

application of good judgement to human conduct.[57] The Aristotelian distinction between theoretical and practical knowledge has proved immensely significant for the development of thinking about knowledge and science, and about law. On the one side of the spectrum, the natural sciences, epitomizing modern science, emphasize theoretical knowledge with universality, objectivity, certainty and rationality as their guiding principles. On the other side, the sciences that emphasize practical wisdom rather than theoretical knowledge are focused on attention to particularity, intersubjectivity, sustainability, and argumentation. Knowledge in practical matters such as law is never general knowledge, but always dependent on the context. It is therefore always provisional. Here is the root of the problems involved in interdisciplinary ventures between law and an exact science.

If we accept the premise that *technē* is revealing, we can come to the heart of the matter, for

> [...] what is decisive in *technē* does not lie at all in making and manipulating nor in the using of means, but rather in the aforementioned revealing. It is as revealing and not as manufacturing, that *technē* is a bringing-forth [QCT, 13].

For my point here, that implies that if we make the mistake of thinking of technology purely as production and application, we disregard the way in which the four aspects of causality work together to "bring forth" the correctness of an idea, i.e. the truth of technology. Now, one might respond that this is all very well for the Greeks in their time, but it does not fit modern technology, based as it is on the exact, the natural sciences. And while this may be true, to Heidegger, "[t]he decisive question still remains: of what essence is modern technology that it happens to think of putting exact science to use?" [QCT, 14].

An exclusive focus on the *causa finalis* rather than on the combined effort of all aspects of causality and, more specifically, given the importance of our choice and awareness of perspective, on the human activity in the sense of "making", *poiēsis*, involved here, leads to the mistake of looking upon technology as production only, as an object at hand, rather than trying to bring forth the correctness of the idea, the truth.[58] This also affects man in that he

---

57 *The Nicomachean Ethics*, trans. D. Ross (Oxford: Oxford University Press, 1991), II, 2. See also my, "Law in Context, Law, Equity, and the Realm of Human Affairs", in *Practising Equity, Addressing Law. Equity in Law and Literature,* ed. D. Carpi (Heidelberg: Winter, 2008), 33–70.

58 See also Heidegger, "The Question Concerning Technology" that "everywhere everything is ordered to stand by, to be immediately at hand, indeed to stand there just so that it may be on call for a further ordering" [17].

too can and will be objectified, while he thinks that he is the maker who is in charge.[59] The latter illusion, in turn, gives rise to one final delusion:

> [i]t seems as though man everywhere and always encounters only himself. [...] *In truth, however, precisely nowhere does man today any longer encounter himself, i.e., his essence* (emphasis in the original). [...] [he] fails in every way to hear in what respect he ek-sists [QCT, 27].

In short, the result of modern technology as we have come to use it, is that we ourselves have been reduced to "being at hand" only, objects that stand reserve for purposes of further ordering. If we apply this line of argument to the case forensic DNA technology that we discussed above obviously Heidegger has a very good point here. The reduction of the human subject to an object up for grabs, that is ordered by law and that can literally be plundered for bodily materials without giving enough attention to the question of the "what" of technology, thereby runs the severe risk of compromising traditionally respected legal values of contemporary polities.[60] What technology *does* in the sense of the result of its application does not coincide 100% with what it *is*, hence the need to remain critical with regards to the question whether technology should be put to use for the simple reason that it is available.

So we must not accept what is held before our eyes as the inevitable effect of science and technology but rather challenge forensic technology, both in the Heideggerian and the everyday meaning of the word. Since obviously

---

[59] Heidegger, "The Question Concerning Technology": "[...] he [man] comes to the point where he himself will have to be taken as standing-reserve. Meanwhile man, precisely as the one so threatened, exalts himself to the posture of the lord of the earth. In this way the impression comes to prevail that everything man encounters exists only insofar as it is his construct". [27].

[60] In the context of genetic modification, see D. A. Kirby and L. A. Gaither, "Genetic Coming of Age: Genomics, Enhancement, and Identity in Film", *New Literary History* 36 (2005): 263–282, for the view that a disruption of the individual's freedom as far as life choices are concerned is caused by genetic modification. They refer to J. Habermas, *The Future of Human Nature* (Cambridge: Polity Press, 2003), 72, for the consequences at the juridico-political level of human equality. The objectification of the self as a "designed product" ["Genetic Coming of Age: Genomics, Enhancement, and Identity in Film", 273] is prominent in the 1995 film *Judge Dredd* (directed by Danny Cannon, featuring Sylvester Stallone as Judge Dredd), an example that struck me as my worst possible nightmare, because this judge not only has to struggle with the understanding that he is genetically modified but that the government created him as a law enforcement tool ["Genetic Coming of Age: Genomics, Enhancement, and Identity in Film", 275].

man is the one who accomplishes the act of revealing, the question will then be: "to what extent is man capable of such a revealing?" [QCT, 18]. In order to be able to do things right, we should return to the arts, Heidegger advises, for it is there that we can find inspiration for our current task of accomplishing the revealing of the what of technology in the form of a methodology for reflection:

> [t]here was a time when it was not technology alone that bore the name *technē*. Once that revealing that brings forth truth into the splendor of radiant appearing also was called *technē*. Once there was a time when the bringing-forth of the true into the beautiful was called *technē*. And the *poiēsis* of the fine arts also was called *technē*. [...] What, then, was art – perhaps only for that brief but magnificent time? Why did art bear the modest name *technē*? Because it was a revealing that brought forth and hither, and therefore belonged within *poiēsis* [QCT, 34].

In short, we must turn to the humanities to learn about poetic revealing in its original sense in order to reveal by reflection the core of contemporary technologies in their applications so that we may be able to ask the right questions with respect to all aspects of their causality. Or, in Heidegger's prophetic words,

> [b]ecause the essence of technology is nothing technological, essential reflection upon technology and decisive confrontation with it must happen in a realm that is, on the one hand, akin to the essence of technology and, on the other, fundamentally different from it. Such a realm is art. But certainly only if reflection on art, for its part, does not shut its eyes to the constellation of truth after which we are *questioning*. [...] For questioning is the piety of thought [QCT, 35].

Now that the rise of modern science and its concomitant disputes on what may rightfully be called "science", on whether in the humanities methods should focus on explaining human behaviour in terms of the natural science model or in hermeneutic manner,[61] have divorced *phusis* from *technē* and *poiēsis*, and have made them into direct opposites, *thesis*, positing, ordinance, a.k.a. *nomos*, law in the sense of what is generally accepted as an ordered whole, has become predominant. Applied to our subject, the result of this divorce is a proclivity to accept at face value what is held before us as forensic technology, and to reduce our ethical thought on the subject to a discussion of the current *mores* rather than foreground the truly moral side of humanity.

---

[61] See also Gaakeer, "Law in Context, Law, Equity, and the Realm of Human Affairs", 62–65.

So we must keep questioning (forensic) technology and realize that its language too is a perspective, albeit a forceful one, with residue and loss. To do so, we need the humanities. In his interview for the German periodical *Der Spiegel*, "Only a God Can Save Us", Heidegger not only returned to the subject of technology, this time in the sense of the supremacy of cybernetics, and speaks of it as an example of disenchantment in the Weberian sense, but he also brings back to our attention and vigourously promotes the idea of *Bildung*, a broad, humanist education, as a goal to stem the tide of instrumentalism.[62]

The Just: the Case for Law and Literature

To me, this suggests and includes a return to a discussion of morality in the original sense before it came to be polluted by a rule-bound model of reasoning, a return to a reflection on man as a moral being. Moral, that is, in the sense of the human ethos of searching for and analyzing values.[63] This is even more pertinent given the Enlightenment values underlying law as discussed above with their penchant for inherent(ly) human rights and human dignity.

As Charles Taylor, already in 1991, urged us to do, it is time to reconsider the primacy of instrumental reason in modernity, "the kind of rationality we draw on when we calculate the most economical application of means to a given end", because of the grave consequences of instrumental reason for our political lives, one example of these being that

> [...] the institutions and structures of industrial-technological society severely restrict our choices, [that] they force societies as well as individuals to give a weight to instrumental reason that in serious moral deliberation we would never do, and which may even be highly destructive.[64]

---

[62] M. Heidegger, "Only a God Can Save Us" (The *Der Spiegel* interview), *Philosophy Today*, XX.4/4 (1976): 267–285.

[63] For a comparable view on the moral aspect as far as literature is concerned, see J. Gardner, *On Moral Fiction* (New York: Basic Books, 1978).

[64] Instrumental reason is one of the (three) malaises of our times (the other two being individualism and loss of freedom) that Charles Taylor identifies in *The Ethics of Authenticity* (Cambridge, Mass.: Harvard University Press, 1991), here cited 5 and 8.

So the question we have to ask is whether we dominate technology or technology dominates us. The risk that Charles Taylor perceives is that instrumental reason becomes enframed in a project of domination that seriously
affects our freedom in the sense of our capacity to remake the conditions of
our existence. That, I would argue, is precisely what the matter is in forensic
technology. Or, as Taylor elaborates upon it, together with instrumental reason we witness the growth of, "a disengaged model of the human subject",
which

> [...] offers an ideal picture of a human thinking that has disengaged from its messy
> embedding in our bodily constitution, our dialogical situation, our emotions, and
> our traditional life forms in order to be pure, self-verifying rationality.[65]

For this development we have to thank Descartes, of course, as we already
saw above, who gave us the idea that "we are pure mind, distinct from body,
and our normal way of seeing ourselves is a regrettable confusion".[66] To Taylor, technology in the service of an ethic of benevolence towards real flesh
and blood people should be our subject, and that means that we should look
for alternative framings of technology.[67]

Taken together with Heidegger's view on revealing, Taylor's suggestions
open the case for literature as poetic revealing as I would claim Houellebecq's novel already shows if we read well, i.e. with attention and concern.
Here, the epistemological position that Martha Nussbaum takes is of great
importance. She says:

> [my] aim is to establish that certain literary texts [...] are indispensable to a philo
> sophical inquiry in the ethical sphere; not by any means sufficient, but sources of
> insight without which the enquiry cannot be complete

---

[65] See Taylor, *The Ethics of Authenticity*, 101–102.

[66] See Taylor, *The Ethics of Authenticity*, 102. In *Sources of the Self, the Making of the Modern
Identity* (Cambridge: Cambridge University Press, 1989), esp. ch 8, "Descartes' disengaged reason", Taylor elaborates upon the influence of Descartes on the concept of rationality in modernity. Andrew Marvell's "Dialogue between the Soul
and the Body" is a poem on the theological implications.

[67] See Taylor, *The Ethics of Authenticity*, 106–107, "Instead of seeing it purely in the
context of an enterprise of ever-increasing control, of an ever-receding frontier of
resistant nature, perhaps animated by a sense of power and freedom, we have to
come to understand it as well in the moral frame of practical benevolence [...]",
with this benevolence also placed, "in turn in the framework of a proper understanding of human agency, not in relation to the disembodied ghost of disengaged
reason, inhabiting an objectified machine. We have to relate technology as well to
this very ideal of disengaged reason, but now as an ideal, rather than as a distorted
picture of the human essence".

and that we need literature, "that talks of human lives and choices as if they matter to us all".[68] This claim that literature's capacity of making valuable ethical and social contributions when incorporated into the professional lives of lawyers – for that is Nussbaum's context here – makes it an indispensable medium not only to learn about law but also to bring in the moral perspective as defined above in our reflection on the forensic role of science, is especially acute. Not in the least because science divorced from morality will lead to the production of monsters, figuratively and literally, with Mary Shelley's *Frankenstein* as an early case in point.[69]

Furthermore, in the interdisciplinary field of *Law and Literature* we have an eminent way of letting metaphysics (back) in, after the sociologist and realist jurisprudence of prominent jurists such as Oliver Wendell Holmes jr. and Karl Llewellyn, not to mention Richard Posner's contemporary form of pragmatism which more often than not epitomizes instrumental reason. Law's dealing with technology in a positivist manner, trying to draw boundaries by means of rules without questioning on beyond technology as *datum* of scientific applicability desperately needs supplementing with, if not replacement by, a moral engagement in forensic technological issues. Narrative fiction is eminently suited to the task in that it offers us an immediate confrontation with what happens to others.

The importance of DNA as the biological source of human identity *per se* needs no questioning, but its discourse or discursive perspective in the Heideggerian sense when it comes to answering the question of our authenticity, our uniqueness, will benefit if we return to that other source of the self, literature.[70]

---

[68] See M. C. Nussbaum, *Love's Knowledge* (Oxford: Oxford University Press, 1990), 23–24, 171. See also M. C. Nussbaum, *Poetic Justice, the Literary Imagination in Public Life* (Boston: Beacon Press, 1995). For the origin of that line of thought in her works, see M. C. Nussbaum, *The Fragility of Goodness* (Cambridge: Cambridge University Press, 1986), in which human ethical philosophy along Aristotelian lines is combined with the suggestion to study the narrative and emotional structures of novels. A literary exemplification of that admonition can be found in M. C. Nussbaum, "Little C" in M. C. Nussbaum and C. R. Sunstein, *Clones and Clones: Facts and Fantasies about Human Cloning* (New York: W. W. Norton & Co., 1998), 338–346, a story about human relations and cloning.

[69] See also J. McLarren Caldwell, *Literature and Medicine in Nineteenth-Century Britain from Mary Shelley to George Eliot* (Cambridge: University Press, 2004), 29.

[70] The sciences need the arts in another way too. Recent research suggests that through the arts we can develop cognitive competencies and experience cognitive growth, such as using the imagination to determine action, making decision in the absence of a "rule", understanding that problem solving can benefit from a change of perspective; see E. Eisner, *The Arts and the Creation of Mind* (New Haven: Yale University Press, 2002).

If (bio)technology changes our perceptions of who we are,[71] and in doing so develops its discursive idiosyncrasies, as recent research in cultural studies already prominently shows,[72] then we should further the discussion by questioning the moral and cultural assumptions behind the discursive *loci* of (bio)technology. We must consider these carefully, also because of their importance for interdisciplinary studies in law as the example of the cooperation with behavioral sciences above shows, lest sterile attentiveness that does not result in meaningful action or experience becomes the key word.[73]

---

[71] See also Weinbaum, "Racial Aura: Walter Benjamin and the Work of Art in a Biotechnological Age", for an application of Walter Benjamin's view as laid down in his essay, "The Work of Art in the Age of Mechanical Reproduction" to an analysis of contemporary biotechnological developments. On the view, and central thesis that, "changes in the mode of production are manifest in transformations in cultural production and, in turn, in human sense perception" [214], "the powerful biotechnologies [...] those which, through use transform our perception of biological life itself" [215–216] bring with them, "the liberatory potential of mechanical reproduction" while at the same time the concomitant aesthetics leads to fascism in political life, as Walter Benjamin argued for art itself in the age of mechanical reproduction.

[72] See, for example, M. Bal, "Scared to Death" in Bal and Boer, *The Point of Theory: Practices of Cultural Analysis*, 32–47, on the subject of the strength and/or surplus value of cognitive metaphors with an elaboration with respect to the relation of science and language in the discourse with which the developers of DNA presented their research to a larger audience: they spoke of "the mother-molecule", "the secret of life" etc. and thus made use of a long tradition of metaphors. Kirby and Gaither, "Genetic Coming of Age: Genomics, Enhancement, and Identity in Film", 277, point to the problem that in the films on the subject of genetic modification, "they (i.e. the protagonists) still have to come to grips with the question of *which* identities will actually represent their 'authentic' selves". To them, the search for authenticity has become a "moral ideal" in contemporary societies. See also W. G. Sebald, *After Nature*, trans. M. Hamburger (New York: Random House, 2002), 62, on the steam engine described by Chamisso, on a whale expedition, as, "the first warm-blooded animal created by mankind", while later on when technology wins, man is spoken of as a machine. See also P. Wald, "Future Perfect: Grammar, Genes, and Geography", *New Literary History* 31 (2000): 681–708, 697, on the anxious expectation that mapping the human genome will indeed lead to "biological determinism, the idea that characters and destinies of human beings are inscribed in our bodies. Although geneticists routinely insist on their rejection of biological determinism, in doing so they often need to work against the assertions embedded in their language and in their maps".

[73] For a detailed discussion of this idea, see R. Rochlitz, *The Disenchantment of Art, the Philosophy of Walter Benjamin*, trans. J. M. Todd (New York: The Guildford Press, 1996). See also on Walter Benjamin, S. Almog, "Windows and Windows: Reflections on Law and Literature in the Digital Age", *University of Toronto Law Journal* 57

In bioethics, the importance of literature has already been highlighted these past years. As early as 1982, Stephen Toulmin observed how the rise of bioethics could be associated with the importance of cases.[74] More recently, this notion is elaborated upon for bioethics by Tod Chambers who has argued that an engagement with narrative is a new and fruitful direction in the field,

> it is because cases are pivotal to the task of bioethics and because cases are a narrative genre that I argue that the tools of narrative theory are central to the rhetoric of the discipline.[75]

Here is also an opening for a comparison with the narrative strand in *Law and Literature*. Lawyers, from their first day in law school to their last day of practice, solve cases by first comparing imaginary cases before moving on to the case at hand. Furthermore, the subject of the stories told in law is always the discrepancy between what happened and what was expected to happen. The core business of the legal profession is the translation of the brute force of client stories into a narrative that can be mediated by law.[76]

Transposed to the broader plane of forensic (bio)technology, this suggests a more disruptive reading of Houellebecq's novel, especially of its ending, because it would seem that Hubczejak as Michel Djerzinski's biographer misreads or misunderstands his subject's contribution to science and humanity in that as the die-hard of positivist science that he himself is, he accepts on the surface of it the critical connection in Michel's ideas. Michel accepts as irrefutable the position that because of quantummechanics "Man no longer needed God nor the idea of an underlying reality" [A, 359], for the very reason that

> in experiments, it is possible to get a group of observers to agree on reasonable intersubjectivity; these experiments are linked by theories, which should, as far as possible, be succinct and which must, by definition, be refutable. This is a perceived world, a world of feelings, a human world [A, 360].

(2007): 755–780. Also of interest as far as the interrelation of science, literature, and law is concerned is R. R. M. Verchick, "Steinbeck's Holism: Science, Literature, and Environmental Law", *Stanford Environmental Law Journal* 22 (2003): 1–41.

[74] S. Toulmin, "How Medicine Saved the Life of Ethics", *Perspectives of Biology and Medicine* 25 (1982): 637–750.

[75] See T. Chambers, *The Fiction of Bioethics, Cases as Literary Texts* (New York and London: Routledge, 1999), 13. See also *Being Human: Readings from the President's Council on Bioethics,* an extensive anthology of works of literature that speak to bioethical dilemmas, http://www.bioethics.gov/bookshelf/reader/table_of_contents.html. Also of interest is F. Meulenberg and I. de Beaufort eds., *Science, Fiction, and Science-Fiction, the Role of Fiction in Public Debates on Medical Ethical Issues and in Medical Education* (Overveen, Netherlands: Belvédère Publishers, 2006).

[76] See also my "(Con)temporary Law", *European Journal of English Studies* 11.1 (2007): 29–46.

He does ask himself, though, whether "[…] the need to find meaning [is] simply a childish defect of the human mind" [A, 360]. Yet, Hubczejak also mentions that the decisive moment in the evolution of Djerzinski's ideas was when he first saw *The Book of Kells*, the study of which caused scientific intuitions that in retrospect – remember the year is 2079 – proved correct. A childish defect? That Houellebecq has Hubczejak derail in positivism is the ironic twist that sows the seed of doubt at the end of the novel. Hubczejak is the prime example of the effect of the third metaphysical mutation, a global transformation in the values to which the majority subscribes in Djerzinski's terms, combined with an absolute belief in what Djerzinski told about a "consistent Griffiths' history". What the literal reader Hubczejak lacks perhaps is what Anthony Cunningham calls and we, I would claim, are right to do, and that for once literally, is reading for life.[77]

On this view, Houellebecq's depictions of Bruno and Michel's lives and feelings should be read for what they hide as much as for what they reveal. And on this view too, literature is no mere counterphenomenon to law and science, no mere critique, but a dissenting opinion connected as it is to law via the perspective of cases and narratives which is the locus for a reflection and, if necessary, opposition to the language of science and technology when that is incorporated into law instrumentally. So ends my second reading of Houellebecq.

A fine example supportive of this claim can be found in a literary work, Michael Cunningham's novel, *Specimen Days*, with the last part entitled "Like Beauty", set in a scene 150 years from now, in a situation in which the difference between biological humans and biomechanical, i.e. constructed humans, can only be discerned if you look closely into their eyes. The latter are commodified, or as Simon, one of the biotechnological products, says,

---

[77] See A. Cunningham, *The Heart of What Matters, the Role for Literature in Moral Philosophy* (Berkeley: University of California Press, 2001), 3, "[b]y providing detailed depictions of the complex interior life of fictional characters embroiled in the messy business of living, fine literature directs our attention to the subtleties and nuances of what should rightly command our attention. By reading the right kinds of novels in the right way, we can literally read for life, thereby honing our capacity to see clearly and choose wisely when it comes to real life". See also A. Z. Newton, *Narrative Ethics* (Cambridge, Mass.: Harvard University Press, 1995), and M. Dirda, *Book by Book, Notes on Reading and Life* (New York: Henry Holt and Co., 2005), e.g. at xv, "[…] the humanities encourage the self-development of our own humanity. They are instruments of self-exploration".

[a] couple of years ago the Council identified all artificials as stolen property, because the whole debate about natural versus engineered life just went on and on.
We were monsters and abominations. Or we were the innocent victims of science,
and deserved protection. [...] Somebody in Texas invented and patented a soul-
measuring apparatus, but the courts disallowed it. [...] Because we were manufactured [...] We had essentially stolen ourselves.

Simon, however, was secretly given a poetry chip by his maker Emory Lowell
for the purpose of regulating his behaviour, and to instil into him, "[...]
some moral sense as well", or as Lowell confesses, "I thought that if you
were programmed with the work of great poets, you'd be better able to appreciate the consequences of your actions".[78]

If what the engineers of the human body create is "unbelievable" to us
given our current conceptual legal universe, and if they threaten to take over
or erase the moral discourse, then writers and others involved in the humanities would do well to resume the thread of moral discourse as engineers
of the human soul and that not as Stalinist indoctrination would have it, of
course,[79] but in the classical Greek sense of being the *poietai*, the makers.

There is good reason for doing so because scientific explanations and
humanistic understandings are both forms of story-telling, related though
distinct. Furthermore, in case we should need a more practical argument for
the legal profession to legitimate the enterprise: for judges who are unaccus-
tomed to the specifics of a discipline other than their own, it can be most
helpful to gain insight in the way in which other forms of cognition function
when they have to decide cases in which the clash of disciplines is obvious.[80]
We do need all the guidance that we can get to inform us about how to live
and act well, in short, how to be human. Literature can also help us understand how distortions can occur in our perception of (forensic) technology –

---

[78] See M. Cunningham, *Specimen Days* (London: Fourth Estate, 2005), 245 and 281.
To mind also come Jeffrey Eugenides' *Middlesex*, Zadie Smith's *White Teeth*, Margaret Atwood's *Oryx and Crake*, and David Mitchell's *Cloud Atlas*, not to mention
Franz Kafka's *The Metamorphosis* and Edward Bulwer-Lyton's *The Coming Race*
(1871).

[79] Joseph Stalin used the term engineer of human souls for writers in a speech in the
house of Maxim Gorki, 26 October 1932, at a meeting of Soviet writers. A satirical
novel on the Stalinist use of the term is Josef Skvorecky's *The Engineer of Human
Souls* (1979).

[80] For a plea for literature as a conceptual wrench, see C. Biet, "Judicial Fiction and
Literary Fiction: the Example of the Factum", *Law & Literature* 20.3 (2008):
403–422, 420, "The function of literature is, then, to question the law with
particular circumstances that bother logic or that show that it does not, in itself,
have the disposition that would permit it to treat correctly the case narrated or
represented".

just think of Ira Levin's *The Boys from Brazil* – and, as I argued above, it can draw our attention to underlying assumptions about science, cultural as well more specifically legal ones too, in order to better appreciate the ethical dilemmas.[81]

I think it would be wrong to already change the question, "What is Man?", to "What does it mean to be posthuman?", i.e. to be genetically enhanced or modified, if we have not yet been fully successful in our efforts of defining, or redefining for that matter, the concepts that lie at the foundation of our common humanity. If, therefore, the project of the invention of the human is still in full swing, as I would claim it is, our ongoing concern and task should be questioning the stories that are told to us, as much as questioning the stories we tell in science and technology.

I have raised a lot of questions and I realize that my questions outnumber my answers, but that is the task of philosophy. Since *épistèmè* and *phronèsis*, theory and practice are indissolubly connected in legal practice and legal theory, on the premise that human identity is a relational affair, and that this notion is important for a forensic environment of technology as well as for law itself, I turn, by way of conclusion, to the philosophical underpinnings of justice as Paul Ricoeur developed them in *The Just*.

Ricoeur's most famous contribution to moral philosophy on the subject of human relations is *Oneself as Another* in which he distinguished between identity as *idem*, e.g. the correspondence between who we say we are and our bodily evidence in the form of DNA, and identity as *ipse*, as human authenticity and uniqueness.[82] He continued his search in *The Just*,[83] which is a return to the subject matter of the earlier work when that is applied to law and justice. One such line, or axis, as Ricoeur calls it, is that of the dialogical constitution of the self, of identity when that applies to selfhood. The second axis is the philosophical inquiry into the way in which the predicates that qualify human actions in terms of morality are hierarchically constituted. The just, then, is situated at the intersection, and this means that, as Ricoeur points out,

---

81 See also P. Wald and J. Clayton, "Editors' Preface: Genomics in Literature, Visual Arts, and Culture", *Literature and Medicine* 26.1 (2007): vi–xvi, and www.literature-andgenetics.org. The development of cinematography on the subject of DNA technology is traced in great detail by D. A. Kirby, "The Devil in Our DNA: A Brief History of Eugenics in Science Fiction Films", *Literature and Medicine* 26.1 (2007): 83–108.

82 P. Ricoeur, *Oneself as Another* [1992], trans. K. Blamey (Chicago: University of Chicago Press, 1994).

83 P. Ricoeur, *The Just* [1995], trans. D. Pellauer (University of Chicago Press, Chicago, 2000).

[...] the self only constitutes its identity through a relational structure that places the dialogical dimension above the monological one inherited from the great tradition of reflective philosophy.[84]

Here what Heidegger and Habermas taught us on the subject of choice in speech and dialogue again shows its immense value. "The virtue of justice is based on a relation of distance from the other",[85] and it is the institution that mediates the relation to the other: in the dialogue of friends, the other is "you", on the plane of justice it is "anyone", as can be seen in the Roman jurist Ulpian's foundational principle for law, found in the opening passage of the emperor Justinian's *Digests*, justice is the *constans et perpetua voluntas ius suum cuique tribuere*. For justice as distribution in a societal setting, a hierarchy of institutions is needed and the incarnation of that is the figure of the judge who is given authority and institutional power to fulfill his task of impartial third party: "[...] the judge is to the juridical what the teacher of justice is to moral thought and what the prince, or any other personalized figure of sovereign power is to the political".[86] To Ricoeur, this means that on the deontological plane of obligation, the just is identified with the legal, with rules. On the plane of legal theory, especially when we speak of the subject of human agency, capacity, imputability or accountability, it is, I would claim, connected to foundational concepts such as volition. On the plane of practical reason, the just to Ricoeur is no longer the good in the sense of the good life, and neither is it the legal, but it is the equitable:

> [t]he equitable is the figure that clothes the idea of the just in situations of incertitude and of conflict, or, to put it in a better way, in the ordinary – or extraordinary – realm of the tragic dimension of action.[87]

In short, for the judge to fulfill the task of making the just happen, his equitable side must be informed by the variety of answers that literature offers to our question, "What is man?", significantly the question behind the whole enterprise of law, literature and *Law and Literature*.

---

[84] Ricoeur, *The Just*, xiii.

[85] Ricoeur, *The Just*, xiii.

[86] Ricoeur, *The Just*, xiii.

[87] Ricoeur, *The Just*, xxiv. See also my "Law in Context, Law, Equity, and the Realm of Human Affairs" [34], where I argue that the relationship of equity to law in practice, i. e in the sense of legal conflict solving in actual cases, has its parallels in the relationship of *Law and Literature* – as an interdisciplinary strand in legal theory – to traditional doctrinal legal thought, "[...] equity is to law, what *Law and Literature* is to law as science in the double sense of that term, both as a one-sided focus on a positivist rigidity of black letter, doctrinal law, and as an intellectually restricted, theoretical approach to law as an autonomous, merely rule-oriented system".

Here, as far as a feasible methodology is concerned, Ricoeur's model of mimesis from his seminal work *Time and Narrative*[88] can be of great importance since law and literature both need insight in the way in which story-telling works. Ricoeur distinguishes three stages. The first is prefiguration, or mimesis$_1$, a term denoting the temporality of the world of action. It is followed by configuration, or mimesis$_2$, a term denoting the world of the narrative emplotment of events that has the following prerequisites,

> [...] the composition of the plot is grounded in a pre-understanding of the world of action, its meaningful structures, its symbolic resources, and its temporal character.

This means that

> [...] an event [...] gets its definition from its contribution to the development of a plot. A story, too, must be more than just an enumeration of events in a serial order; it must organize them into an intelligible whole, of a sort that we can always ask what is the "thought" of this story. In short, emplotment is the operation that draws a configuration out of a simple succession.[89]

On this view, the plot, equated already by Aristotle with the configuration of opposite views, can be fruitfully compared to legal debate.[90] The latter being dialogically structured, emblematic of law itself, it is obvious that when it comes to the phase of applying law which is the stage of refiguration or mimesis$_3$ in Ricoeur's terminology, when the worlds of mimesis$_1$ and mimesis$_2$ interact and influence one another, this third form of mimesis is essentially a reflexive application, after having carefully weighed all the evidence. When then in forensic technologies we are confronted with the situations of incertitude and of conflict that Ricoeur points to, we need not dismiss science, but neither should we cease from our questioning and our exploration, in the manner that philosophical hermeneutics teaches us to, of the ways in which law interrelates with the other arts and sciences.

---

[88] P. Ricoeur, *Time and Narrative* [1984], trans. Kathleen McLaughlin and David Pellauer (Chicago: University of Chicago Press, 1988), vols. 1–3.

[89] Ricoeur, *Time and Narrative*, vol. 1, 54 and 64.

[90] For another inquiry into the concept of mimesis, see my "O negócio da Lei e da Literatura: Criar uma ordem, imaginar o homem", in *Direito e Literatura. Mundos em Diálogo*, eds. H. Buescu, C. Trabuco, and S. Ribeiro (Coimbea: Edições Almedina, 2010) 13–47, to be reprinted in D. Carpi, J. Gaakeer eds., *The Liminal Discourses of Law* (London and New York: De Gruyter, forthcoming 2012).

Ian Ward
University of Newcastle

Ghostly Presences:
The Case of Bertha Mason

Writing in the *Journal of Mental Science* in 1858, John Bucknill bemoaned the prevalence of a fanciful journalistic genre that pedalled accounts of supposedly "sane people confined in lunatic asylums". They were, he concluded, more often than not the "ghosts of newspaper readings", spectres raised by calculating writers and editors, designed to haunt the febrile cultural imagination of mid-Victorian England.[1] One of the greatest novels in the English canon was written precisely with this same conjuration in mind. It may not have been written for an immediately journalistic audience, unlike for example Wilkie Collins's *Woman in White*, but it was written with the same audience in mind, and the same aspiration. This novel was Charlotte Brontë's *Jane Eyre*. The spectral presence was, of course, Bertha Mason, the first Mrs. Rochester. The purpose of this essay is to revisit the case of Bertha Mason, in the process uncovering the often latent, occasionally more overt, jurisprudence which it engages. The first part will discuss the necessary tensions which defined the discourse of madness in mid-nineteenth century English. The second part will then consider the Mason case which Charlotte Brontë wrote into the heart of the novel.

The Discourses of Madness

Madness was a subject of considerable discursive contestation in mid-Victorian England. For some, the occurrence of madness could be most readily

---

[1] For discussions of these panics, see P. McCandless, "Dangerous to Themselves and Others: the Victorian Debate over the Prevention of Wrongful Confinement", *Journal of British Studies* 23 (1983): 84–104, and also "Liberty and Lunacy: The Victorians and Wrongful Confinement" in *Madhouses, Mad-Doctors and Madmen: The Social History of Psychiatry in The Victorian Era*, ed. A. Scull (Philadelphia: University of Pennsylvania Press, 1981), 339–357, discussing the particular ambiguities that riddled Victorian responses to supposedly improper confinement, 340–341, and the role of the media, 356–357.

explained in terms of moral failings. Such arguments often assumed a more obviously cultural aspect. For others, madness was something best understood in terms of physiology and psychology. Science challenged morality, and at a remove theology; a challenge which, of course, was writ large across much of mid and late nineteenth century intellectual life. And somewhere amidst this contest could be found another voice, altogether quieter, more reticent; the voice of jurisprudence. Whilst the moralist and the medic had much to say on the mooted causes of madness, its manifestations and the best way of addressing it, the lawyer, it seemed, had very little to say at all.

Reticent or not, the law had to engage the question of madness. Indeed, most obviously in the matter of criminal jurisdiction, it had presumed a role in controlling lunatics for centuries.[2] But in matters of civil jurisdiction the law was altogether less coherent. Here there were two principal issues; the question of the insanity of an alleged lunatic, and the collateral issue of their confinement. In both cases, recourse to the law was largely voluntary.

With regard to proceedings for securing certificates of lunacy, families could petition the Lord Chancellor's Masters in Lunacy for an "inquisition" or Commission in Lunacy in order to gauge whether an alleged lunatic was "a proper person to be taken charge of and detained under care and treatment".[3] The reasons why they might chose to do so were various. Primary was the desire to secure entailed estates. The care of a committed lunatic passed to a "guardian", whilst the management of his or her estates was vested in a "committee". Very often they were one and the same, more often still they were appointed from within the committed lunatic's family.[4]

The possibilities for abuse and corruption were patent; something which led, in time, to the establishment of an Alleged Lunatics Friends Society in 1845.[5] Up until 1853 lunacy certificates could be issued on the evidence of just two medical experts, commonly known as "alienists", and most notably, perhaps, without any need for a personal examination of the alleged

---

[2] P. Bartlett, "Legal Madness in the Nineteenth Century", *Social History of Medicine* 14 (2001): 107.

[3] See A. Suzuki, *Madness at Home: The Psychiatrist, the Patient and the Family in England 1820–1860* (Berkeley: University of California Press, 2006), 19–20, and also Bartlett, "Legal Madness in the Nineteenth Century", 109, citing the 1853 Act, 16 & 17 Vic.c.97 sch. F (1) & (3).

[4] For an overview of the procedure, see Bartlett, "Legal Madness in the Nineteenth Century", 119–21.

[5] McCandless, "Dangerous to Themselves and Others", 86–88.

lunatic.[6] This latter fact caused considerable disquiet, even within the medical profession. In his *Enquiry Concerning the Indications of Lunacy*, published in 1830, John Conolly had loudly adverted to the dangers inherent in a system that seemed so dependent on often semi-detached medical opinions and so ready to confirm the incarceration of family members who might, in reality, be more inconvenient than mad.[7] Connolly's view was further supported by other prominent psychiatrists such as Henry Monro, and increasingly, and perhaps most importantly, by successive editors of the *Lancet*.[8]

At the same time, however, counter-polemics, such as James Prichard's *Treatise on Insanity*, published five years after Conolly's *Enquiry*, just as vigorously defended the notion that families were best placed to adjudge the nature and extent of any "moral" lunacy that might be found in their midst, just as it also recognised their complementary capacity to ascertain the precise threat that such a lunatic might pose to their property and estates.[9] Such justification, however, did little to allay popular suspicions regarding the relation of mutual dependency that existed between families who wanted relatives committed and "alienists" who wanted to be hired to provide such certificates.[10]

---

[6] The term "alienist" is slightly ambiguous, and to a degree pejorative. It was often used to denote a mad-doctor from a fully qualified doctor or psychiatrist; with the latter category understandably keen to have the distinction widely appreciated. More generally, however, any doctor, professionally qualified or not, who gained a reputation for providing certificates of lunacy for payment became known as an "alienist". For a discussion of the distinction, see Scull, *The Most Solitary of Afflictions: Madness and Society in Britain 1700–1900* (New Haven and London: Yale UP, 1993), 249–251.

[7] Suzuki, *Madness at Home*, 73–77. For a concise commentary on Conolly's work, noting its inculcation of patriarchal, if otherwise apparently progressive, principles of treatment and care, see E. Showalter, *The Female Malady: Women, Madness, and English Culture, 1830–1980* (New York: Pantheon Books, 1985), chapter 1.

[8] Suzuki, *Madness at Home*, 171, and also McCandless, "Dangerous to Themselves and Others", 90–91.

[9] See Scull, "The Social History of Psychiatry in the Victorian Era" in *Madhouses, Mad-Doctors and Madmen*, 7–10, and also Suzuki, *Madness at Home*, 54–59, 81–84, discussing the particular view of the controversial early Victorian psychiatrist, George Mann Burrows, who argued that given the "moral" dimension to lunacy, it was not just sufficient for an examination to be conducted through family members, but preferable that a professional opinion was not corrupted by personal examination. He also charts the notoriety of the *Davies* case, in which Lord Brougham, hired to defend the alleged lunatic, presented a devastating critique of Mann Burrows and his refusal to undertake personal examinations. Brougham's critique played a major part in campaigns for the reform of this particular anomaly.

[10] The 1889 Lunacy Acts Amendment Act would tighten the procedure for the issuing of committal orders, requiring them to be signed by magistrates who enjoyed

The attitude of the courts, particularly in prominent cases of appeal from Lunacy Commissions, was far from consistent; oscillating between periods during which it articulated a greater concern for individual liberty and periods when it was clearly inclined to defer to medical opinion. Whilst late eighteenth century courts had occasionally sought to protect individual liberty, as in *Lord Donegal's Case*, by the early nineteenth century, however, courts seemed altogether more sensitive to the concerns of anxious families keen to secure their estates. Lord Eldon's notorious judgement in *Ridgway v Darwin* in 1802, that in assessing mental capacity, a court or a Commission should first and foremost look to an ability to properly manage an estate, was confirmed four years later by Lord Erskine in *Cranmer*, who agreed that an inability to properly conduct business as well as personal "affairs" was itself evidence of an "unsound mind".[11]

Half a century on, certain judges were prepared to articulate a rather greater scepticism, particularly in the matter of the supposed sanctity of medical evidence. In *Nottidge* in 1849, Baron Pollock commented that courts should reject the "notion" that "any person may be confined in a lunatic asylum or a madhouse who has any absurd or even mad opinion upon any religious subject", whilst the Lunacy Commissioners should "liberate every person who is not dangerous to himself or others".[12] In the main, however, judges remained willing to countenance medical evidence which was often couched in nothing more solid than vague notions of "love-madness" or an "unsound mind".[13] For men such as Forbes Winslow, writing in the *Journal of Psychological Medicine*, the court and the family retained a dual responsibility to

---

the power to order public inquiries; a strategy which, whilst it was intended to guard against abuses, may well have simply encouraged families to internalise supposed lunatics. See Suzuki, *Madness at Home*, 65–66, 91–92, and also McCandless, "Dangerous to Themselves and Others", 99.

[11] See *Lord Donegal's Case* 2 Ves. Sen. 408, *Ridgway v Darwin* 8 Ves. 65–66, and *Ex parte Cranmer* 12 Ves. Jun. 445–456.

[12] For a discussion of *Nottidge*, see McCandless, "Dangerous to Themselves and Others", 92–93. For further accounts of judicial disquiet, particularly in criminal cases, see J. Eigen, *Unconscious Crime: Mental Absence and Criminal Responsibility in Victorian London* (Baltimore NJ: Johns Hopkins UP, 2003), 115–116, 127.

[13] See Suzuki, *Madness at Home*, 156–158, and also M. Williams, *Secrets and Laws: Collected Essays in Law, Lives and Literature* (London: UCL Press, 2005), particularly 132–137. As late as 1935, in *Quarry*, Chief Justice Hewart could still articulate his frustration at medical evidence that was not just vague but very often little more than "sentimental nonsense". Cited in T. Ward, "Law, Common Sense and the Authority of Science: Expert Witnesses and Criminal Insanity in England c. 1840–1940", *Social and Legal Studies* 6 (1997): 336.

protect the "insane person" from "himself and others whatever may be the degree of his mental disturbance".[14] For the most part, then, in matters of lunacy certification, whilst courts and Commission retained a nominal jurisdiction, the dominant discourse was fashioned by the alienists and the families who hired them.[15]

The second, necessarily collateral, issue was that of confinement. And it was here, in the matter of supposedly "wrongful confinement", that public concerns were greatest. Having secured a certificate, a guardian might then seek to have their ward placed in a public or private asylum; neither of which, up until 1845, were subject to any coherent system of public inspection.[16] There again they might not; and there was no legal requirement that they should. Strategies of "domestic psychiatry" had an obvious attraction.[17] For one, they evaded public embarrassment; and few ailments caused Victorian's greater concern than familial insanity.[18] In 1860, a Parliamentary Select Committee on the Care and Treatment of Lunatics sympathised that "Insanity under any shape is so fearful a malady, that the desire to withdraw it from the observation of the world is both natural and commendable".[19] Equally, and rather more positively, there was a strong evangelical literature which supported strategies of "unbossoming"; a confessional as well as redemptive process which was presumed to work most effectively within the supposedly more receptive and nurturing environment of the private family.[20] Thackeray chose to pursue this approach in the case of his intermittently psychotic wife.[21]

---

[14] For Winslow's response, see McCandless, "Dangerous to Themselves and Others", 93–94.

[15] Suzuki, *Madness at Home*, 21–25, 66–67.

[16] Private asylums were subject to a licensing regime from the time of the 1774 Madhouse Act, but investing the responsibility in local magistrates meant that they were rarely inspected in any formal or consistent manner. From 1828, this licensing regime was administered by the Metropolitan Commissioners in Lunacy, renamed the Commissioners in Lunacy from 1845. For a discussion of the inauguration of formal inspections from 1845, see McCandless, "Dangerous to Themselves and Others", 84–85, 89–90.

[17] Suzuki, *Madness at Home*, 114, 138, 151, 174–175. For a similar thesis, see E. Showalter, *The Female Malady*, 28 discussing the "domestication of insanity".

[18] See Suzuki, *Madness at Home*, 121–122, and also 139 suggesting that it was the "exposure" of familial lunacy which really troubled Victorians most. For Victorian anxieties regarding such apparent disabilities, see also McCandless, "Dangerous to Themselves", 86, and also C. Kaplan, *Victoriana: Histories, Fictions, Criticism* (Edinburgh: Edinburgh UP, 2007), 141–143.

[19] Quoted in McCandless, "Dangerous to Themselves and Others", 96.

[20] Suzuki, *Madness at Home*, 111–117, 119, 180–182.

[21] Showalter, *The Female Malady*, 53.

In such circumstances, families would very often seek to employ the services of professional "keepers", generally men or women who presented themselves as having had experience of dealing with lunatics in public or private asylums.[22] All in all, as Sir Robert Peel concluded, for reasons of both principle and practice, it was always "preferable to leave" lunatics "in the custody of their relations, than to lock them up in madhouses".[23] It was a solution that chimed, moreover, with received perceptions of the sanctuary, and indeed sanctity, of the family in middle-class Victorian England. As Lord Brougham confirmed, it was "better to trust to the relatives, wives, husbands, or children of persons unhappily afflicted" than to the Lunacy Commissioners.[24] Except that, here again, the opportunities for abuse were patent. Accounts of abused relatives incarcerated, as John Conolly advised, "in garrets, or in the secluded wings of country mansions", were frequently uncovered.[25] And here again, courts were loath to intervene; no matter how great the alleged abuse might be.[26] The common law was evasive, the civil jurisdiction particularly so; and that evasion was reciprocal. Many Victorians preferred not to engage the law and it, by and large, preferred not to have to engage their lunatic relatives.

If the courtrooms were quieter than might have been expected, the living rooms of mid-Victorian England were not. Madness then, as now, fascinated. There were, on the one hand, the claims of medical science. To an extent this emergent discourse appeared to lend a degree of reassurance

---

[22] Suzuki, *Madness at Home*, 110–111.

[23] Quoted in Suzuki, *Madness at Home*, 174.

[24] Quoted in Suzuki, *Madness at Home*, 174.

[25] In his essay "Residences for the Insane", published in the *Transactions of the National Association for the Promotion of Social Science* in 1858, and quoted in McCandless, "Dangerous to Themselves and Others", 97.

[26] In 1855, the Criminal Court of Appeal confirmed that in instances of domestic "custody of a lunatic" the ordinary common law did not necessarily apply. See *Rundle's Case* in *English Reports*, Deras, 482, and also Suzuki, *Madness at Home*, 128–129, 146–150 and 175. In her *The Female Malady*, 107, Showalter cites the contemporary writings of Furneaux Jordan, who argued that some women had a genetic propensity to encourage violence in their spouses. These "congenital impulses of character", he opined, could be categorised as a form of madness. Of course, the issue of domestic violence was to come to the fore just two years later with the enactment of the Divorce Act, which opened the eyes of the public to an issue which had hitherto remained largely behind closed doors. For contemporary testament to this, see Frances Power Cobbe's essay "Celibacy v Marriage" in *Criminals, Idiots, Women and Minors: Victorian Writing by Women on Women*, ed. S. Hamilton (Peterborough, Ontario: Broadview Press, 1995), 81–82.

which the law appeared unable or unwilling to furnish.[27] There again, to just as many, it was this very confidence which was troubling. Most obviously, despite the copious writings on the subjects, in newspapers and learned journals, no one seemed to agree what constituted insanity, still less how it might be best treated.[28] Such concerns, particularly when crystallized by accounts of "wrongful confinement", generated considerable debate. In *On Liberty*, John Stuart Mill articulated the common fear that anyone who fails to do "what everybody does" ran the constant risk "of a commission *de lunatico*, and of having their property taken away from them and given to their relations".[29] The extent to which necessarily "elastic" measures of what constituted lunatic behaviour were dependent on the perceived transgression of presumed cultural norms is patent; especially so in the context of debates which oscillated round the original idea of "moral" insanity and its later pseudo-medical mutations, such as hysteria and neurasthenia.[30]

Of course the fear and the fascination were mutually sustaining. As every newspaper editor knew, nothing sold better than lurid accounts of madness, particularly amongst the well-to-do, and more particularly still if they involved any titillating sexual perversions. The case of Rosa Bagster, whose family alleged that an unhealthy sexual proclivity had led their daughter to elope and make an unsuitable marriage, was a notorious example.[31] The Bagster case engaged a particular species of madness that was very obviously the construct of social prejudice as much as scientific veracity: "moral mad-

---

[27] See Scull, "The Social History of Psychiatry in the Victorian Era", 17 and Suzuki, *Madness at Home*, 2–3. For a representative, and influential, collection of essays on the associated rise of professional psychiatry and institutional confinement, see Scull, *Madhouses, Mad-Doctors and Madmen*, and Scull, *Museums of Madness: The Social Organisation of Insanity in Nineteenth Century England* (London: Penguin, 1982). Scull's work remains perhaps the most impressive and substantial attempt to investigate the role of professionalization and institutionalism in Victorian responses to madness. The overarching thesis finds support in Helen Small and Elaine Showalter's more particular studies of women and madness in this period. See H. Small, *Love's Madness: Medicine, the Novel and Female Insanity* (Oxford: Oxford UP, 1996), 22–24, and Showalter, *The Female Malady*, 24–25.

[28] See Suzuki, *Madness at Home*, 44–46 and 50–54, discussing the controversy surrounding such "pseudo-certificates", and also 91 discussing the absence of a uniform definition of madness. See also Williams, *Secrets and Laws: Collected Essays in Law, Lives and Literature*, 126, on the effect of the absence of a uniform definition of insanity.

[29] J. Mill, *On Liberty* (Chicago: Chicago UP, 1955), 99–100.

[30] Suzuki, *Madness at Home*, 138–139.

[31] For accounts of these cases see Suzuki, *Madness at Home*, 12–18, 122–127.

ness".[32] In his rather florid, but highly influential, definition of this kind of insanity, James Prichard depicted "moral madness" as a "morbid perversion of the natural feelings, affections, inclinations, temper, habit, moral dispositions, and natural impulses, without any remarkable disorder or defect of the intellect".[33] It was, Prichard insinuated, also the kind of madness to which young women like Rosa Bagster might be peculiarly susceptible. In his *Outlines of Lectures on Mental Diseases*, Alexander Morison concurred. An excessive sensibility to "hopeless love" could be categorised as "a form of madness".[34] It was often caused, moreover, by an excessive exposure to romantic novels.

The science and the culture cannot be readily distinguished. And, as Elaine Showalter famously confirmed, this is all the more apparent in the context of gender.[35] The authority of science, as expounded by the likes of Prichard and Morison, lent a veneer of legitimacy to a darker misogyny.[36]

---

[32] For further discussion of "moral madness", and its growing popularity in early and mid-nineteenth century discourses regarding varieties of presumed insanity, see Scull, "The Social History of Psychiatry in the Victorian Era", 7–10, and P. McCandless, "Liberty and Lunacy: The Victorians and Wrongful Confinement", 354–356, both in Scull, *Madhouses, Mad-doctors and Madmen*, as well as Suzuki, *Madness at Home*, 80–87.

[33] In Showalter, *The Female Malady*, 29. For a broad discussion of "moral madness" and its portrayal in alternative medical and cultural discourses, see S. Shuttleworth, *Charlotte Brontë and Victorian Psychology*, (Cambridge: Cambridge UP, 1996), 49–56.

[34] See Small, *Love's Madness: Medicine, the Novel and Female Insanity*, 33, 36–38.

[35] See Showalter, *The Female Malady*, 3–4, 18–19, 29–31. Showalter's study remains seminal. Its interpretation of statistical resources has been subject to criticism. See, for example, J. Busfield, "The Female Malady? Men, Women and Madness in Nineteenth Century England", *Sociology* 28 (1994): 259–277, arguing that statistics of asylum committals and the like are too uncertain to permit broader conclusions, particularly those which impute a gender bias into medical, and indeed legal, determinants of insanity. But Showalter's overarching theses have assumed a canonical status. See J. Ainsley, "Some Mysterious Agency: Women, Violent Crime, and the Insanity Acquittal in the Victorian Courtroom", *Canadian Journal of History* 35 (2000): 38–40, and also the broad approval of Helen Small in *Love's Madness: Medicine, The Novel and Female Insanity*, as well as anticipatory affirmation in Sandra Gilbert's and Susan Gubar's hugely influential literary criticism of the subject, *The Madwoman in the Attic: The Woman Writer and the Nineteenth Century Imagination* (New Haven: Yale UP, 1979). Showalter's general assertion that public discourses of madness inhere gender prejudices has also received support from contemporary sociologists such as Janet Homshaw and Sheila Hillier in their "Gender and Culture: A Sociological Perspective to Mental Health Problems in Women" in *Women and Mental Health*, ed. D. Kohen (London: Routledge, 2000), 39.

[36] See Small, *Love's Madness: Medicine, the Novel and Female Insanity*, 15–19, Shuttleworth, *Charlotte Brontë and Victorian Psychology*, 4–6, and P. Ingham, *The Brontës* (Oxford: Oxford UP, 2006), 65–66.

Thus leading figures on the medical establishment such as John Cheyne and Henry Maudsley propounded the "fact" that most instances of perceived female mental disorder might be traced back to improper sexual behaviour, most readily forms of sexual self-abuse, but also including complications arising from irregular menstruation. The latter affliction, it was confirmed, was the primary cause of "nymphomania", the most acute form of "moral madness".[37] The doctors, like the lawyers, wrote from within a particular, and particularly gendered, culture; and it was not one from which they could be reasonably expected to detach themselves.

Against this misogyny, later Victorian feminists such as Josephine Butler and Mona Caird argued the case for writing a female literature, a "literature of our own".[38] A distinctive genre of female literary testament spoke to precisely this need. And it might well be argued that a novel such as Charlotte Brontë's *Jane Eyre* is a canonical example of this genre. But there was, of course, an irreducible paradox. Poetic confession, particularly where it oscillates around sexual frustration, might just as easily be castigated for nurturing "love-madness" amongst its readership. Even so, there is no doubting the vitality of this genre. Earlier precursors of the genre could be found in Mary Wollstonecraft's semi-autobiographical novellas *Maria or the Wrongs of Woman* and *Mary* and Mary Hay's *Victim of Prejudice*.[39] At the heart of the Victorian canon were Edward Bulwer Lytton's *Lucretia*, Mary Elizabeth Brad-

---

[37] See Shuttleworth, *Charlotte Brontë and Victorian Psychology*, 75–82 and 85–94, concluding at 92, that "Behind the careful regulation of Victorian girls' lives lay the ever-present fear of their promiscuous libidinal behaviour". See also Showalter, *The Female Malady*, 55–59, 74–78, and also Busfield, "The Female Malady? Men, Women and Madness in Nineteenth Century England", 272. For a broader commentary on contemporary anxieties regarding female "passion" and sexuality, see M. Poovey, *Uneven Developments: The Ideological Work of Gender in Mid-Victorian England* (London: Virago, 1989), 4–11. For examples of criminal courts accepting evidence which imputed a relation between irregular menstrual cycles and instances of female insanity, see Eigen, *Unconscious Crime: Mental Absence and Criminal Responsibility in Victorian London*, 58, 71–75.

[38] Butler's comments can be found in Susan Hamilton's Introduction to her collection *Criminals, Idiots, Women and Minors: Victorian Writing by Women on Women*, 9. Caird's are also found in Hamilton's *Criminals, Idiots, Women and Minors* in her essay "A Defence of the So-Called Wild Women", 287–304. In their *The Madwoman in the Attic*, 52, Gilbert and Gubar term such strategies a "kind of metaphorical germ warfare".

[39] See Showalter, *The Female Malady*, 1–2, 10–17 and also Small, *Love's Madness: Medicine, the Novel and Female Insanity*, 28–32, suggesting that all of Wollstonecraft's novels can be characterised by their particular interest in the mental condition of their female protagonists, and also 123–138.

don's *Lady Audley's Secret* and Wilkie Collins's *The Woman in White* as well as Charles Reade's *Hard Cash* and Sheridan LeFanu's *The Rose and the Key*.[40]

But by the middle of the nineteenth century, as Gilbert and Gubar confirm, the gendered image of madness and confinement had become culturally "all-pervasive". The "madwoman in the attic", the "mad double", was everywhere:

> Images of enclosure and escape, fantasies in which maddened doubles functioned as asocial surrogates for docile selves, metaphors of physical discomfort manifested in frozen landscapes and fiery interiors – such patterns recurred throughout this tradition, along with obsessive depictions of diseases like anorexia, agoraphobia, and claustrophobia.[41]

And one semi-autobiographical testament reigned supreme in the genre of "female malady" or "wrongful confinement" literature, both in terms of contemporary audience and critical reputation.[42]

The Case of Bertha Mason

Critics have long noted the extent of Charlotte Brontë's interest in associated aspects of mental illness, phrenology, mesmerism and the like.[43] She read

---

[40] Even though it is largely forgotten today, Reade's novel, published in 1863, had an enormous impact on public opinion, as did LeFanu's. See McCandless, "Liberty and Lunacy: The Victorians and Wrongful Confinement", 346–7; Showalter, *The Female Malady*, 71–73, 102–103, and Small, *Love's Madness: Medicine, the Novel and Female Insanity*, 142–154 and 179, 191, 207.

[41] See Gilbert and Gubar, *The Madwoman in the Attic*, xi, and also 77–85.

[42] The full-title of the novel is *Jane Eyre: An Autobiography*. See Small, *Love's Madness: Medicine, the Novel and Female Insanity*, 140, 154, Ingham, *The Brontës*, 172–173, Kaplan, *Victoriana: Histories, Fictions, Criticism*, 7, H. Glen, *Charlotte Brontë: The Imagination in History* (Oxford: Oxford UP, 2002), 25, and also L. Miller, *The Brontë Myth* (London: Jonathan Cape, 2001), 13, noting the importance of the semi-autobiographical form of Charlotte's testament. For the seminal importance of *Jane Eyre*, see Gilbert and Gubar, *The Madwoman in the Attic*, concluding at 440 that its author was a "phenomenologist – attacking the discrepancy between reason and imagination, insisting on the subjectivity of the objective work of art, choosing as the subject of her fiction the victims of objectification, inviting her readers to experience with her the interiority of the Other".

[43] Shuttleworth, *Charlotte Brontë and Victorian Psychology*, especially chapters 2 and 4, investigating the extent to which these subjects were discussed, at length, in the contemporary journals received at Haworth parsonage, as well as at the Keighley Mechanics Institute, which various members of the family often visited. See also Ingham, *The Brontës*, 157–158 and 167–169, and Gilbert and Gubar, *The Madwoman in the Attic*, 311–312, observing, possibly with a degree of overstatement, that Charlotte was "essentially a trance-writer".

widely on the subject, and later in life even visited the notorious Bethlem Mental Hospital.[44] It has also been suggested that Brontë was familiar with various local cases of insanity in the Haworth area, most of which were dealt with behind closed doors. The story of the curate of Oakworth incarcerating his mad wife in his parsonage was well-known at the time. The same is true of the Sidgwick family of Norton Conyers.[45] On a more personal level, Charlotte also knew of the committal of her friend Ellen Nussey's brother to a private asylum near York. It is also possible that Charlotte might have been aware, however dimly, of the travails of her literary hero, William Makepeace Thackeray, whose own wife was subject to increasingly common fits of *post-partum* insanity.[46] Altogether closer to home was the sorry demise of brother Branwell, whose final months were characterised by alcohol-induced moments of often violent delirium, and then the collapse in health of sister Emily, whose intermittent bouts of depression and anorexia nervosa were chronicled by an anxious Charlotte.[47]

---

[44] She visited in 1853. Critics have tended to assume that her experience was translated into various accounts of nervous disorder and insanity in her later novel *Villette*. See Showalter, *The Female Malady*, 69–70.

[45] See J. Barker, *The Brontës* (London: Weidenfeld and Nicolson, 1994), 511–512.

[46] Brontë always denied knowing of Mrs. Thackeray's illness. Contemporary critics, however, were not so sure. The second edition of *Jane Eyre*, in December 1847, included a laudatory Prefatory dedication; an act which inheres an obvious irony. See C. Brontë, *Jane Eyre* (Oxford: Oxford UP, 2000), 3–5 [hereafter JE]. Contemporaries mused on the possibility that the anonymous author of *Jane Eyre* might be the governess of the Thackeray children, even the "keeper" of Mrs Thackeray. "Well may it be said", Charlotte was said to have commented, when later told of the condition of Thackeray's wife, "that fact is often stranger than fiction". The "coincidence", she added, "struck me as equally unfortunate and extraordinary". Thackeray, who was an avowed admirer of the novel, remained haunted by the assumed association, long after the true author of *Jane Eyre* was revealed. It is recorded that when one inquisitive American admirer asked, at a dinner party in 1860, whether it was "true, the dreadful story about you and Currer Bell", Thackeray responded "Alas Madam, it is all too true. And the fruits of that unhallowed intimacy were six children. I slew them all with my own hand", in Small, *Love's Madness: Medicine, the Novel and Female Insanity*, 179. For evidence of Thackeray's admiration, see his letter to Charlotte written in gratitude for receipt of an early copy of *Jane Eyre*, in M. Allott, *The Brontës: The Critical Heritage* (London: RKP, 1974), 70. For further commentary on the Thackeray context, see L. Miller, *The Brontë Myth*, 21, Barker, *The Brontës*, 541–542, and also R. Kaye, "A Good Woman on Five Thousand Pounds: *Jane Eyre*, *Vanity Fair*, and Literary Rivalry", *Studies in English Literature* 35 (1995): 723–739.

[47] See Ingham, *The Brontës*, 13–14. For the supposition that Cathy Earnshaw's apparent eating disorders in *Wuthering Heights* had an autobiographical root, see Gilbert and Gubar, *The Madwoman in the Attic*, 284–285.

It is generally thought that Charlotte herself suffered from similar nervous disorders.[48] Private correspondence from 1846 and 1847 confirms Charlotte's concerns, much of it following the contemporary assumption that there was a necessary relation between manifestations of physical and psychological dysfunction and wider social dislocation; a common experience of "spasms, cramps and frenzy-fits", as she put it.[49] Another contemporary trope which Charlotte appears to have adopted in her writings from around 1847 is that which supposed an author to be a literary surgeon, responsible for diagnosing and treating social ills. Surgical and psychopathic metaphors saturate *Jane Eyre*, as they do the later *Villette*. Everyone surveys everyone else. Everyone diagnoses. Everyone presumes to treat and to cure.[50] Interestingly, the closer depiction of insanity in *Jane Eyre* tended to evade contemporary reviews. Not so, however, modern critics; most of whom have focussed on both immediate concerns raised in the novel, the nature of Bertha Mason's affliction, and the nature of her confinement.[51]

The nature of Bertha's malady is left to the inference of the reader. Bertha remains almost ephemeral, a physical force that is more "sensed than seen".[52] Various suppositions are intimated. Rochester, clearly sympathetic to the broader notions of "moral insanity", ascribes his wife's condition to genetic disease. Bertha's mother was "both a madwoman and a drunkard", whilst her daughter descended into a state that was "intemperate and unchaste", possessed of apparently monstrous sexual "propensities" [JE, 291–292,

---

48 Again in private correspondence, looking back on her earlier years at Roe Head School, Charlotte bore testimony to the "concentrated anguish of certain insufferable moments and the heavy gloom of many long hours" from which "the morbid nerves can know neither peace nor enjoyment – whatever touches – pierces them" in Ingham, *The Brontës*, 177. See also Barker, *The Brontës*, 249 and 289.

49 See Small, *Love's Madness: Medicine, the Novel and Female Insanity*, 155, and also Shuttleworth, *Charlotte Brontë and Victorian Psychology*, 6, 12–13, 32, 150–152, and also 182 and 244–245, stressing the extent that this particular appreciation of the relation of social and psych-physical decay creates a necessary tension between the supposed realms of the public and the private in Charlotte's fiction.

50 Shuttleworth, *Charlotte Brontë and Victorian Psychology*, 15–17, 39–45, suggesting, at 39, that Charlotte's fiction "circles obsessively" around the related questions of surveillance and confinement, and also 57 and 71, arguing, not just that Charlotte was determined to engage a wider social responsibility, but to do so within identifiable mid-Victorian "discourses of femininity", and also at 173 and 219–221 emphasising the degree to which the presumption of diagnosis and the practice of surveillance is pervasive in Charlotte's final two novels.

51 Gilbert and Gubar, *The Madwoman in the Attic*, 339.

52 A. Rich, "Jane Eyre: The Temptations of a Motherless Woman", in her *On Lies, Secrets and Silence: Selected Prose 1966–1978* (New York: Norton, 1979), 99.

305–306].[53] To this extent, Rochester's depiction chimes with contemporary presumptions, seemingly shared by Charlotte, with regard to sexual excess leading indubitably to mental instability.[54] Her "vices", Rochester affirms to Jane, "sprung up fast and rank", the "germs of insanity" already evident in her rampant sexual energy on their catastrophic wedding-night [JE, 306–307].[55]

In private correspondence alluding to complaints by some critics that her portrayal of Bertha Mason was just too hideous, Charlotte affirmed:

> The character [of Bertha] is shocking, but I know that it is but too natural. There is a phase of insanity which may be called moral madness, in which all that is good or even human seems to disappear from the mind and a fiend-like nature replaces it. The sole aim and desire of the being thus possessed is to exasperate, to molest, to destroy, and preternatural ingenuity and energy are often exercised to that dreadful end. The aspect in such cases assimilates with the disposition; all seem demonised. It is true that profound pity ought to be the only sentiment elicited by the view of such degradation, and equally true is it that I have not sufficiently dwelt on that feeling; I have erred in making horror too predominant. Mrs. Rochester indeed lived a sinful life before she was insane, but sin is itself a species of insanity: the truly good behold and compassionate it as such.[56]

The concession, that the depiction might be too brutal, and too horrific, is striking.

Certainly Bertha appears to suffer from a peculiarly intense form of "moral madness", far removed from the gentler disturbances which tended to afflict a Jane Austen heroine, for example.[57] The description which Char-

---

[53] For these critical conclusions, see Small, *Love's Madness: Medicine, the Novel and Female Insanity*, 163–164 and Shuttleworth, *Charlotte Brontë and Victorian Psychology*, 166–167.

[54] See Shuttleworth, *Charlotte Brontë and Victorian Psychology*, 11–12.

[55] For Rich's commentary on Rochester's view of his wife, emphasising his particular horror of her sexual violence, see "Jane Eyre: The Temptations of a Motherless Woman", 99–100.

[56] Letter to her publishing reader at Smith, Elder & Co, W. S. Williams, dated 4 January 1848, and printed in *The Brontës: Their Lives, Friendships and Correspondence*, eds. T. Wise and J. Symington (Oxford: Basic Blackwell, Shakespeare Head Brontë, 1932), ii, 173–174. For an affirmation that Charlotte clearly intended to portray a species of "moral madness", see P. Grudin, "Jane and the Other Mrs Rochester: Excess and Restraint in *Jane Eyre*", *Novel* 10 (1977): 149–50.

[57] For a discussion of Charlotte's apparent deployment of contemporary ideas of "moral madness" in the depiction, not just of her Bertha Mason, but also her brother Richard Mason, see Shuttleworth, *Charlotte Brontë and Victorian Psychology*, 14–15, 52–55. See also Rich, "Jane Eyre: The Temptations of a Motherless Woman", 95–96, emphasising the studious resistance to romanticism which characterises all of Charlotte Brontë's protagonists. For a slightly darker commentary on presentations of female instability in Austen, see Gilbert and Gubar, *The Madwoman in the Attic*, chapters 4 and 5.

lotte provides of Jane's first glimpse of Bertha is designed to perplex and to horrify:

> In the deep shade, at the furthest end of the room, a figure ran backwards and forwards. What it was, whether beast or human being, one could not, at first sight, tell: it grovelled, seemingly, on all fours; it snatched and growled like some strange wild animal: but it was covered with clothing; and a quantity of dark, grizzled hair, wild as a mane, hid its head and face [JE, 293].

There is something animalistic about the "horror" of Bertha Mason. She is a "strange wild animal", a "wolfish thing", a "clothed hyena" [JE, 293–294].[58] Nancy Armstrong argues that such a rampantly aggressive sexuality, so obviously capable of destroying the domestic harmony of the family home, represented a critique, albeit perhaps unconscious on the part of its author, of the fragility of the early Victorian domestic idyll.[59] The physicality is intense.[60] Corporeal and particularly sanguinary images are prevalent; allowing Charlotte to conjure up familiar Gothic images of vampyric demons, whilst also attaching contemporary associations between female insanity and disturbed menstrual cycles.[61]

Curiously, perhaps, Jane's perception of Bertha, which might be expected to confirm her creator's preconceptions, is in fact slightly more evasive; the product on the one hand of Jane's own status as an outsider and, on the other, of an ambiguity which reaches across Charlotte's corpus, between the

---

[58] For commentaries on the animalistic metaphors, see S. Thomas, "The Tropical Extravagance of Bertha Mason", *Victorian Literature and Culture* (1999): 7; Shuttleworth, *Charlotte Brontë and Victorian Psychology*, 165–166, and Small, *Love's Madness: Medicine, the Novel and Female Insanity*, 158–159, focussing particularly upon the hyena allusion which carried a particular contemporary familiarity in cultural depictions of mad women. According to Jenny Sharpe, the Bertha portrayed by Charlotte Brontë is "Calibanesque". See her *Allegories of Empire: The Figure of Woman in the Colonial Text* (Minneapolis: University of Minnesota Press, 1993), 43.

[59] N. Armstrong, *Desire and Domestic Fiction: A Political History of the Novel* (Oxford: Oxford UP, 1987), 164, 174.

[60] And original too, according to Sally Shuttleworth, at least in the context of mid-nineteenth century literature. See her *Charlotte Brontë and Victorian Psychology*, 164.

[61] See, in particular [JE, 208–214], describing the aftermath to Bertha's attack on her brother Richard. "She sucked my blood", Richard records, "she said she'd drain my heart". Jane makes the same allusion, to the "foul German spectre – the Vampyre" [JE, 284]. Even the medicine that Mason is given is "crimson". For commentaries on the prevalence of images of red and blood, see Showalter, *The Female Malady*, 67 and Shuttleworth, *Charlotte Brontë and Victorian Psychology*, 166–167.

alternative lures of sense and sensibility.[62] Initial encounters with Bertha are written to tempt, to entice her heroine into a more Gothic sensibility, one which might lure the "romantic reader", but which Jane must resist; the "demoniac laugh" from one of the rooms off the long passage "with its two rows of small black doors all shut, like a corridor in some Bluebeard's castle"; Mason's allusion to being attacked by a vampire; Jane's own recollection of Bertha's "fearful and ghastly" face as she leant over her bed two nights before her intended marriage to Rochester, a recollection which, she confesses, momentarily disrupted her own reason, rendering her "insensible from terror" [JE, 147–148, 213, 282–285].[63] All quite "creepy" as Queen Victoria famously observed.[64]

It has been suggested that this Gothic sensibility might have been impelled by Charlotte's desire, conscious or sub-conscious, to reproduce a "fairy-tale" romance, replete, not just with a putative princess rising from her place at the hearth, but with its monster incarcerated at the top of the castle. The fact that Rochester might also, in this paradigm, be depicted as a "beast" himself, against Jane's "beauty", merely reinforces the thought that there is a deeper psychological and sexual contest in play; the pilgrim Jane destroying the monster which has cast a spell over her "beast" and his enchanted home.[65] There is, of course, a different fairy-tale tradition which critics have

---

[62] For commentaries on Jane as outsider, and on the sense and sensibility theme, see Shuttleworth, *Charlotte Brontë and Victorian Psychology*, 148–155, and also C. Christ, "Imaginative constraint, feminine duty and the form of Charlotte Brontë's fiction", *Women's Studies* 6 (1979): 292–293.

[63] For a discussion of the Gothic, even Byronic, depiction of Bertha, see Small, *Love's Madness: Medicine, the Novel and Female Insanity*, 161–163, and also Grudin, "Jane and the Other Mrs Rochester: Excess and Restraint in *Jane Eyre*", 147–148.

[64] She also thought it a "wonderful book", the "description of the mysterious maniac's nightly appearances awfully thrilling". See her diary entries, for May 21, 1858, and undated 1880, in Allott, *The Brontës: The Critical Heritage*, 389–390.

[65] Critics often seize upon the predominance of fairy-tale images in conversations between Jane and Rochester. Their first encounter, at [JE, 113], is commonly cited. For a commentary on this psycho-sexual interpretation of a reinvested fairy-tale in *Jane Eyre*, focussing particularly on the resonances with *Beauty and the Beast*, see K. Rowe, "Fairy-born and human-bred: Jane Eyre's Education in Romance" in *The Voyage In: Fictions of Female Development*, eds. E. Abel, M. Hirsch and E. Land-land (Hanover and London: New England UP, 1983), 69–89. See also Glen, *Charlotte Brontë: The Imagination in History*, 58–61 and 66 preferring to perceive an allusion to the story of *Cinderella*, and noting Charlotte's teasing recourse to fairy-tales in Rochester's attempt to lure Jane first into marriage and then adultery, with the promise, "I will clasp the bracelets on these fine wrists, and load these fairy-like fingers with rings" [JE, 259]; a juxtaposition of images that aligns bondage with

subsequently projected into the fate of Bertha Mason: that of the sexually alluring concubine and the voracious oriental despot. Here again, Jane's allusion to Bluebeard's castle would have had an unmistakable resonance.[66]

But the plainer "truth" must out; the fairy-tale must give way to the personal testament.[67] There is no "ghostliness", no "curious cachinnation", even if the "clamorous peal" of laughter, as "tragic, as preternatural a laugh as any I ever heard", seemed to "echo in every lonely chamber" [JE, 107]. Silenced she might be, but Bertha's voice ultimately reinforces her animality. At the moment of her death, as the innkeeper recalls, she "yelled, and gave a spring, and the next minute she lay smashed on the pavement", as "dead as the stones on which her brains and blood were scattered" [JE, 427–428].[68] Bertha Mason may indeed appear to be a "shocking", even fantastical lunatic, at least at first glance. But she is also, as Charlotte readily admitted in her letter to her publisher, an ordinary and familiar one, a "natural" lunatic.

The second issue is that of confinement. The theme of "constraint", physical, emotional, intensely gendered, dominates *Jane Eyre*; the drama of social interaction and constraint, as Sally Shuttleworth has suggested, "played out [...] on the terrain of the female body".[69] And critics have long presumed that the depiction of Bertha's confinement was intended as a contribution to a current debate of likely interest to readers. H. F. Chorley, reviewing the novel for the *Athenaeum*, observed that the "mystery" of Thornfield was no "exaggeration", for "We, ourselves, know of a large mansion-house in a distant county where, for many years, a miscreant was kept in close confinement"; an observation which revisits the supposition that

---

fantasy. For a further comment noting the pervasive presence, alongside vulnerable princes, of evil madwomen in the culture of European fairy-tales, see Gilbert and Gubar, *The Madwoman in the Attic*, 36 and 351.

[66] The allusion to Bluebeard is complemented by Jane's later condemnation, at [JE, 269–70], of Rochester's proposed affair as the kind of thing more suited to a sultan's seraglio. For a commentary, see Ingham, *The Brontës*, 144–145.

[67] See Rowe, "Fairy-born and human-bred: Jane Eyre's Education in Romance", 89, and T. Eagleton, *Myths of Power: A Marxist Study of the Brontës* (Basingstoke: Macmillan, 1988), 17–20 and 75–78, imputing in the tension between the Gothic Jane and the pragmatic, an essential intellectual struggle between the rational and sentimental which pervaded, not just Charlotte Brontë's novels, but nineteenth century English literature as a whole.

[68] As critics such as Helen Small have noted, it is certainly altogether more brutal than that upon which it most obviously drew as a literary progenitor, the death of Ulrica in Sir Walter Scott's *Ivanhoe*. See Small, *Love's Madness: Medicine, the Novel and Female Insanity*, 157–158.

[69] See Shuttleworth, *Charlotte Brontë and Victorian Psychology*, 148, and also Grudin, "Jane and the Other Mrs Rochester: Excess and Restraint in *Jane Eyre*", 145.

Charlotte and her audience would probably have been quite familiar with such apparently dubious incarcerations.[70]

Aware that the common law would not countenance his divorce from Bertha, Rochester confesses that this solution appeared to be his only credible strategy; a supposition which, as we have again noted, would have chimed with many readers.[71] Rochester even contemplates his strategy in terms of a moral injunction, convincing himself, "Let her identity, her connection with yourself, be buried in oblivion: you are bound to impart them to no living being" [JE, 309]. Incarceration at Thornfield has the virtue, above all, of being discrete; allowing Rochester not merely to hide his wife from public view, but himself too. Rochester, as Mrs Fairfax confirms, withdrew from his own family and his own "society" on discovering the predicament in which he was placed by the terms of his father's estate settlement.[72] Thornfield is a house full of different inmates, all enduring different forms of confinement and alienation.[73]

Indeed, there is much in Rochester's strategy of confinement that would have chimed with contemporary attitudes, not least the recruitment of an experienced "keeper" Grace Poole, to whom he is more than willing to convey responsibility for the care of his first wife.[74] The same might be said of the apparent reluctance, wherever possible, to use physical restraints, which was commended by Conolly in his *Treatment of the Insane without Mechanical Restraint,* published just a few years later in 1856; though again, in the context of Bertha's final acts of destruction, the wisdom of this strategy is rendered rather less certain.[75] Interestingly, Conolly actually cited the "case" of Bertha Mason in his treatise, precisely to warn against such domestic confinement. Whilst the relationship between the Rochesters, he observed, became ever more "fierce and unnatural", Thornfield itself was rendered:

---

[70] In Allott, *The Brontës: The Critical Heritage,* 72.

[71] JE, 306–7.

[72] JE, 127–8. In essence Rochester resents being left to secure his own estate by locating a rich heiress. It is notable that Rochester's first recourse, on encountering any setback in the novel, is to retreat, to shut himself away from society. He confesses to Jane that he has even contemplated suicide in the past.

[73] See JE, 141–142 for a particularly suggestive conversation on this theme between Jane and Rochester, and also 146 for Jane's musing on Rochester's "alienation".

[74] JE, 306–309.

[75] See Showalter, *The Female Malady,* 68. Grace Poole does take the precaution of having rope always available, in order to restrain Bertha whenever necessary. See JE, 293, and also 309–10, commenting that Grace has proved to be, in general, a competent keeper.

Awful by the presence of a deranged creature under the same roof: her voice, her sudden and violent efforts to destroy things or persons; her vehement rushings to fire and window; her very tread and stamp in her dark and disordered and remote-chamber, have seemed to penetrate the whole house; and, assailed by her wild energy, the very walls and roof have appeared unsafe, and capable of partial demo-lition.[76]

It seems reasonable to suppose that Charlotte intended the destruction of Thornfield to articulate her own doubts in regard to strategies of internal or familial confinement.[77] Jane's casting of Rochester as Bluebeard reinforces this thought.[78] Later, in a slightly different perspective, Rochester confesses to Jane that

Concealing the mad-woman's neighbourhood from you, however, was something like covering a child with a cloak, and laying down near a upas-tree: the demon's vicinage is poisoned, and always was [JE, 300].

There again it might be argued that Rochester's strategy appears to have been effective, at least for a while; so much so that the local clergyman subsequently confesses himself astounded by the thought that there might have been a wife lurking in the attic for years.[79] Moreover, the audience is clearly supposed to sympathise with the incarcerator's predicament; dealing with a species of insanity conspicuous in its violence. Small wonder, perhaps, that Rochester's first inclination was to hide a wife who, not just in her apparent illness, but also in her original physiognomic condition, might have appeared to his contemporaries to be semi-barbarian. It was, after all, what Brougham and Peel and many others would have chosen to do.

In its presentation of Bertha Mason, *Jane Eyre* is ultimately evasive; both in terms of the nature of Bertha's madness and in its strategic interiorization in both Thornfield and in the novel. At a critical level, this evasion resonates very obviously with Irigaray's famous alignment of madness and the denial of voice.[80] It also resonates with Virginia Woolf's identification of a "twisted" rage in the novel; a temptation to self-immolation which must be

---

[76] J. Conolly, *The Treatment of the Insane without Mechanical Restraints* (1856), 149–150, quoted in Small, *Love's Madness: Medicine, the Novel and Female Insanity*, 54–55.

[77] This conclusion is reached by Sally Shuttleworth, in *Charlotte Brontë and Victorian Psychology*, 37–39 and 160.

[78] JE, 107.

[79] JE, 291, for the clergyman's response, when Bertha's existence is presented at the abortive marriage ceremony for Jane and Rochester.

[80] See Small, *Love's Madness: Medicine, the Novel and Female Insanity*, 25–27.

resisted.[81] It has been argued that Bertha's inarticulate rage might express
Brontë's own deeper sexual frustrations and anxieties; her own determination to resist the lures of "moral" insanity which might have been inferred
if she had succumbed to Rochester's invitation to adultery.[82] In this dominant interpretive paradigm, the two questions, of the nature of Bertha's madness, and the manner of confinement, come together; even as they articulate
a vivid and empowering testimony of silence and disempowerment.[83]

The Appeal of Bertha Mason

At the same time, of course, it must be admitted that Charlotte Brontë did
not write *Jane Eyre*, at least not primarily, as a strategic contribution to the
discourse of madness; even if she did consciously exploit its contemporary
resonance. The fate of Bertha Mason was shaped by the demands of the plot.
She dies because it is convenient, for Rochester, for Jane, and for Charlotte.
It is what Charlotte knew that her audience would expect.[84] Yet, the nature of
her death, like that of her confinement, cannot but make a statement. There
was, in short, in the presentation of Bertha Mason pretty much everything
that the incipiently neurotic mid-Victorian audience might have expected to
find when invited to consider the phenomenon of the madwoman in the
attic. Moreover, in stimulating debate, of course, in bringing before the public gaze precisely the kind of narrative or testament that was otherwise missing, it also challenged all the same prejudices.

---

[81] V. Woolf, *A Room of One's Own* (Oxford: Oxford UP, 1992), 90–95, further depicting a Charlotte at "war with her lot", and writing a novel infected with "anger".
For commentaries on Woolf's determination to deflect simplistic assumptions
that the "rage" of Charlotte Brontë, still less that of an entire sex, might be read
into the particular circumstances of either Jane Eyre or Bertha Mason, see Kaplan,
Victoriana: Histories, Fictions, Criticism, 18–19, Small, *Love's Madness: Medicine, the
Novel and Female Insanity*, 160 and Miller, *The Brontë Myth*, 159.

[82] On the sexual, Byronic nature of Rochester's threat to Jane, see H. Moglen, *Charlotte Brontë: The Self Conceived* (Madison, Wisconsin: University of Wisconsin Press,
1984), 118–20 and 126–128, and also M. Blom, "*Jane Eyre*: Mind as a Law unto
Itself", *Criticism* 15 (1975): 359–361, and C. Senf, "*Jane Eyre*: The Prison-House of
Victorian Marriage", *Journal of Women's Studies in Literature* 1 (1979): 354–356. For a
discussion of the broader parallel between the sexual surveillance to which the
two women in the novel are exposed, see Shuttleworth, *Charlotte Brontë and Victorian Psychology*, 151–152.

[83] See Small, *Love's Madness: Medicine, the Novel and Female Insanity*, 160–161, 169–170.

[84] See Gilbert and Gubar, *The Madwoman in the Attic*, 362, commenting on the necessary "liberating" function of Jane's death.

The after-life of *Jane Eyre* is relatively easy to trace. Famously, prospective publishers initially shied away from publishing a novel that oscillated with such intensity around issues of bigamy and adultery, as well as familial insanity.[85] Certainly contemporary critics were troubled.[86] And the depiction, and fate, of Bertha Mason, was particularly exercising; the primary "defect", according to G. H. Lewes, in *Fraser's Magazine*.[87] A number, most famously perhaps Elizabeth Rigby in the *Quarterly Review*, were simply appalled; a book of "great coarseness" and "horrid taste", its protagonist "unregenerate and undisciplined".[88] There was, the reviewer for the *Spectator* agreed, a rather "low tone of behaviour" abroad in the novel.[89] At the same time, despite, or perhaps because of, this reception, *Jane Eyre* rapidly became a publishing sensation, selling 2500 copies in the first couple of months of early autumn 1847. In terms of its audience, the importance of the novel is unarguable.[90] It is here that concerns, such as those expressed by John Bucknill, resonate. In like terms, Edward Seymour bemoaned a contemporary predilection for sensationalist novels such as *Jane Eyre*:

> Still the feeling fostered by novel writers (who never, by the way, as far as I know, really depict a lunatic case), the feeling for absolute secrecy which pervades society, the idea that where there is secrecy there is the opportunity for injustices – all these operate on the public mind to decry similar institutions.[91]

---

85 In correspondence with the critic G. H. Lewes, Charlotte confessed "It is an autobiography – not perhaps, in the naked facts and circumstances, but in the actual suffering and experience". At the same time, she readily conceded that "It has no learning, no research, it discusses no subject of public interest". Quoted in Miller, *The Brontë Myth*, 14 and Glen, *Charlotte Brontë: The Imagination in History*, 65. For an overview of this criticism, importing into it an unsurprising gender prejudice, see Gilbert and Gubar, *The Madwoman in the Attic*, 337–338, and also Kaplan, *Victoriana: Histories, Fictions, Criticism*, 21–22, musing on the peculiar "intensity" of the novel.

86 For comments on the nature of this contemporary critical reception, and the perceived subversion in the novel, see S. Foster, *Women's Fiction: Marriage, Freedom and the Individual* (Helm: Croom, 1985), 84–85 and also T. Winnifrith, "Charlotte and Emily Brontë: A Study in the Rise and Fall of Literary Reputations", *Yearbook of English Studies* 26 (1996): 15–16.

87 So did Eugene Forcade, reviewing the novel for *Revue des deux mondes*. For both reviews see Allott, *The Brontës: The Critical Heritage*, 84–85 and 100–101.

88 In Allott, *The Brontës: The Critical Heritage*, 106–11.

89 In Allott, *The Brontës: The Critical Heritage*, 75.

90 See Glen, *Charlotte Brontë: The Imagination in History*, 109–110, 134–136 and Ingham, *The Brontës*, 26.

91 In his *Letter to the Right Honourable the Earl of Shaftesbury, on the Laws Which Regulate Private Lunatic Asylums*, quoted in Suzuki, *Madness at Home*, 176.

Joseph Mason Cox agreed, revisiting the thought that such literature actually rendered young impressionable women more susceptible to "love-madness".[92]

The situation of the novel at the vanguard of an emergent genre of testamentary feminist literature was, however, quickly appreciated.[93] Margaret Oliphant, writing as soon as 1855, noted that "the most alarming revolution of modern times has followed the invasion of *Jane Eyre*", continuing perceptively that:

> Nobody perceived that it was the new generation nailing its colours to the mast. No one would understand that this furious love-making was but a wild declaration of the "Rights of Women" in a new aspect [...] Here is your true revolution. France is but one of the western powers; woman is the half the world [...] And this new Bellona steps forth in armour, throws down her glove and defies you – to conquer her if you can.[94]

The prescience is striking; for the reputation of *Jane Eyre* in this particular canon has become almost hagiographic. Elaine Showalter showers praise on a novel which represents a critical "evolution from Romantic stereotypes of female insanity to a brilliant interrogation of the meaning of madness in women's daily lives".[95] Sandra Gilbert and Susan Gubar deploy its "madwoman in the attic", Bertha Mason, as a paradigmatic figure, representative of an entire genre of feminist literature fashioned by the repressed creative anxiety of its authors.[96] Heather Glen likewise attests to the importance of *Jane Eyre* as an original "story of a woman's struggle for independence"; one that necessarily challenged all the received cultural prejudices which were levelled against the notion that autonomy was anyway in the better interests of

---

[92] See Small, *Love's Madness: Medicine, the Novel and Female Insanity*, 6–9, 44–48 and 56–62. Jane Austen famously seized upon this particular concern, regarding the impact of romance literature on young women, and deployed it to comic effect in *Northanger Abbey*. For a commentary on the "subversive" impact of such novels, and contemporary annuals of women's literature, see Glen, *Charlotte Brontë: The Imagination in History*, 116–125 and 134–143. In the circumstances, there is a further paradox in the recommendation of some psychiatrists that young women should read literature as a suitable form of treatment. The story of Ophelia was thought to be especially suitable, and sobering, and commonly prescribed.

[93] Ingham, *The Brontës*, 216–311.

[94] In Allott, *The Brontës: The Critical Heritage*, 312–313.

[95] Showalter, *The Female Malady*, 66.

[96] For commentaries on this paradigm, each of which is sceptical of overly simplistic interpretations of Jane as feminist crusader, see Kaplan, *Victoriana: Histories, Fictions, Criticism*, 17–19, Showalter, *The Female Malady*, 68 and also Miller, *The Brontë Myth*, 163–164.

women.[97] Speaking to the broader canonical import of *Jane Eyre*, Adrienne Rich confirms the novel's "special force and survival value". "The wind that blows through this novel", she concludes, "is the wind of sexual equality – spiritual and practical".[98] *Jane Eyre* has, in short, become a "modern myth".[99] And so has Bertha Mason.[100] The suppression of Bertha Mason was, of course, a critical element in Elizabeth Gaskell's attempt to purify Charlotte's reputation.[101] By the mid-twentieth century, however, Bertha had exacted her revenge, not just entrancing critics and casting her marital successor into a relative critical darkness, but also reasserting her identity and changing her name.[102]

A pivotal moment in this literary redemption was the publication of Jean Rhys' novel *Wide Sargasso Sea* in 1966; and then its interrogation two decades later by Gayatri Spivak, determined to restore the particular historical and colonial context of *Jane Eyre* against those critics who preferred to focus on the text as speaking across cultures and generations. This is not the place to investigate more closely the re-casting of Bertha Mason in Rhys' prequel.[103] But is the place to note Spivak's more particular comments on the underlying

---

[97] Glen, *Charlotte Brontë: The Imagination in History*, 25–26, 52.

[98] A. Rich, "Jane Eyre: The Temptations of a Motherless Woman", in her *On Lies, Secrets and Silence: Selected Prose 1966–1978* (New York: Norton, 1979), 89, 105.

[99] See Kaplan, *Victoriana: Histories, Fictions, Criticism*, 7, 15 and 31, emphasising its continued relevance in the context of modern debates regarding the domestic condition of women. According to Kaplan, *Jane Eyre* is a "mnemonic" of female "desire and rage", a text that is remembered through the generations. For similar attestations see also Miller, *The Brontë Myth*, 13–14, and Ingham, *The Brontës*, 223–225 and 232–244.

[100] In her *The Brontë Myth*, 164, Lucasta Miller presents a striking, if slightly odd, testament to the mythic status of Bertha Mason, in Roy Jenkins' description of Margaret Thatcher as the "great incubus of John Major's premiership, comparable with Mr Rochester's mad wife in *Jane Eyre*".

[101] See Miller, *The Brontë Myth*, 72 and 89, noting the apparent success of the strategy, as recorded by the daughter of the feminist critic Elizabeth Malleson, who recalled that her mother had read her the novel as a child "entirely omitting Rochester's mad wife, and so skilfully that we noticed nothing amiss with the plot".

[102] For an overview of this critical strategy, within which re-writings of Bertha Mason can be placed, see E. Baer, "The Sisterhood of Jane Eyre and Antoinette Cosway" in Abel, Hirsch and Langland, *The Voyage In*, particularly 131–133, 147–148. For a variant, which prefers to retain the received assumption that Bertha and Jane are bound by their "shared experiences" in Rhys' novel, just as they are by allusion in Charlotte Brontë's, see M. Emery, *Jean Rhys at "Worlds End": Novels of Colonial and Sexual Exile* (Austin: Texas UP, 1990), 56–57.

[103] See I. Ward, "The Rochester Wives", *Law and Humanities* 2 (2008): 123–128 for a closer discussion of this re-casting.

jurisprudence of Bertha Mason's case; a jurisprudence which is all the more apparent in *Wide Sargasso Sea*. Alongside the deep racism that Spivak reads into *Jane Eyre*, through Rhys' re-writing, there is the facilitative power of the law. This relation, of law and power and female subjugation was, of course, famously noted by Frances Power Cobbe, in her ironic conflation of "criminal, idiots, women and minors"; whose conjoined fate was enshrined in the "common law of England".[104]

Where Charlotte imputes the "force" of law, Rhys foregrounds it. Her Antoinette is confined by this force. It is because of the law, because her dowry will resuscitate the ailing Rochester estate, that she has been married. It is the law which conveys her property, and her person, to Rochester; even if the latter inclines to view the transaction more in terms of contamination than resuscitation.[105] Antoinette/Bertha is a chattel; passed by one man, her brother, to another, her husband. "She's mad", Rochester later agonises, "but mine, mine" [WSS, 106]. There are repeated allusions to her brother's failure to arrange a protective legal settlement for Antoinette.[106] "I am not rich now", the newly married Antoinette/Bertha informs an uncomprehending Christophine, "I have no money of my own at all, everything I had belongs to him". "That", she confirms, "is English Law" [WSS, 69]. The parallel is an obvious one. The sequestration of an estate through marriage in nineteenth century Jamaica, or indeed nineteenth century England, is no different in substance than the purchasing of a slave; "The Letter of the Law. Same thing as slavery. They got magistrate. They got fine, they got jail house and chain gang. They got tread machine to mash up people's feet".[107] When Antoinette/Bertha later appeals to her brother to redeem himself and to save her, Christophine's deeper suspicions are again vindicated. "I cannot interfere legally between your husband and yourself", Richard Mason replies. Bertha, an uncomprehending Jane records, then "flew at him". As Spivak notes, it was not the simple matter of confinement that triggered the rage. It is the "dissimulation" that Bertha "discerns in the word 'legally'."[108]

---

[104] In Hamilton, *Criminals, Idiots, Women and Minors: Victorian Writing by Women on Women*, 108–128.
[105] J. Rhys, *Wide Sargasso Sea* (Harmondsworth: Penguin, 2000) [hereafter WSS], 101.
[106] See WSS, 69, 72.
[107] See P. Le Gallez, *The Rhys Woman* (London: Macmillan, 1990), 142–146 for a sophisticated account of the legal intricacies Rhys excavates in the Rochester-Mason-Cosway marriage settlement, and also Thomas, "The Tropical Extravagance of Bertha Mason", 10–11, exploring the more immediate allusions to colonial law.
[108] G. Spivak, "Three Women's Texts and a Critique of Imperialism", *Critical Inquiry* 12 (1985): 249–251.

Rhys' Rochester becomes the physical embodiment of the laws of patriarchy, cultural and jurisprudential, all their various dissimulations, all the "masquerading". It is the "letter of the Law" which permits Rochester to return to England with his hated wife, and her estate; disregarding Christophine's desperate plea that he might at least leave his wife behind, with some remnants of her estate, and her freedom.[109] Christophine has anticipated Rochester's likely strategy: "It is in your mind to pretend she is mad. I know it. The doctors say what you tell them to say" [WSS, 103]. And it is the law, as Christophine also appreciates, that precisely allows him, even encourages him, to confine Antoinette to the third floor of Thornfield.[110] The very day following his final altercation with Christophine, Rochester is writing home, instructing his lawyers to engage a "discreet" staff and to expect his return.[111] As critics such as John Hearne have noted, Rochester needs the law, not just to secure his estate, or to allow him to hide his mad wife, but to justify the violence he inflicts on her, when he makes love to her, and when he later incarcerates her. It furnishes another set of collateral excuses for the original sin into which he fell, providing an emotional prophylactic for the paranoid self-loathing into which he is, thereafter, prone to falling.[112] The laws of patriarchy make men like Rochester feel a bit better about themselves; though not always, as Rhys imputes, that much better.

It is the enduring truth of this inference which continues to ensure the place of both *Jane Eyre* and indeed *Wide Sargasso Sea* in the feminist literary canon; and more particularly still that which engages the relation of law and power with the myriad "cultures" of madness. More generally, the nature of madness, and more prosaically how best to engage it, remains a matter of considerable, often heated, debate. As one contemporary historian observes

> We, like the Victorians, are still baffled by the nature of insanity and the best way to deal with it. Moreover, we, like them, are still sceptical of the ability of medicine to diagnose and treat it effectively. As long as these conditions exist, the fear of wrongful confinement is likely to continue.[113]

---

[109] WSS, 101–102.

[110] Spivak, "Three Women's Texts and a Critique of Imperialism", 249, 254.

[111] WSS, 105.

[112] J. Hearne, "The Wide Sargasso Sea: A West Indian Reflection" in *Critical Perspectives on Jean Rhys*, ed. P. Frickey (Washington DC: Three Continents Press, 1990), 190–191. The same conclusion is also reached, in the more immediate context of *Jane Eyre* by Sally Shuttleworth, in her *Charlotte Brontë and Victorian Psychology*, 169, where she concludes that Rochester, like so many of his contemporaries, has come to realise that his "legal dominance" has been "purchased at the cost of self-hatred". For instances of violence and ill-treatment in WSS, see 96–97.

[113] McCandless, "Dangerous to Themselves and Others", 104.

The gender consonance, meanwhile, is just as marked. The "demonstrable elements of bias" in the dominant Freudian paradigms of insanity remain, whilst the madwoman, particularly the violent madwoman, remains an especial totem of deviance; one cherished by newspaper editors and thriller writers alike.[114]

In this context, the case for recovering testaments of madness becomes urgent. The public, invariably patriarchal, narratives of the medical and legal professions must be challenged by the intensely private testaments, autobiographical or semi-autobiographical, of those who experienced the sharper end of successive Victorian madness and wrongful confinement "scares".[115] It is for this reason that the case of Bertha Mason matters. On the one hand it serves chronicle contemporary debates regarding insanity and the most appropriate strategies for its containment, whilst also confirming the root of present prejudices in modern legal and medical discourses of insanity, whilst on the other it engages an altogether larger jurisprudential enterprise. As Maria Aristodemou observes, writing to this grander aspiration, and deploying a metaphor of peculiar pertinence in the context of this article:

> While man was running away, or hiding in his own self-created prisons, in labyrinths variously called law, reason, knowledge, God, woman has travelled the world. From the prison he imposed on her called home she has been on a journey that does not return back to oneself, or lead to death, but reaches out towards the other.[116]

In the context of the case of Bertha Mason the allusion could not be more prescient, and the same is true of the attendant supposition.

---

[114] See M. Williams, *Empty Justice: One Hundred Years of Law, Literature and Philosophy* (London: Cavendish, 2002), 76, and also K. Kendall, "Beyond Reason: Social Constructions of Mentally Disordered Female Offenders" in *Women, Madness and the Law*, eds. W. Chan, D. Chunn and R. Menzies (London: Glasshouse Press, 2005), particularly 41–51, 56–57.

[115] See Williams, *Empty Justice: One Hundred Years of Law, Literature and Philosophy*, 214–215, urging the need to recover testaments of "passivity" as a counter to the narratives of "total agency" pronounced "on the part of the masculine".

[116] M. Aristodemou, *Law and Literature: Journeys from her to eternity* (Oxford: Oxford UP, 2000), 270–271.

Gary Watt
University of Warwick

The Case of Conjoined Twins:
Medical Dilemma in Law and Literature

When King Solomon was called upon to allocate a child to one of the two women claiming to be its mother, his task was to distinguish the right outcome from the wrong outcome.[1] The case of *Re A (Separation of Conjoined Twins)*,[2] a decision of the Court of Appeal of England and Wales, presented the court with a problem for which there was no right answer – a choice between a rock and a hard place. The dilemma was to leave baby twins conjoined and facing a high probability of death within a short timescale; or to separate them, thereby producing the almost immediate death of one whilst securing a good likelihood of normal lifespan for the other. The court of first instance and the Court of Appeal were asked to make a choice and, rightly or wrongly, made one; and their decision in favour of separation may or may not have been the least wrong one. I am not called upon to determine the dilemma, even in theory, so I will resist the temptation to favour one outcome over another on the specific facts of *Re A*. The purpose of this essay is to recommend improvements to the form and content of the conversations that ensue when officials and professionals are called upon to comment and decide upon a dilemma of this sort, whichever way they decide. It is difficult to escape the linguistic norms to which we are subject. How can we express our means of escape without employing the very language we mean to escape from? When we discuss the case of *Re A*, linguistic norms compel us to refer to "twins" (plural) and "one" and "other"; terms which imply nothing of the high degree of physical connection or contiguity between the baby girls. There is an individualistic bias to our social discourse (and certainly to our legal and scientific discourse) which is hard to avoid. I do not underestimate

---

[1] I am grateful to my friends and colleagues Marc Stauch and Jane Bryan for their insightful comments on this paper, in response to which I made a number of changes and should probably have made a number more. 1 Kings 3,16–28.
[2] In *Re A (Children) (Conjoined Twins: Surgical Separation)* [2001] Fam 147, Court of Appeal. Throughout the remainder of this paper quotes from this case are followed by references (RA, page number in parenthesis) to the relevant page number of this report.

the linguistic conventions that constrain medical and legal practitioners, nor do I underestimate the lengths to which the judges in the case of *Re A* undoubtedly went to wrestle with their decision, but I do argue that individualistic linguistic norms have obscured alternative modes of thought and expression which might have the capacity to confer metaphysical meaning on instances of physical confusion.[3] We will discover that alternatives are to be found in non-legal literature, but that they are also to be found concealed in legal literature, where they supply an internal critique of the law's own dominant norms of speech. We will also discover that alternatives have the power to enter legal language through the exercise of creative literary imagination.

The case of *Re A* began when a pregnant mother, carrying conjoined twin girls, travelled with her husband from their home on the Maltese island of Gozo to the United Kingdom. They had travelled because the medical profession in the UK has an expertise in delivering conjoined twins that could not be found on their rural island. They also came to consult on the possibility of surgical separation. When the children were born it became apparent that one of the twins, called "Mary" for the purpose of the subsequent legal proceedings, was wholly dependent upon the cardio-vascular system located within the skeletal structure of the other twin, "Jodie". As Brooke LJ stated in his summing up, Mary's heart, brain and lungs were "for all practical purposes useless" and "nobody would have even tried to extend her life artificially if she had not, fortuitously, been deriving oxygenated blood from her sister's bloodstream" [RA, 239]. In the words of Robert Walker LJ, "the awful paradox at the centre of this case" lay in the fact that Mary was "alive as a distinct personality" but was not "viable as a separate human being" [RA, 242]. The medics saw that the strain on Jodie's heart would cause cardiac arrest and the death of both twins within a period of a few years at best, but more likely much sooner. They therefore applied to court for permission to separate the babies despite the parents' objection, based on their Roman Catholic religious beliefs, that they would prefer both babies to die naturally rather than kill one to save (and inevitably scar, possibly debilitate) the other. The hospital had received the family for the purpose of delivering the babies and advising on the possibility of separation, but when the parents objected to the proposed separation, the hospital sought to *enforce* it through the courts. Thus the hospital turned from host to hostage-taker. That might

---

[3] Compare A. Domurat Dreger, who advances the argument that conjoined twins are prejudiced by norms of body image perpetuated by singletons (her term for people who are physically un-conjoined): *One of Us: Conjoined Twins and the Future of Normal* (Cambridge, Mass.: Harvard University Press, 2004).

sound like hyperbole (after all, the parents at no stage bore any personal ani-
mosity to the individual medics charged with care of the twins), but it is
sobering to note that a modern hospital can, without any hint of official cen-
sure, commit that species of treachery by host to guest which the ancients
considered to be a breach of the fundamental law of *xenia* ("hospitality")
and, as such, to be deserving of the Furies' fierce revenges. The hospital's
decision to seek separation against the parents' express wishes was the first
of many evil means employed in the course of pursuing the medics' conviction that surgical separation of the twins would be the right and legitimate
end. It might be thought that in the case of a true dilemma, where both ends
are bad, no end should be pursued where there is patent evil in the means by
which it is achieved. Ward LJ decided, however, that the medics had "a
proper interest" in asking the court to decide; even going so far as to state
that to decide questions of life and death is "surely and pre-eminently a
matter for a court of law to judge. That is what courts are here for" [RA,
174]. His lordship was painfully aware that this was a case with no right
answer, but "after anxious thought" he concluded "that the court cannot ab-
dicate responsibility and simply say it is too difficult to decide" [RA, 192].

The possibility of leaving the terrible choice to the parents did not com-
mend itself to the court. Perhaps the court's intervention was motivated by a
paternalistic concern to spare the parents the ordeal of the decision. That
would hardly be an excuse, but it would at least be an explanation. Without it,
it is hard to see what better qualification than the parents the legal and medical professionals had to determine on this dilemma. The court had, after all,
acknowledged that "[o]ther medical teams may well have accepted the parents' decision" and that, had the medics in this case done so "there could not
have been the slightest criticism of them for letting nature take its course in
accordance with the parents' wishes" [RA, 173]. It is only because the issues
in this case were framed in terms of the competing best interests of two individuals that the court could claim a special competence to hear the matter, but
it was the medics and the lawyers between them who framed the issues in
those terms. The court assumed competency to hear the case on the basis of
a language constructed on its own individualistic terms. There were alternatives. The court might have disclaimed competence on the ground that the
dilemma concerned a choice between wrongs as opposed to a choice between
rights, or, better still, the court could have framed the dilemma in this case as a
dilemma faced by both twins, as a shared problem.[4] Instead it framed the di-

---

4 M. Q. Bratton and S. B. Chetwynd, "One Into Two Will Not Go: Conceptualising
Conjoined Twins", *Journal of Medical Ethics* 30 (2004): 279–285.

lemma in terms of a competition between them, as is painfully apparent from the following part of the judgment of Ward LJ, which should be quoted in full:

> In this unique case it is, in my judgment, impossible not to put in the scales of each child the manner in which they are individually able to exercise their right to life. Mary may have a right to life, but she has little right to be alive. She is alive because and only because, to put it bluntly, but none the less accurately, she sucks the life-blood of Jodie and she sucks the lifeblood out of Jodie. She will survive only so long as Jodie survives. Jodie will not survive long because constitutionally she will not be able to cope. Mary's parasitic living will be the cause of Jodie's ceasing to live. If Jodie could speak, she would surely protest, "Stop it, Mary, you're killing me". Mary would have no answer to that. Into my scales of fairness and justice between the children goes the fact that nobody but the doctors can help Jodie. Mary is beyond help.
>
> Hence I am in no doubt at all that the scales come down heavily in Jodie's favour. The best interests of the twins is to give the chance of life to the child whose actual bodily condition is capable of accepting the chance to her advantage even if that has to be at the cost of the sacrifice of the life which is so unnaturally supported. I am wholly satisfied that the least detrimental choice, balancing the interests of Mary against Jodie and Jodie against Mary, is to permit the operation to be performed [RA, 197].

Towards the conclusion of his judgment, his lordship even went so far as to state that he could "see no difference in essence between" the legitimacy of killing a "six-year-old boy" who was "indiscriminately shooting all and sundry in the school playground" and the legitimacy of killing Mary in defence of Jodie [RA, 204]. That the judge does not acknowledge any essential difference between these two cases does not so much reflect upon the judge himself as upon the law's narrowly abstract view of what counts as essential.

Having framed the issues in terms of competition between the twins, it was almost inevitable that the court would decide to approve the medical procedure to separate them.[5] The process of separating the twins in thought and word prepared them for separation by scalpel. There was still a huge hurdle to overcome, of course, for this would be the first case in which, contrary to parents' expressed wishes, any court had sanctioned the killing of an innocent child to save the life of another. If their lordships imagined that they were searching for the legally right or wrong solution to the problem before them, they were searching in vain. There could be no legally "correct" outcome to the conundrum of the conjoined twins – certainly there was no binding precedent for the case, and their lordships made it clear that they did not intend their decision to stand as a precedent for future cases. Robert Walker LJ rightly acknowledged that there was "no helpful analogy or parallel to the situation" [RA, 255] which the court had to consider in *Re A*, but the case

---

[5] See, further, Bratton and Chetwynd, "One Into Two Will Not Go".

nevertheless became a battleground for analogy and metaphor. If ever a case proved the importance of a literary analysis of legal language, this was it. The case did not entail a search for governing law, but a search for a governing story by which their lordships could convince themselves and their audience. Their lordships' choice of metaphor should concern us, for forms of official speech always have substantial significance. When the court assumed, or accepted, the task of speaking for the voiceless baby girls it also assumed or accepted the task of providing a fitting narrative, and ought to have been mindful that in the case of Mary it would be the lasting official account of her life. Their lordships rightly rejected the analogy, which was used at first instance, that depicted Jodie as the life-support machine for Mary; as indeed they had to, for taken to its natural conclusion the analogy would lead to the illogical proposition that the child Mary be "switched off" to save "the machine" Jodie. The metaphor of the child as machine clearly failed to do justice to Jodie, but the descriptions employed by Ward LJ in the passage quoted earlier surely failed to do justice to Mary. In that passage we are told that she "sucks the lifeblood out of Jodie", that her living is "parasitic" and that she is "killing" Jodie. If such descriptions appear to add insult to the injury, his lordship might argue that he depicted Mary as an assailant in order to justify a finding that the medics had acted in Jodie's defence. He might argue, therefore, that he added the insult to *avoid* finding an injury. If so, that argument fails to convince. To kill an innocent child is always injurious. It is not only morally wrong, but legally wrong. That it might, morally and legally speaking, be the lesser of two injuries does not change the fact that it is an injury. At another point in his judgment, Ward LJ compared the parents' choice to a choice that might have been presented to parents at the gates of a concentration camp:

> If a family at the gates of a concentration camp were told they might free one of their children but if no choice were made both would die, compassionate parents with equal love for their twins would elect to save the stronger and see the weak one destined for death pass through the gates. [RA, 196]

The concentration camp analogy, which appears to be inspired by the story of *Sophie's Choice*,[6] was doubtless a well-intentioned attempt to express the insolubility of the parents' dilemma. It should not be taken literally, but it de-

---

6 W. Styron, *Sophie's Choice* (New York: Random House, 1979). The novel was made into a film (Alan J. Pakula, 1982) in which Meryl Streep played the part of Sophie, for which she was awarded an Academy Award for Best Actress. The fictional story of *Sophie's Choice* is a great deal more circumspect on the merits of making the choice. In Sophie's case she "decided" to save her son and allow her daughter, the younger sibling, to be taken to her death. As things transpired, Sophie was tortured by her choice and committed suicide in later life.

mands to be taken seriously; and if it is taken seriously, it inevitably prompts one to search the facts of *Re A* for equivalents to "the parents", "the children" and "the murderous officials". It is by no means straightforward to find them. To identify the children with the twins is unproblematic. It is also relatively straightforward to identify the officials with the hospital and the court, for they were the officious agencies in this case, and, "to put it bluntly, but none the less accurately", the "killers"; even allowing for the argument that to kill Mary was less evil than to let both twins die. Certainly the parents of Jodie and Mary cannot be accused of killing, even if they might be accused of choosing not to save. We might assume, therefore, that the parents of Jodie and Mary are equivalent to the parents in the concentration camp analogy; but that is decidedly not the case, for the parents of Jodie and Mary were offered no real power over the fate of their daughters. It turns out that the judge has imagined himself in the position of the parents in the concentration camp analogy, acting as *parens patriae*. The image therefore deflects attention from the judge's official role as superintendent over life and death, whilst subtly excising the parents of Jodie and Mary from the picture.

There are rare cases, and this was one, in which the typical language of law, and specifically the language of competing individual rights, is bound to fail (despite the judge's claim to be "in no doubt at all"). What is then required is to employ another language; a language of communal wisdom. A family can provide this for itself, but judges should, perhaps, recognise that they also have capacity to provide it, even if they have to admit that, in cases beyond the law's official remit, they are speaking as State-appointed wise men and women; speaking lore, not sitting as judges of law. So, what narrative might the court have supplied to explain (perhaps to justify) the surgical separation in this case? What is required is a narrative that will capture the essentials of the dilemma as closely as possible but which removes any suggestion of competition between the twins and any sense that the weaker twin is assaulting the stronger. If the narrative can supply a worthy epitaph to the weaker twin and at the same time a story to comfort and inspire the survivor, so much the better. The analogy of mountaineers, roped together, was mentioned briefly in *Re A*; suitably extended, it might serve:

> These baby girls are like expert mountaineers, also twin sisters, who have overcome immense challenges together and now face their most daunting challenge to date – to scale the north face of the Eiger. It is not for sport that they undertake this challenge, but because the very nature of their whole life together depends upon meeting and overcoming such challenges as they present themselves. Together they are undaunted. They begin their ascent, roped together, as always they are. Tragically, when far from the ground, but still far short of the summit, an accident occurs. An avalanche of snow and stone falls down. The stones strike the

trailing twin, knocking her unconscious and causing her to fall and break her legs so badly that loss of blood will kill her before she has any chance of being hauled to the summit or helped to safety. The conscious twin now clings on to the mountain for dear life, but the single piton by which the rope is held cannot bear the combined weight of both twins for very much longer. Even if the piton holds, exposure to the elements will, before long, take both their lives. The conscious twin has no means to cut the rope, and even if she had the means she would by no means cut it. Perhaps the law would excuse her if she did, perhaps the law would even say that she has the right to cut the rope,[7] but she cannot in conscience kill her twin to save herself. The unconscious twin has no such dilemma, for she is deep in unknown dreams, and she can never wake. Now suppose that in a dream an offer comes to her by an unseen hand: to cut the rope and send her sleeping to a peaceful death. Will she accept? Would she sacrifice herself to save her sister? We think she surely would. Today, in the case of Mary and Jodie, we make that offer and hear that reply.

This, I suggest, provides a much more appropriate account of the dilemma and a more fitting epitaph for Mary than the implication that she is a "blood-sucker", "parasite" and "sister-killer". We have seen, though, that those labels fit with the law's demand that there be an assault on Jodie before it can be asserted that Mary was justifiably killed in Jodie's defence. Thus the language of the law compels a distortion of the language of life. It might have distorted it further, for Ward LJ was not insensitive to the need for humane language. His lordship deliberately resisted the option of labelling Mary "an unjust aggressor", noting that this "American terminology" would be "wholly inappropriate language for the sad and helpless position in which Mary finds herself" [RA, 203]. His lordship was right, of course, but one wonders, with respect, whether he might have done better to avoid the suggestion that Mary's "parasitic living", "sucks the lifeblood out of Jodie" and that Jodie would have said "stop it, Mary, you're killing me". Even if the law demands a finding of assault it is nevertheless possible, with imagination and sensitivity, to accommodate such a finding within the story of the mountaineers. Thus the story could be amended to include the following coda:

> But the law will not let us hear Mary's request and it will not let us act on her reply. So we must return to her and explain that we must show that Jodie has acted in self-defence, and that for this we must allege that Mary is a threat to Jodie. And so we imagine that we have asked Mary for permission to say that she is a threat to Jodie, and we imagine that she would reply "if it will save my sister, you may call me whatever you like".

A narrative along these lines might, in fact, have been in the minds of the judges in *Re A*. If so, it is regrettable that they did not feel able to express

---

[7] C. Oakes Finkelstein, "Two Men and a Plank", *Legal Theory* 7.3 (2001): 279–306.

them for the record. There is no denying that legal categories are, and inevitably will be, abstracted from life; but there are cases in which the process of abstraction should be expressly acknowledged and a fuller account provided. When the court accepted jurisdiction over the dilemma in *Re A*, it framed the question in the empirical language of scales (an ingrained, but nevertheless inadequate, metaphor for justice) and thereby committed itself to disposing of the question in terms of conflict and cold calculation. An issue such as this ought to have been convened in a court of pathetic conversation. In his novel *Hard Times*, Dickens satirises empirical utilitarianism in the figure of Thomas Gradgrind,[8] who is introduced in the second chapter (entitled "Murdering the Innocents") in the following terms:

> THOMAS GRADGRIND, sir. A man of realities. A man of facts and calculations. A man who proceeds upon the principle that two and two are four, and nothing over, and who is not to be talked into allowing for anything over. Thomas Gradgrind, sir – peremptorily Thomas – Thomas Gradgrind. With a rule and a pair of scales, and the multiplication table always in his pocket, sir, ready to weigh and measure any parcel of human nature, and tell you exactly what it comes to. It is a mere question of figures, a case of simple arithmetic.[9]

Gradgrind is a friend to facts and a foe to "fancy". Is there something transcendental, almost ineffably beautiful, in the power and grace of a horse? Not for Gradgrind. When Sissy Jupe (he calls her "girl number twenty") cannot define a horse, he calls on another child to perform:

> "Bitzer", said Thomas Gradgrind. "Your definition of a horse".
> "Quadruped. Graminivorous. Forty teeth, namely twenty-four grinders, four eye-teeth, and twelve incisive. Sheds coat in the spring; in marshy countries, sheds hoofs, too. Hoofs hard, but requiring to be shod with iron. Age known by marks in mouth".
> Thus (and much more) Bitzer.
> "Now girl number twenty", said Mr. Gradgrind. "You know what a horse is".[10]

The satire of Gradgrind is effective as a critique of the empiricist brand of judgment in *Re A*, right down to the application of "a rule and a pair of scales". The authors of the excellent critical analysis of *Re A* entitled "One Into Two Will Not Go",[11] are alert to the inadequacy of simplistic mathematical formulae and figures when it comes to appreciating the complexity of the human figure in a conjoined state. It is not possible justly to dispose of a case like *Re A* through a formulaic allocation of body parts, still less justifi-

---

8 Commonly thought to represent Jeremy Bentham.
9 C. Dickens, *Hard Times* [1854] (London: Penguin Popular Classics, 1994), 2.
10 Dickens, *Hard Times*, 4.
11 Bratton and Chetwynd, "One Into Two Will Not Go".

able to conceive of the case as a competition between the twins. It is never a "mere question of figures". It is not about numerical counting, but about producing a just account. The officials are bound to fail in their attempt to produce a just outcome in a case such as this, but they must not compound that failure by producing an unjust narrative.

One way to challenge the assumption that an empirical "head count" approach to problems of medical ethics is inevitable is to show that ideas of physical and numerical confusion are deeply embedded in some of our legal, cultural and political conceptions of the beautiful and worthwhile. In 2005, a girl who was born with four arms and four legs was named Lakshmi by her Hindu parents; an acknowledgment of her resemblance to the four-armed Hindu goddess. This illustrates the potential to ascribe a positive story to an atypical physical form. It does not mean that the form will remain unaltered (in 2007, Lakshmi underwent surgery to leave her with two arms and two legs), but it does provide an alternative to the generally negative account of conjoined physicality that is produced by the dominant discourse.[12] It is trite to observe that unusual physiology is not the norm, but a mistake that is often made is to identify the norm (the normal) with the ideal. If there is an "ideal" human form, a regular template of the human body, it is a type that can never correspond to human reality in every detail; like a type font that matches no person's individual handwriting. I am not suggesting for one moment that a conjoined physical state is something that one (or two) might aspire to, but I am suggesting that where it exists one ought to apply one's imagination to the appreciation of its positive as well as its negative aspects. It follows that, in the event of separation surgery, one ought to acknowledge, as Bratton and Chetwynd urge us to acknowledge, that something significant is lost in the process of separation, even as something is gained.[13] One of the things that is lost through separation surgery is a degree of human closeness and contiguity that the vast majority of us will never experience. The embrace of family and friends is close, but not as close; even the physical intimacy of lovers is not so fundamentally confused. The pathetic sight of conjoined twin babies depicts utter closeness and interdependence and therefore signifies something of what it means to be fully human. In Plato's *The Symposium*, the comic character of Aristophanes employs the image of a conjoined "Androgyne" to account for the universal human drive to find utmost intimacy with another person. It may be nothing more than a satire on the whole genre of "origin myth", but at least it provides a broadly positive image of physically conjoined persons:

---

12 On this, see further, Dreger, *One of Us: Conjoined Twins and the Future of Normal.*
13 Bratton and Chetwynd, "One Into Two Will Not Go".

[…] the original human nature was not like the present, but different. The sexes were not two as they are now, but originally three in number; there was man, woman, and the union of the two, having a name corresponding to this double nature, which had once a real existence, but is now lost, and the word "Androgynous" is only preserved as a term of reproach. In the second place, the primeval man was round, his back and sides forming a circle; and he had four hands and four feet, one head with two faces, looking opposite ways, set on a round neck and precisely alike; also four ears, two privy members, and the remainder to correspond. He could walk upright as men now do, backwards or forwards as he pleased, and he could also roll over and over at a great pace, turning on his four hands and four feet, eight in all, like tumblers going over and over with their legs in the air; this was when he wanted to run fast […] Terrible was their might and strength, and the thoughts of their hearts were great, and they made an attack upon the gods […] At last, after a good deal of reflection, Zeus […] said: "Methinks I have a plan which will humble their pride and improve their manners; men shall continue to exist, but I will cut them in two and then they will be diminished in strength and increased in numbers; this will have the advantage of making them more profitable to us. They shall walk upright on two legs, and if they continue insolent and will not be quiet, I will split them again and they shall hop about on a single leg". He spoke and cut men in two, like a sorb-apple which is halved for pickling, or as you might divide an egg with a hair; and as he cut them one after another, he bade Apollo give the face and the half of the neck a turn in order that the man might contemplate the section of himself: he would thus learn a lesson of humility […] After the division the two parts of man, each desiring his other half, came together, and throwing their arms about one another, entwined in mutual embraces, longing to grow into one […].[14]

Modern science confirms that there is an element of ambiguity, even androgyny, in our physical makeup as we develop in our mother's womb. It is trite to say that the original diploid cell, the zygote (from the Greek for "yoke") is made up of male and female gametes (the male sperm and the female egg) yoked together, but even at the foetal stage males and females share the same form (phenotype) despite differences in their genetic makeup (genotype). It is not until several weeks after conception that a foetus with an "x" and a "y" chromosome starts to develop into a male (a foetus with two "x" chromosomes continues to develop as a female). Plato's account captures this very basic sense that two have derived from one. It also captures the fact that monozygotic twins are formed, "two out of one", from a single primal cluster of cells. This is the process which, if not performed completely, produces conjoined twins. The story of Adam and Eve, fabulous as it seems to modern minds, also captures something of the genetic science. The orthodox translation tells us that God took a rib from Adam and made woman from it, but the word "rib" can also be translated as "side". We now know

---

[14] Plato, *The Symposium* (Charleston: Forgotten Books, 2008), 22–24.

that, generally speaking, so far the sex chromosomes are concerned, the male has two different "sides" – "x" and "y" – and that the female has only the "x" side. Feminists have even argued that the original Adam was an Hermaphrodite,[15] having female and male sides to his/her nature in a manner not so very different to Plato's Androgyne. Of course, the Biblical and Platonic stories were not solely concerned to provide an account of human origins. They were also concerned to provide a narrative to explain the reality of human relationships, and in particular to account for a human being's felt need for a companion or lover. The story of the one who became two is not offered as a factual reality on its own account, but as a way of accounting for the fact that true companions or lovers have the desire and capacity to make "two become one". This theme of companions and lovers as halves of a whole is, of course, a familiar one in romantic literature. It is a theme that Shakespeare explores with typical accomplishment in a poem he wrote for a 1601 collection on the "phoenix and the turtle-dove" (the phoenix representing the Queen and the turtle-dove a doting courtier).[16] It has been described as "the first great published metaphysical poem".[17] Since 1807, the poem has been known as "The Phoenix And Turtle" or by its first line "Let the Bird of Loudest Lay". The following are stanzas 7 to 10:

> So they loved, as love in twain
> Had the essence but in one;
> Two distincts, division none:
> Number there in love was slain.
> Hearts remote, yet not asunder;
> Distance, and no space was seen
> 'Twixt this turtle and his queen:
> But in them it were a wonder.
> So between them love did shine,
> That the turtle saw his right
> Flaming in the phoenix' sight;
> Either was the other's mine.
> Property was thus appalled,
> That the self was not the same;
> Single nature's double name
> Neither two nor one was called.

---

[15] A. T. Reisenberger, "The creation of Adam as hermaphrodite – and its implications for feminist theology", *Judaism: A Quarterly Journal of Jewish Life and Thought* 42 (1993): 447–452.

[16] *Love's Martyr: or Rosalins Complaint. Allegorically shadowing the truth of Loue, in the constant Fate of the Phoenix and Turtle etc* [1601].

[17] P. G. Cheney, *The Cambridge Companion to Shakespeare's Poetry* (Cambridge: Cambridge University Press, 2007), 117.

When Shakespeare writes that number was "slain" and that property was "appalled", he is acknowledging, even celebrating, the fact that the paradox of "two as one" cannot be expressed, still less appreciated, in terms of mathematical and empirical science.

Shakespeare's play *Twelfth Night: Or What You Will*, which is thought to have been written in 1601, and is known to have been performed in December 1602 to an audience of lawyers at Middle Temple (one of the Inns of Court), culminates in a moving reunion of brother and sister twins in which Shakespeare alludes to Plato's myth of the Androgyne and, specifically, to the metaphor of the halved apple.[18] Confronted with the spectacle of Sebastian alongside his sister Viola (disguised as a man), Sebastian's friend Antonio asks:

> "How have you made division of yourself?
> An apple, cleft in two, is not more twin
> Than these two creatures [...]" (5.1.207–9)

The scene is made all the more moving by the fact that Shakespeare was himself the father of boy and girl twins, Hamnet and Judith, and the fact that Hamnet had died in 1596, aged 11. It is not unreasonable to speculate that *Twelfth Night*, which depicts the tragic separation of twins and their fantastic reunion, was at some level written with his own children in mind.

In Early Modern Europe, the metaphysics of two as one was as significant to politics as it was to poetry. Indeed, it was hardly possible to separate the body politic from the body poetic at the time. (The theme of the Phoenix and Turtledove was, as already mentioned, employed to describe relationships at court, rather than purely private love.) It is not surprising, therefore, that in *Calvin's Case* (1608),[19] Sir Edward Coke, the Lord Chief Justice of England, drew upon the contemporary (and classical) image of the conjoined body, to describe the union between England and Scotland that had been brought about by the accession of James I (James VI of Scotland) to the English throne in 1603. The case concerned the question whether a Scot (in this case Robert Calvin) who was born after the accession of King James I could own land in England, hence the case is also known as *The Case of the Postnati*:

---

[18] J. Bate and E. Rasmussen, *The RSC Shakespeare: Complete Works* (London: Macmillan, 2007), 645–646. In this paper, all quotations from the works of Shakespeare are taken from this edition.

[19] *Calvin's Case*, or the *Case of the Postnati* [1608] 7 Coke's Reports 1a; 77 ER 377. This text, including translations of the original Latin sections, is taken from ed. S. Sheppard, *The Selected Writings of Sir Edward Coke*, vol. I (Indianapolis, Indiana: Liberty Fund, 2003), 200–201.

no man will say, that now the King of England can make warr or league with the King of Scotland, *et sic de caeteris*: [and thus the union] and so in case of an alien born, you must of necessity have two several ligeances to two several persons. And to conclude this point concerning laws, *Non adservatur diversitas regnor' sed regnant', non patriarum, sed patrum patriar', non coronarum, sed coronatorum, non legum municipalium, sed regum majestatum.* [a distinction is not to be made of realms, but of rulers; not of countries, but of fathers of countries; not of crowns, but of the crowned; not of municipal laws, but of king's majesties] And therefore thus were directly and clearly answered, as well the objections drawn from the severalty of the kingdoms, seeing there is but one head of both, and the *Postnati* and us joyned in ligeance to that one head, which is *copula et tanquam oculus* [a coupling, and, as it were, an eye] of this case; as also the distinction of the Laws, seeing that ligeance of the subjects of both kingdoms, is due to their Sovereign by one law, and that is the Law of nature [...][20]

Later in the passage, the Lord Chief Justice repeats that the union of Scotland and England under one king is "a uniting of the hearts of the subjects of both kingdoms one to another, under one head and sovereign".[21] How poignant, and ironic, that the law of England should struggle to conceptualise the conjoined physical constitution of two human beings, when one of its greatest judges had conceptualised the constitutional union of England and Scotland in terms of two bodies united by a single heart and head. Sir Edward Coke even criticised arguments which *"disjungere conjungenda"* ("separate things which ought to be conjoined").[22] His Lordship also noted that the conjoined nature of the newly united nations was reflected in the official political imagery of heraldry, for the coat of arms of King James I included, in Coke's words, a "union and conjunction [...] of the three Lions of England, and that one of Scotland, united and quartered in one escutcheon". Even today the shield (escutcheon) of the Royal Coat of Arms is quartered. When used in England, two quarters bear the three lions of England and one quarter the solitary lion of Scotland; when used in Scotland, two quarters bear the solitary lion of Scotland and one quarter bears the three lions of England. In each case the fourth quarter bears the harp of Ireland. Shakespeare also refers to heraldry as a symbol of friendship that is close to the point of being conjoined, as in the following speech from *A Midsummer Night's Dream*, where Helena eulogises her friendship to Hermia:

---

[20] 7 Coke's Reports 1a, 14b.
[21] 7 Coke's Reports 1a, 15a.
[22] 7 Coke's Reports 1a, 15a.

> We, Hermia, like two artificial gods,
> Have with our needles created both one flower,
> Both on one sampler, sitting on one cushion,
> Both warbling of one song, both in one key,
> As if our hands, our sides, voices and minds,
> Had been incorporate. So we grow together,
> Like to a double cherry, seeming parted,
> But yet an union in partition,
> Two lovely berries moulded on one stem,
> So with two seeming bodies but one heart,
> Two of the first, like coats in heraldry,
> Due but to one and crownèd with one crest.
> (3.2.204–215)[23]

Writing some two and half centuries after Shakespeare and Coke, the artist John Ruskin reflected on the aesthetic and metaphysical appeal of the conjunction of colour in heraldic composition, which he considered to be exemplary of "an eternal and universal" principle that is found, not only in art, but in "human life" and even in the political union of "nations that are unlike". Ruskin called it the principle of "reciprocal interference":

> Whether, indeed, derived from the quarterings of the knights' shields, or from what other source, I know not; but there is one magnificent attribute of the colouring of the late twelfth, the whole thirteenth, and the early fourteenth century, which I do not find definitely in any previous work, nor afterwards in general art, though constantly, and necessarily, in that of great colourists, namely, the union of one colour with another by reciprocal interference: that is to say, if a mass of red is to be set beside a mass of blue, a piece of the red will be carried into the blue, and a piece of the blue carried into the red; sometimes in nearly equal portions, as in a shield divided into four quarters, of which the uppermost on one side will be of the same colour as the lowermost on the other; sometimes in smaller fragments, but, in the periods above named, always definitely and grandly, though in a thousand various ways. And I call it a magnificent principle, for it is an eternal and universal one, not in art only, but in human life. It is the great principle of Brotherhood, not by equality, nor by likeness, but by giving and receiving; the souls that are unlike, and the nations that are unlike, and the natures that are unlike, being bound into one noble whole by each receiving something from, and of, the others' gifts and the others' glory".[24]

Conscious that he was trespassing "upon too high ground", Ruskin even went so far as to suggest that there is no greater Divine ordinance than that "the most lovely and perfect unity shall be obtained by the taking of one na-

---

[23] Compare Lysander's romantic overture to Hermia: "[…] my heart unto yours is knit / So that but one heart we can make of it". (2. 2. 47–8.).

[24] J. Ruskin, *The Stones of Venice* (London: The Folio Society, 2001), 282–283. *The Stones of Venice* was first published in three volumes between 1851 and 1853.

ture into another". He ventured this to be a "vast" law that "has rule over the smallest things". Amongst the "smallest things" we might find "conjoined baby twins", the flesh of one committed to the other through "reciprocal interference", connected to the other not by the equality of scales nor by individual likeness (in resources, strength or anything that can be measured), but "by giving and receiving". We might even find that two may be "bound into one noble whole".

Michele Sesta
University of Bologna

Vida Interminable:
Patients and Family Members Between the Right to Live and the Obligation Not to Die

Isabel Allende's short story *Vida Interminable* (in *Cuentos de Eva Luna* or *The Stories of Eva Luna*, 1989) tells the story of two spouses who lived their entire lives together, "so close that over the years they had come to look like brother and sister".[1]

Roberto was a doctor who had developed the belief that every man had the right to a peaceful (*apacible*) death from his clinical experience, during which he had nurtured a great compassion for the fragile bodies of patients who were shackled to life-machines and tormented by needles and tubes. In his opinion, science denied these patients a worthy decease under the pretext that one must be kept breathing at all costs. As a doctor, he suffered because he could not help them depart this world and was compelled to hold them languishing in their beds against their will.

On pondering over these topics Roberto, while believing that a physician's work must consist of facilitating death rather than contributing to what he deemed was its bothersome bureaucracy,[2] realised that such a decision could not depend exclusively on the medical staff's evaluation or on the com-

---

[1] I. Allende, "Interminable Life" in *The Stories of Eva Luna*, trans. Margaret Sayers Penden (New York: Atheneum, Macmillan Publishing Company, 1991), 237.

[2] As far as this matter is concerned, the one who interprets the norm cannot avoid addressing law no. 587/1993, under *Norme per l'accertamento e la certificazione di morte*. On the subject see Campione, *Commento alla legge 29 dicembre 1993, n. 578*, in *Codice della famiglia* [2nd ed.], ed. M. Sesta (Milan: Giuffré, 2009), 3520ff., where the process of dying has been made into a procedure, as well as App. Milano, 18 December, 2006, in *Foro it.* I (2007), 571 with comments by Casaburi, who claimed that the request for an emergency provision put forward by the tutor of a *non compos mentis* person in a persistent vegetative mental state to forcefully interrupt his nourishment, thus shortly bringing about his death, is groundless. Only by ascertaining the subject's cerebral death – to which the evoked law no. 587/1993 refers to – is it possible to incontrovertibly establish the limit of therapeutic activity directed towards extending the patient's life. This decision was later overruled by the Court of Cassation's intervention (see Cass., 16 October, 2007, no. 21748, cit. and Court of Appeal, Milan, 9, July, 2008 in *Guida al dir.* (2008), f. 30, 62).

passion of the patient's family members, but had to conform to legal criteria. This was a keen observation, for it seeks to surpass a merely subjective perspective where the termination of life is concerned. Such an issue involves not only the individual, but also society in its entirety: it therefore turns to the juridical world for an answer.

In Allende's narration, Roberto's experience is not limited to theoretical considerations, but clashes with a real choice which presents itself when his wife falls terminally ill with cancer. The spouses, bound by a life-long familiarity, had nurtured the desire to die together, preferring to face death together, hand in hand, so that "at the fleeting instant in which the spirit disengages, they would not run the risk of losing each other in some warp in the vast universe".[3]

This results in Roberto's decision to inject his consenting wife with a lethal poison and console her with the promise that they would reunite a few minutes later. However, after refilling the syringe, Roberto cannot inject the poison into himself; therefore he asks a friend to help him die.

This short story and its open ending is also open to the reader's conjectures, and grasps the true Gordian knot which binds the ethical and juridical issues embedded in decisions to terminate life. This "knot" concerns the absolute or "balanceable" character of the right to live, whose unconditional safeguarding at times seems to cross into the obligation to not die. More generally, it presents the problem, which is common to all ethically relevant issues, of the relationship between ethical evaluations and juridical regulations, which are particularly complex in ethically pluralist societies such as European ones.

In Italy, the problems deriving from such a relationship have been drawing attention for the last forty years since the Italian legal system, in renovating the pre-existing juridical regulation, allowed people options that were previously prohibited by law. Among these, for instance – to state the most ethically and socially debated situations – the introduction of divorce laws in 1970, that of laws concerning termination of pregnancy in 1978, the 2004 regulation on medically assisted pregnancy, as well as the ongoing debate on living wills.[4]

In all of these cases, though with different results, opposing ethical perspectives have faced each other: one inspired by a fixed conception of family

---

[3] Allende, "Interminable Life", 249

[4] To study the matter in depth, see M. Sesta, *Riflessioni sul testamento biologico* in *Fam. e dir.* (2008), 407ff; G. Franzoni, *Testamento biologico, autodeterminazione e responsabilità* in *Resp. civ.* (2008), 581ff.; Various Authors, *Testamento biologico. Riflessioni di dieci giuristi*, eds. la Fondazione Veronesi (Milan: Bruno Mondadori, 2006); L. Balestra, *Efficacia del testamento biologico e ruolo del medico* in *Familia* (2006), 435ff.

or human life to the point of subjecting the individual's interests and will to it; the other, on the contrary, prone to carrying out a single person's prerogatives to their extremes, even when they are deemed incoercible by more or less shared moral visions.

As a result, to a certain idea of marriage corresponds the refusal of its solvability; to a determined idea that grants equal dignity to the unborn child and its mother's life and right to health, corresponds the prohibition (ratified by criminal law) to interrupt pregnancy. In the same manner, a determined consideration of the unborn child is at the basis of the limits reinforced by law no. 40/2004 of the various practices of medically assisted reproduction. Moreover, the principle of life's sacredness is founded on announced dispositions in regards to matters of informed consent and advanced directives on health care, as summarized in the act which was approved by the Senate of the Italian Republic[5] in clear disregard of the disposition adopted by the Italian magistracy during the Englaro case.

Therefore, the real issue that Allende's story ponders and from which we started – which will later be considered with great delicacy in the autobiographical novel *Paula* – is that concerning the legitimacy of actions or omissions that are consistent with the patient's will and suitable to produce or accelerate his or her death. It is indeed necessary to point out that on both ethical and juridical levels, positive actions and omissions do not necessarily undergo the same treatment. Summarizing, it is interesting to observe how on one hand, the legitimacy of a conscious refusal to life-saving treatments[6]

---

[5] See Act no. 2350 of the Chamber of Deputies, *Proposta di legge approvata in un testo unificato dal Senato della Repubblica il 26 marzo 2009.*

[6] The civil, administrative and penal courts have reached an agreement on the issue by now (among the many others, see the penal judgement as stated by the Court of Cassation on 10 October, 2001, no. 36519 in *Riv. It. Med. Leg.* (2002), 573 with footnote by De Matteis; T.A.R. Milan Lombardy, 26 January, 2009, no. 214 in *Giust. Civ.* I (2009), 788, Cass., 16 October 2007, no. 21748, in *Fam. E dir.*, with a footnote by Campione), as well as on most of the jurisprudence (see, in the last reference, G. Cocco, *Un punto sul diritto di libertà di rifiutare terapie mediche anche salva vita (con qualche considerazione penalistica)* in *Resp. civ. prev.* (2009), 493). On the matter, also see the Comitato Nazionale per la Bioetica's opinion in the document entitled *Rifiuto e rinuncia consapevole al trattamento sanitario nella relazione medico – paziente* which was approved on 24 October, 2008. Here the legitimacy, on a juridical level, of an autonomous and competent patient's refusal and renunciation of therapy is confirmed, 7. But now it is time to consider the bill that is currently in course of approval: see Act no. 2350 of the Chamber of Deputies, *Proposta di legge approvata in un testo unificato dal Senato della Repubblica il 26 marzo 2009.*

is indubitable to this day, whereas on the other hand, at least on an ethical level,[7] the lawfulness of the active and not merely passive behaviour of a physician who – before a patient who is *compos mentis* but physically unable to evade the treatment in course – makes the patient's wish to not be cured any longer effective (withdrawal of consent to go through with the treatment) by interrupting nourishment treatments and artificial respiration.

Contemporary regulations provide rather diversified, when not antithetic, answers to this problem.

As far as Europe is concerned, it is sufficient, for instance, to point out that Holland has already equipped itself with a law [*Termination of Life on Request and Assisted Suicide (Review Procedures) Act*] that, in specific cases, legalised euthanasia and assisted suicide,[8] as has Belgium [cfr. *The Belgian Act on Eutanasia of May 28, 2002*], whereas in other countries (for example in Switzerland) only the latter case is admitted. On the contrary, a second group of regulations – though with some blurring – assumed a closed attitude towards euthanasia practices and assisted suicide. Such is, for example, the case in France, a country where the law, though allowing life-saving treatments to be interrupted on the patient's conscious request [cfr. *Loi n. 2005/370 du 22 avril 2005 relative aux droits des malades et à la fin de vie*], sanctions any physician's

---

[7] As far as the juridical profile is concerned, Italian jurisprudence has in fact decreed that the fundamental right to refuse life – supporting medical treatment that – in the light of national and international law – can be neither compulsorily imposed on a dissident patient nor carried through against his will except in the case of the consent's withdrawal. As a result, the behaviour of a physician who, in the case of a patient's physical incapacity, made his or her right to not continue therapeutic treatment effective by physically acting, e.g. switching the patient's automatic ventilator off, thus determining a fatal respiratory failure, is not to be considered *contra legem* (Roman Court of Law, 6 March, 2007 in *Riv. pen.* 2007, 548, with note by Musco)

[8] In particular, the law does not consider such behaviour unlawful when enacted by a physician who:
"a. holds the conviction that the request by the patient was voluntary and well considered,
b. holds the conviction that the patient's suffering was lasting and unbearable,
c. has informed the patient about the situation he was in and about his prospects (his/her situation)
d. and the patient held the conviction that there was no other reasonable solution for the situation he was in,
e. has consulted at least one other independent physician who has seen the patient and has given his written opinion on the requirements of due care, referred to in parts a – d,
f. has terminated a life or assisted in a suicide with due care" (art. 2, comma 1).

conduct towards causing a patient's decease (see art. 221–1 *Code Penal*: *"Le fait de donner volontairement la mort à autrui constitue un meurtre. Il est puni de trente ans de réclusion criminelle"*). Italy substantially follows France's line of action, in that its jurisprudence (see the *Welby* case[9]) recognises the validity of an informed refusal of life-saving treatments made by a patient who is of sound mind; nevertheless euthanasia and assisted suicide are considered absolutely illicit.

In our country, the matter of life-terminating choices and euthanasia practices recently became a subject of debate because of a judiciary event, i.e. the Englaro case, which led to a significant, as well as controversial and criticised, judicial sentence on legitimacy.[10] In 2007 in fact, the Court of Cassation asserted the principle by which the judge, on a legal agent's request, may authorize the deactivation of a vegetative patient's life-saving treatment if it can be proved that such an appeal, based on clear, univocal and convincing evidence, is an expression of the patient's will. The latter must derive from his or her previous statements, or from his or her personality, life style and convictions, which correspond to his or her way of conceiving, prior to lapsing into a state of unconsciousness, the very idea of a person's dignity.

This decision is particularly relevant because it sustains that a legal system founded upon the pluralism of values which centres the relation between patient and physician upon the principle of self-determination and freedom of choice must respect the decision of a person who believes it absolutely against his or her own convictions to be compelled to survive indefinitely in a state of life devoid of perceiving the outside world. In such cases he or she – through a legal delegate – may demand the deactivation of all medical treatments, including hydration and artificial nourishment by nasal-gastric tube.

The judgement which has been mentioned here, moreover, pauses to consider the position that, in choices concerning termination of life, people other than the directly involved subject (in this case the patient's legal delegate and father) hold. In this aspect too this decision can be related to Isabel Allende's story (its protagonist, who is the patient's husband, has in fact a decisive role), which can therefore provide interesting starting points for discussions, with reference to the configurability and limits of the patient's relatives' right to partake in life-terminating choices.

---

[9] For more on the matter, see S. Seminara, *Le sentenze sul caso Englaro e sul caso Welby: una prima lettura* in *Dir. pen. proc.* (2007):1561ff.; R. Campione, *Caso Welby: il rifiuto di cure tra ambiguità legislative ed elaborazione degli interpreti* in *Fam. e dir.* 3 (2007): 296ff.; G. Iadecola, *Qualche riflessione sul piano giuridico e deontologico a margine del caso Welby* in *Giur. merito* 4 (2007): 1002ff.
[10] Cass., 16 October, 2007, no. 21748, cit.

The Court of Appeal of Milan, before which the Court of Cassation put the matter, finally authorized (dictating moreover relative accessorial instructions to be followed in its realization phase) the tutor's request to interrupt the artificial life support treatment. In fact, the existence of both conditions required by the legal principle established by the Court of Cassation[11] – i.e. the vegetative state's irreversible nature and the request's correspondence to the "patient's voice" as results from his or her preceding statements, personality and lifestyle on the basis of clear, univocal and convincing proof – had been verified.

As opposed to the Supreme Court's interpretation of the normative data, which was substantially founded on the principle of self-determination as far as medical treatments are concerned as can be interpretatively understood from articles 2, 13, and 32 of the Constitution, the Italian legislator acted through two means. The first consisted of an initiative on the Senate's part in order to relieve the conflict of power with the Court of Cassation before the Constitutional Court;[12] the second of a bill that had already been approved by the Senate of the Republic whose main trait is its reaffirming the prohibition of any form of euthanasia, assistance or help to suicide through its considering medical practice as exclusively directed towards the protection of life and health, as well as relief from suffering. In such a context, the space for the patient's will is notably reduced both as far as the "capable person" is concerned and explicitly as regards advance health care directives. Such directives cannot concern decisions about the nourishment and hydration of an individual who has fallen into a state of incapacity even through the intervention of a legal delegate or so called health fiduciary.

The conflict between opposing juridical solutions reflects the current bioethical debate's opposite poles on advanced health care directives: one centred upon the absolute – and as such inalienable – value of human life, and the other which sees life as an alienable value which can be balanced with others, chiefly the autonomy of the patient's choice. Catholic scholars have stressed this latter value's vulnerability by suggesting that it appears as often decisively open to influence from the interaction between medicine and society, which thus become the background for treatment relation.

The basic issue, i.e. that concerning the boundaries that the law's will encounters while circumscribing individual prerogatives in the name of a determined vision of human life, therefore reemerges and, on a level of positive law, must necessarily resolve itself in the light of constitutional principles

---

[11] Court of Appeal, Milan, 9 July, 2008 in *Guida al dir.* (2008): f. 30, 62.

[12] Constitutional Court, 8 October, 2008, no. 334 in *Giur. Cost.* (2008): 3713.

that seem to impose criteria of reasonable balance between opposing interests. This for instance has occurred in the case of termination of pregnancy
or medically assisted reproduction[13] – although it is evident that such a procedure will *necessarily* imply a *vulnus* for the absolute characteristic of determined values and hence the ethical visions which are inspired by them.

As we can see, Italy's experience in the matter after all seems to have em-
barked on a course which is very distant from that followed by the protagonist of Isabel Allende's story from which these brief considerations have
arisen. These ideas however have been positively confirmed in the above
mentioned foreign legislations. On the other hand, the story's ambiguous
epilogue reveals how hard it is to move from an abstract and theoretical
vision of life-ending issues to practical decisions, just as it is true that the
story's protagonist was incapable of carrying out the decisions he agreed
upon with his dying wife by himself. This difficulty recalls how important it
is to be cautious where the fulfilment of a fully capable and healthy person's
expressed will is concerned in order to regulate the future and abstract future
of the terminally ill.

---

[13] On the matter see, in particular, Constitutional Court, 18 February, 1975, no. 27,
(the text is available at www.cortecostituzionale.it) which, although preceding law
no. 194/1978's coming into force, *Norme per la tutela sociale della maternità e sull'inter-
ruzione volontaria della gravidanza,* balances between the unborn child's protection
and the mother's right to health. See also Constitutional Court, 8 May, 2009,
no. 151, in *Fam. e dir.* (2009): 761, with a note by M. Dogliotti concerning medically
assisted reproduction.

Jane Bryan
University of Warwick

Reading Beyond the *Ratio*: Searching for the Subtext in the "Enforced Caesarean" Cases

Introduction

Law students are taught that "[f]inding the *ratio decidendi* of a case is an important part of the training of a lawyer".[1] Lawyers, therefore, search judicial decisions for the pithy kernel that is held out to be the very essence of the case: the part that, by reason of *stare decisis,* possesses the authority to influence future cases and affect the way in which the law develops. Any other statement in the decision is considered to be *obiter dictum:* a "mere saying by the way, a chance remark".[2] An *obiter dictum* may be of interest but, being non-binding, it lacks the potency of the *ratio.*

However, no reader of literature would be content to rely upon a pithy summary of a novel, a play, a poem. They would not be reassured that this could summarise the complexity of the piece or capture its essence. Far more is said in the lines, and between the lines, of a literary work than can ever be truly captured in a concise précis.

If we are to take seriously the claim that law should be treated *as* literature, then we must train lawyers to read cases as literature: reading beyond the *ratio* to unearth the subtext of the judgment. This is particularly important, it is submitted, when the images, characters and narratives found within the *obiter dicta* undercut the explicit message of the *ratio decidendi.* In the same way that Hollywood tacks a trite happy ending on to the adaptation of a complex novel in the hope of dispelling the darkness of the original, so a judgment may leave a lasting legacy through the rhetoric of the *ratio,* which, for instance, portrays the law as upholding liberal humanist values whilst the very same judgment propagates images and stories that covertly work against these values. By accepting the sound bite flourish of the *ratio* as the essence of the case, the reader of the case is only getting part of the story.

---

[1] G. Williams, *Learning the Law* (London: Stevens and Sons, 1982), 67.
[2] Williams, *Learning the Law,* 77.

To illustrate this point, the judgments of the "enforced caesarean" cases, which came before the English courts in the 1990s, will be analysed to show that narratives expressed or implied therein subvert the official legacy of these cases as encapsulated in their *ratios*. The *ratios*, which continue to be uncritically accepted and widely repeated in textbooks, journals and subsequent cases,[3] assert that the foetus has no legal standing in English law and that a competent pregnant woman can refuse medical treatment even though the consequence may be her death or that of the foetus or both.

Adopting a Derridean analysis, the "margins" of the judgments in the "enforced caesarean" cases will be deconstructed to find narratives and images which appear tangential or extraneous to the *ratio* but which, it is submitted, are the key to the critical interpretation of the cases.[4]

Drawing upon the postmodernist theory that all knowledge is discursively constructed, this analysis will explore the characters drawn and the stories told in the "enforced caesarean" cases. Adopting Foucault's theory of the relationship between knowledge and power,[5] the function of these narratives and images in the operation of power relations will also be explored, in particular, those that exist between the pregnant woman and the medical profession.

It is submitted that the "enforced caesarean" cases, rather than being protective of the autonomy of the pregnant woman, as their *ratios* proclaim, operate, through their subtexts, to expose the pregnant woman to covert non-legal mechanisms of disciplinary control (to use Foucault's terminology), in particular, those found within the current medical management of birth. Further, the subtext supports such mechanisms through images and narratives, telling the story of the pregnant woman as a threat to the foetus and the doctor[6] as its

---

[3] See, for example, M. Ford, "Evans v UK: What Implications for the Jurisprudence of Pregnancy", *Human Rights Law Review* 8.1 (2008): 171–184; S. Christie, "Crimes Against the Foetus: the Rights and Wrongs of Protecting the Unborn", *Medico-Legal Journal of Ireland* 12.2 (2006): 66–76; R. Scott, *Rights, Duties and the Body: Law and Ethics of the Maternal-Fetal Conflict* (Oregon: Hart Publishing, 2002). John Seymour also writes in *Childbirth and the Law* (Oxford: OUP, 2000) that, with regard to the right of a pregnant woman to refuse medical treatment, "[I]n England, the decision of the Court of Appeal in *Re MB* [...] left no room for doubt on the matter", 217.

[4] J. Derrida, *Of Grammatology,* trans. Gayatri Chakravorty Spivak (Baltimore: John Hopkins UP, 1976), 141–164.

[5] M. Foucault, "Two Lectures" in *Power/Knowledge: Selected Interviews and Other Writings, 1972–1977*, ed. C. Gordon (New York: Pantheon, 1980).

[6] This analysis has cast the doctor as male, the inaccuracy of which is accepted at the outset. However, it is defensible, it is submitted, as it reflects the representation of

guardian, conditioning pregnant women to seek out and willingly submit to medical supervision and control as a means of safeguarding the wellbeing of herself and her foetus.[7]

A literary analysis of the subtext of the "enforced caesarean" cases exposes the disingenuousness at their heart. The cases are able to expressly support the autonomy of the pregnant woman and deny the legal standing of the foetus because the subtext of the cases promotes the covert disempowerment of the woman and protection of the foetus. The *ratio* exists in the form it does, *because* of the subtle functioning of the subtext.

The pregnant woman has *de jure* but not *de facto* autonomy; the foetus has *de facto* but not *de jure* protection. The lack of fit between the apparent freedom and autonomy of the pregnant woman as protected by the common law and findings from empirical research[8] which document the sense of disempowerment experienced by pregnant women in their dealings with the maternity services, is not an example of pregnant women failing to avail themselves of the protection the law affords them but is an example of the divergence of the *ratio* and the subtext, and the convergence of legal and non-legal methods of regulation. By paying attention to the stories of the subtext, rather than the rhetoric of the *ratio*, the interplay of legal and non-legal regulatory mechanisms becomes clearer.

As Gavigan argues, if we look only for explicit social control and discrimination (of the sort which, I would argue, is found in the *ratio* of a case), we will miss the subtle ways in which medical and legal discourses are implicated in the subordination of women.[9]

---

doctors within general societal discourse. Senior members of the profession, who are generally male, also lay down the policies and practices that guide the profession as a whole. Also, it is argued that the binary opposition of doctor and patient is rooted in the male-female binary wherein the male term is privileged and the female term is marked. This is so, it is submitted, regardless of the gender of the doctor and patient in the individual encounter. The doctor is constructed as rational, objective, and knowledgeable: all favoured (male) traits. Conversely, the patient is, generally, constructed as subjective, emotional and weak: all marked (female) traits.

[7] See: B. Ehrenriech and D. English, *For Her Own Good: 150 Years of the Experts' Advice to Women* (London: Pluto Press, 1979); A. Witz, *Professions and Patriarchy* (London: Routledge, 1992).

[8] As widely documented: Ehrenriech, *For Her Own Good*; A. Oakley, *Women Confined: Towards a Sociology of Childbirth* (Oxford: Martin Robertson, 1987); S. Inch, *Birthrights* (London: Hutchinson & Co, 1989); S. Kitzinger, *Freedom and Choice in Childbirth* (London: Penguin, 1987); P. Thomas, *Birthrights* (London: Thorsons, 1996).

[9] S. A. M. Gavigan, "Law, Gender, Ideology" in *Legal Theory Meets Legal Practice*, ed. A. Bayefsky (Edmonton: Academic Printing and Publishing, 1988), 257–258.

The "Enforced Caesarean" Cases

As is now well documented, a spate of "enforced caesarean" cases came before the English courts for a time between 1992 and 2002.[10] Each case followed a similar pattern in that a heavily pregnant woman refused to undergo medical treatment[11] against the advice of her doctor, such refusal putting her own life and that of her viable, full-term foetus at risk. In each case, the Health Authority responsible for the care of the woman applied to the High Court in urgent circumstances for a declaration that it would be lawful to treat the woman without her consent. In each case, the court made the declaration, and in all but one case the woman was delivered of a healthy baby and apparently suffered no harm to her own physical health.

Sir Stephen Brown P. presided in the first "enforced caesarean" case, *Re S* (1992). In this case, Mrs. S. was in labour but the foetus was in "transverse lie", positioned across her uterus thus making a vaginal delivery impossible. Medical evidence stated that without a caesarean, the foetus would inevitably die and the life of the woman was at risk. S. refused an operation on religious grounds. Sir Stephen Brown P. saw the issue in terms of the conflict of rights between the foetus and the pregnant woman. He conducted a balancing exercise and determined that the foetus's right to life outweighed the woman's right to refuse medical treatment.

The case was heavily criticised by academics and non-academics alike.[12] Subsequent cases have proceeded on a different basis. Consistent with the approach in cases of non-consensual treatment of incompetent patients in general,[13] cases involving pregnant women have begun by addressing the

---

[10] *Re S (Adult: Refusal of Medical Treatment)*[1992] 4 ALL ER 671; *Norfolk and Norwich (NHS) Trust v W* [1996] 2 FLR 613; *Tameside and Glossop Acute Services Trust v CH* [1996] 1 Fam LR 762; *Rochdale Healthcare (NHS) Trust v C* [1997] 1 FCR 274; *Re L (An adult: non-consensual treatment)* [1997] 1 FCR 609; *Re* MB [1997] 2 FLR 426; *St George's Healthcare NHS Trust v s* [1998] 3 ALL ER 673; *Bolton*.

[11] In the case of *Tameside*, the pregnant woman did consent to medical treatment but in view of her "labile state" (she was diagnosed as schizophrenic) the doctors wished to obtain a declaration to authorise them to act in her best interests should she decide to withhold her consent to treatment.

[12] D. Morgan, "Whatever Happened to Consent" *NZLJ* 142 (1992): 14–18; K. Stern, "Court-Ordered Caesarean Sections: In Whose Interests?", *Modern Law Review* 56 (1993): 238–243; K. de Gama, "A Brave New World? Rights Discourse and the Politics of Reproductive Autonomy", *JLS* 20.1 (1993): 114–130; M. Thomson, "After *Re S*", *Medical Law Review* 2 (1994): 127–148.

[13] *Airedale NHS Trust v Bland* [1993] 1 ALL ER 821; *F V West Berkshire HA (Mental Health Act Commission intervening)* [1989] 2 ALL ER 545.

issue of the woman's competency.[14] If it is established that the woman is not sufficiently competent to make a decision about treatment herself, the court then proceeds to consider whether or not the treatment proposed by the doctor is in her best interests. If found to be so, the court will make a declaration that the proposed treatment may lawfully be carried out without her consent and in the face of her refusals. The Court of Appeal in *Re MB* (1997) confirmed the earlier ruling of the High Court in *Norfolk and Norwich (NHS) Trust v W* (1996) that "reasonable force" can be used to ensure the patient undergoes treatment.

The Issue of Competency

In the "enforced caesarean" cases, the focus on the medical issue of competency rather than the legal issue of rights allowed the lower courts for some time to avoid the issue of foetal rights and their effect, if any, on the pregnant woman's right to self-determination. The women in the cases were deemed not to be competent and therefore the stated right of *competent* pregnant women to refuse treatment was unchallenged.

The Court of Appeal finally addressed the status of the foetus in the case of *Re MB* (1997). In this case, doctors advised a caesarean section to save the life of the foetus but M.B. refused the necessary anaesthetic injection because of her severe needle phobia. It was held that:

> A mentally competent patient has an absolute right to refuse to consent to medical treatment for any reason, rational or irrational, or for no reason at all, even where that decision might lead to his or her death. The only situation in which it was lawful for the doctors to intervene was if it was believed the adult patient lacked the capacity to decide and the treatment was in the patient's best interests. The court did not have the jurisdiction to take into account the interests of the unborn child at risk from the refusal of a competent mother to consent to medical intervention [426 f-g].

This position was supported in the later decision of the Court of Appeal in *St George's Healthcare NHS Trust v S* (1998). The Court of Appeal has explicitly closed the door on any arguments premised upon the rights of the foetus. In future, it appears, only applications for court-ordered treatment where the

---

14 Butler-Sloss L. J. in *Re MB* confirmed the correct test of competency was laid down by Thorpe J. in *Re C (Adult: Refusal of Treatment)* [1994] 1 ALL ER 819: the adult patient must be able to comprehend and retain treatment information, believe it, and weigh it in the balance to arrive at a choice.

competency of the pregnant woman is in issue will be considered to lie within the jurisdiction of the court.

The Legacy of the "Enforced Caesarean" Cases

Over a decade later, it is now accepted as well-established law that the foetus has no legal standing in English law and that a competent pregnant woman can always refuse treatment regardless of medical advice and the consequences of non-treatment for herself and her foetus (*Re MB* (1997); *St George's* (1998)).[15]

Such uncritical acceptance and repetition of the *ratio*s of the "enforced caesarean" cases almost has a mantra-like quality: the very repetition of the phrases inculcating belief. The law is fixed; it is beyond question; its reassuring certainty and appealing simplicity discourages challenge. The *ratio*s also arouse little academic controversy, upholding as they do liberal humanist values: they protect the woman's right of autonomy and limit the power of the medical profession. The underdog has won. Why would that be challenged, unsettled, reconsidered?

The *ratio*s also permit the law to continue to operate atomistically: treating individuals as autonomous and disconnected, and containing the challenge of the pregnant woman, who, by her two-in-one nature, embodies connectivity. The cases manage the anomalous dual character of the foetus and pregnant woman by denial: the foetus is ignored and the pregnant woman treated as if she were not pregnant.

However, by reading the judgments as pieces of literature and deconstructing the images, characters and narratives therein, it becomes clearer that the "enforced caesarean" cases work through their subtext to disable the autonomy of the pregnant woman, entrench the power of the medical profession and protect the foetus in a way denied by the simple rhetoric of the *ratio*.

This is achieved, in part, behind the guise of determining the medical issue of competency rather than by explicit reference to issues of rights. The law is served by the narrative of incompetency: it provides a convenient, palatable story to explain the disregarding of the pregnant woman's treatment refusal. It becomes a game: medicine gives judges the means, the evidence of incompetency, by which they can achieve their objective, the protection of the foetus, whilst maintaining the law's position of neutrality and whilst refusing to acknowledge that the woman's right to self-determination has been

---

[15] See Derrida, *Of Grammatology*.

ignored. The narrative of incompetency allows an exception to the general rule that pregnant women can refuse treatment, whilst still sustaining the ostensible integrity of the rule itself. A win-win situation: the law maintains its apparent respect for the woman's autonomy and the foetus is saved.

The judges' eagerness to tell the story of the pregnant woman's incompetence is seen in the case of *Rochdale Healthcare (NHS) Trust v C* (1997) where Johnson J held the pregnant woman to be incompetent in the absence of any medical evidence in support of this and contrary to the opinion of her consultant obstetrician that she *was* competent. Significantly, in *Rochdale* the judge also appears to tell the story that the very fact that the woman was in labour impaired her mental function, thus rendering her incompetent:

> The patient was in the throes of labour with all that is involved in terms of pain and emotional stress. I concluded that a patient who could, in those circumstances, speak in terms which seemed to accept the inevitability of her own death, was not a patient who was able properly to weigh-up the considerations that arose so as to make any valid decision, about anything of even the most trivial kind, surely still less one which involved her own life [275 g].

Such an approach, which appears to consider all labouring women incompetent, could effectively deny to women the right, during labour, to refuse any treatment advised by their doctor.

The Power of Stories

The explicit message of the *ratio* is easy to resist: its imperative, didactic quality is clear. We may not agree with the rule, but we are clearly aware of its existence and its substance. Images and narratives employed within the judgment, however, have a subliminal power which is harder to detect and therefore to resist. Such stories are easily accepted as fact, as truth, as common sense.

The "enforced caesarean" cases will be deconstructed as if they were dramatic performances in order to unearth the images and narratives therein.

The Central Players in the "Enforced Caesarean" Cases: the Pregnant Woman

It is significant that in every case decided after *Re S* (1992), the pregnant woman who rejects medical advice and who risks harming her unborn child is deemed incompetent. This clearly signals that competent or "sane" women accept medical advice and act altruistically to protect their offspring.

In several of the "enforced caesarean" cases, the women either suffered from or had a history of mental illness. In *Tameside and Glossop Acute Services Trust v CH* (1996) the pregnant woman was schizophrenic; in *Norfolk* (1996) the woman had received psychiatric treatment in the past and at the time of the hearing was in arrested labour but refused to accept that she was even pregnant. In *St George's* (1998) the appellant was diagnosed as suffering from the mental illness of depression and the women in *Re L* (1997) and *Re MB* (1997) both suffered from extreme needle phobia. The women are therefore marginalised and presented as different from the norm at the outset.

This narrative of the incompetent pregnant woman has two effects. First, whilst the courts ostensibly uphold the right of women to reject medical advice, refuse treatment and expose their foetus to harm, the message conveyed by the "enforced caesarean" cases is that the only women who would seek to exercise this right are outside the norm, mentally incompetent, "insane". Judge LJ in *St George's* sums up how "right-thinking" people view the conduct of the women in the cases:

> She is entitled not to be forced to submit to an invasion of her body against her will, whether her own life or that of her unborn child depends on it. Her right is not reduced or diminished merely because *her decision to exercise it may appear morally repugnant* (emphasis added) [692 b].

By using such images and narratives, the law exerts a powerful influence over how the issues are perceived, conditioning pregnant women, and society at large, not to refuse treatment but to submit to the power of the medical profession. The narratives discourage recourse to the very rights the cases pledge to uphold. These narratives are all the more potent because they are buried within the supposedly objective, impartial court judgment: fiction masquerading as fact.

Secondly, the narrative of incompetency also leads to the inference that if a woman *is* exercising her right to refuse treatment, then she is incompetent: incompetent patients refuse treatment, therefore if the patient refuses treatment, they must be incompetent.

Although the cases promise that competent patients can refuse treatment for an irrational reason, the inability to "weigh" information "in the balance" is taken as proof of incompetency under the test laid down in *Re C (Adult: Refusal of Treatment,* 1994). An irrational decision is often used as evidence of an inability to conduct this balancing exercise thus leading to a finding of incompetency.

The first effect of the narrative of incompetency operates to discourage women from exercising their right to refuse medical treatment and the second effect ensures that the women who are not so discouraged are deemed to be incompetent and this therefore allows the medical profession, with the sanction of the court, to override her refusal of treatment. The woman is ensnared within the web-like narrative.

The Utilisation of Anti-Abortion Images and Narratives

The subtext within the "enforced caesarean" cases also reproduces and validates narratives and images which are socially prevalent, albeit often unstated, and which owe much to the anti-abortion stories circulating since the Abortion Act 1967. Sally Sheldon's deconstruction of the female legal subject created by the 1967 Act is apposite.[16] Sheldon argues that the opponents of abortion typically portray the woman who seeks an abortion as a minor in terms of immaturity or underdevelopment with regard to matters of responsibility and morality. Her decision to abort is trivialised and denied rational grounding, being perceived as mere selfishness.[17]

This construction is similar to the construction of the women in the "enforced caesarean" cases. They are often described as young: the woman in *Re MB* was described as "a naïve, not very bright, frightened young woman" [430f]. Their irresponsibility is evidenced by their late presentation for antenatal care. This also suggests a lack of interest in the foetus and an oppositional stance to the medical profession.

The reasons for the refusal of treatment are often trivialised: in *Rochdale* the woman is said to want to die rather than endure backache and pain around the caesarean scar. This is implied to be a childish overreaction. The extreme needle phobia of MB is belittled by Butler-Sloss LJ who repeatedly says of the woman's fears:

> What she refused to accept was not the incision of the surgeon's scalpel *but only* the prick of the anaesthetist's needle [...] *All* that was involved here was the prick of the needle (emphasis added) [437 h and 439 g].

Eileen Fegan notes that abortion legislation rests on the characterisation of women as lacking in moral autonomy and undeserving of "choice".[18] They are not trusted to make the decision to abort the foetus alone but require

---

16 S. Sheldon, "'Who is the Mother to Make the Judgement?': the Constructions of Woman in English Abortion Law", *Feminist Legal Studies* 1.1 (1993): 3–22.

17 Sheldon, "Who is the Mother to Make the Judgement?", 6–7.

18 E. V. Fegan, "'Fathers' Foetuses and Abortion Decision-Making: the Reproduction of Maternal Ideology in Canadian Judicial Discourse", *Social and Legal Studies* 5 (1996): 75–93.

authoritatively sanctioned justification. Likewise, in the "enforced caesarean" cases, the story is told of the pregnant woman who is mentally incompetent and unable to be trusted to make her own decisions. Decision-making is therefore handed over to the reassuringly mature and responsible (male) figure of the doctor and the judge.

The stories found in the "enforced caesarean" cases and the anti-abortion movement both position the doctor as better equipped than the woman to judge what is best for her. The reader of the "enforced caesarean" cases, therefore, is prepared for the transfer of decision-making power from the woman to the doctor/court. If the woman is distraught and irrational, as in the case of *Re L* and *Re MB*, then she is an unsuitable party to make such an important decision as to refuse treatment. Equally, if she is selfish and self-centred, as constructed in *St George's* and *Rochdale*, considering her own needs and giving no weight to other factors, notably the foetus, then she is again constructed as an unsuitable party to make such a decision. She is therefore in need of the normalising control of the doctor to impose either calm and rationality or morality and consideration for others.[19]

By drawing upon the images and narratives of women utilised by the anti-abortion movement to signify the immorality of abortion, the subtexts of the "enforced caesarean" cases tap into the emotion aroused by a socially pervasive story of women as a threat to the vulnerable foetus and of women rejecting motherhood. By making a connection between the story of abortion and of treatment refusal, the cases imply that the women in both instances are motivated by a desire to harm the foetus. This is made explicit in the cases of *St George's* and *Norfolk*, where the women express their lack of concern for the well-being of the foetus [678 j and 680 c and 271 e respectively].

The cases reinforce the story of the woman as threat by repeatedly emphasising that it is her decision to refuse treatment that is putting the foetus in a life-threatening situation. It is not only the woman's decisions that endanger the foetus, it is her body too. The cases portray the body of the woman as malfunctioning or hostile, as if it too is rejecting maternity: variously, labour is arrested or obstructed; the placenta blocks the foetus's path out of the uterus; the baby is at risk of suffocating in the birth canal; the uterus is on the point of rupture.

By implication, the cases portray not just the individual, non-consenting women as a threat to the foetus, but all women. Their bodies threaten the

---

[19] Sally Sheldon makes this point with regard to narratives of women seeking an abortion in "Who is the Mother to Make the Judgement?", 17.

foetus. The cases use a double strategy: stigmatising the woman who acts against the norm of the selfless mother, whilst supporting the story that all women are potentially a threat to the foetus and therefore require surveillance and regulation by the discipline of medicine.

The scene is now set for the story of the pregnant woman as a threat to the foetus, this story being the central drama of the "enforced caesarean" cases.

The Doctor: The Hero

The doctor is portrayed as rational and also compassionate towards the foetus: in short, the hero of the drama. In contrast, the pregnant woman is portrayed as irrational, selfish and uncaring: the foe to be conquered. By placing the "right-thinking" doctor on the side of the foetus, and portraying the woman as the villain of the piece, the reader of the judgment is encouraged to empathise with the doctor, condemn the pregnant woman and sympathise with the position of the foetus.

The increasing visualisation of the foetus as a result of obstetric technologies means it is widely perceived as a separate person. Once a divide has been created between the pregnant woman and the foetus, space is created to insert a third party in the form of the doctor, who guards the foetus against the actions and the physiology of the pregnant woman. Pro-natalist language within the cases (the foetus is a *baby* with a *mother*) also encourages sympathy for the foetus by stressing its proximity to personhood.

The doctor has the means, and the desire, to protect and rescue the foetus both in the individual court case and through medical supervision of pregnant women in general. Captive within her uterus, the foetus is perceived to be at the mercy of the pregnant woman. In the cases, the doctor saves the foetus by removing the obstacle of the woman, both legally and physically: "[T]he reality was that the foetus was a *fully formed child*, capable of normal life *if only it could be delivered from the mother*"(emphasis added) [*Norfolk* 273a].

The court's declaration frees the doctor of the woman's refusal of treatment; the doctor's scalpel frees the foetus from the body of the woman. Once born, the child will acquire legal personhood, thus allowing the law to protect it from its mother as necessary.

Because of the proximity of the woman and foetus, and the foetus's relative inaccessibility before birth, stories are used to convince the pregnant woman, the medical profession and society generally that pregnant women cannot be trusted to protect the foetus by themselves. As a result, pregnant women actively seek the assistance of the medical profession, submitting to their supervision and control. Birth is constructed as a medical event and the medical profession are constructed as holders of expert knowledge.

The Narrative of Medicine as Truth

The power of medical knowledge is constructed in the "enforced caesarean" cases, in part, by the assumption that the doctors' opinions and predictions are undeniable, uncontroversial Truth. The unfailing accuracy of medical knowledge is a story told in the cases: the doctors' proposed treatment is portrayed as the only possible way in which harm to the foetus and the woman can be averted; the doctors' precise time estimates are repeated unquestioningly as fact by the judges[20] despite clear evidence that medicine is not an exact science and these deadlines for action must only be approximations on the part of the doctors. In similar cases in the US, where women have managed to avoid the interventions ordered by the courts, the doctors' dire predictions were not realised (see *Jefferson v Griffin Spalding County HA*, 1981).[21] Such inaccuracy in medical predictions must occur regularly on a less visible scale, yet the judges never contemplate this possibility and instead retell the myth of medical infallibility.

There is no sense of the disputed nature of much medical opinion, although it is this very multiplicity of viewpoints that is acknowledged by the *Bolam*[22] test of the standard of medical care. The medical profession reaps the benefit of the law's contradictory use of stories of medical knowledge: medical knowledge can either be undeniable Truth or mere opinion depending upon the issues before the court.

Rescues and Redemptions

The joint efforts of the male actors: the doctors and lawyers save the life of the foetus by conquering the woman. Her wishes are overridden and she is forced to undergo invasive surgery. The image of the forcibly restrained

---

[20] In *Rochdale*, 275 b: Johnson J. repeated the opinion of the consultant obstetrician that "a caesarean section would need to be carried out by *about 5.30pm* [...]"; in *Norfolk*, 270 g, Johnson J. repeated medical evidence that "unless the foetus was delivered by *6pm* [...]" (emphasis not in originals).

[21] 274 SE 2d 457 (Sup Ct Georgia).

[22] *Bolam v Friern Hospital Management Committee* [1957] 2 ALL ER 118.

woman facing the anaesthetist's needle and the surgeon's knife forms a tableau of the physical and legal overpowering and punishment of the woman. The power of the medical profession and the law is reaffirmed. The woman is forced, literally and legally, back into her "natural" role of compliant woman and patient.

The reader of the judgment is familiar with the course of the drama: reared as we are on tales of "goodies" and "baddies", of foes vanquished, and on occasion, repentant and reformed. Once introduced to the players in the drama, readers can readily predict the outcome of the case – the pregnant woman will be overruled and the foetus rescued. As argued by Alan M. Dershowitz,[23] Chekhov's theory of purposive narratives can be applied to court cases: the readers or the jury are primed to interpret facts teleologically, drawing upon familiar fictional patterns. Chekhov told the writer S. S. Schovkin that, "[i]f in the first chapter you say that a gun hung on the wall, in the second or third chapter it must without fail be discharged".[24] In the "enforced caesarean" cases, the pregnant woman is portrayed as the foe to be conquered. Therefore, drawing upon familiar stories of justice, the reader interprets the case in a particular way and therefore experiences no shock when forced medical treatment is ordered. On the contrary, it would be more shocking, given the way in which the characters are drawn in the judgments, if the refusal of treatment had been allowed to stand and the foetus had not been saved.

The judgments usually conclude with a reference to the health of the mother and the baby as a result of the enforced caesarean. The healthy baby is seen as a product of the actions of the men. It could be argued that in the cases the men act out the birth process and take it over. The urgency of the cases; the flurry of activity on the part of the male actors; their struggle against the resistant pregnant woman all tells the story of birth itself. The cases therefore perform the male fantasy of giving birth. In *Re MB* Butler-Sloss L. J. ignores the role of the mother in the birth of the baby and gives full credit to the obstetrician: "Mr N. delivered a healthy baby by caesarean section" [430 b].

The reference to the health of the mother and baby, and often to the woman's subsequent gratitude to the court and medical profession operates as a type of "all's well that ends well" justification for the preceding court action:

---

23 A. M. Dershowitz, "Life is Not a Dramatic Narrative" in *Law's Stories: Narrative and Rhetoric in the Law*, eds. P. Brooks and P. Gewirtz (New Haven: Yale University Press, 1996).

24 *Anton Tchekhov: Literary and Theatrical Reminiscences*, ed. and trans. S. S. Koteliansky (1974) as quoted in Dershowitz, "Life is Not a Dramatic Narrative".

Since making this declaration I have been told that the baby was safely delivered and that the mother and baby are doing well [*Norfolk* 273 b]

The operation was successful and both are doing well [*Rochdale* 276 b]

Before delivering this judgment today I was informed [...] that L had been delivered of a healthy baby, that "*she was delighted with the outcome, and that she expressed apology that she had caused many people so much trouble*" (emphasis added) [*Re L* 612 h].

Such statements seek to justify the violation of the woman's autonomy by stating that the woman got what she wanted in the end, a healthy baby, despite the fact that for a time she did not realise she wanted it. It was the perceptiveness of the male doctors and lawyers that prevented the woman from acting against her own interests by exercising her right of autonomy.

This story undermines the validity of the woman's refusal and retells the story that women cannot be trusted to know their own minds. It also recalls narratives circulating in relation to the issue of rape that the woman may say no but she will not mean it or she will welcome having her refusal overridden. This story of a woman's long-term quiescence, her happiness "ever after", may well create endemic judicial and societal distrust of a woman's short-term resistance.

The gratitude expressed by the woman for her enforced treatment is also the fitting conclusion to this cautionary tale, which has shown the dire consequences that befall pregnant women who reject their "natural" roles of submissive patient and altruistic mother. The drama of the cases depicts the woman's punishment and her eventual redemption and delivers a strong social message warning of the hazards of challenging medical authority, and in a wider sense, patriarchy and the natural role for women.

In the cases, the curtain falls on the woman, embracing her natural role as compliant patient, submissive woman and altruistic mother. She no longer presents a challenge to the social order, which has been re-established and reaffirmed: the woman is compliant; men (in particular, the medical profession) are dominant.

Different Models of Legal Control

My argument that the *ratio* and the subtext of a case effect social control in different ways is, in effect, a literary argument. It follows, as I will demonstrate in this section, that my argument can be illuminated through attention to non-legal literary texts. I will focus in particular upon Shakespeare's exploration, in *A Midsummer Night's Dream,*[25]of differing models of behaviour control.

Set in Ancient Athens, the drama begins with Egeus asking Theseus, the Duke of Athens, to enforce the "ancient privilege of Athens" [1.1.42] that a father can put to death a daughter who refuses to marry the man he has chosen for her. Theseus urges Hermia to obey her father and marry Demetrius:

> Look you arm yourself
> *To fit your fancies to your father's will;*
> Or else the law of Athens yields you up-
> Which by no means we may extenuate-
> To death, or to a vow of single life.
> (emphasis added) [1.1.119–123]

Hermia must consciously adapt her choices and her conduct in order to bring them into line with her father's commands: she must force her fancies to fit his will, or must face the legal consequences. Her father's legal power is clear, as is Hermia's subjugation. Theseus's model conceives of law as a blunt tool of compulsion.

Hermia, however, posits an alternative model: "I wish my father lookt but with my eyes". [1.1.53] Hermia imagines changing Egeus's behaviour without compulsion, by changing his perceptions. If his perceptions could be altered, then he would freely choose to behave differently (by accepting Hermia's choice of husband and not seeking to enforce his "ancient privilege"). She hoped to achieve her will through subtle manipulation of his perspectives.

The different approaches of Theseus and Hermia can be matched to the differing operation of the *ratio* and the subtext in the "enforced caesarean" cases. The *ratio* states the rule that a doctor must respect the right of a pregnant woman to refuse medical treatment. If the doctor fails to obey this rule, then he must face the legal consequences (liability for the civil wrong of battery and the criminal offence of assault). The rule compels him to adapt his behaviour. The subtext, however, employs narratives and images which subtly manipulate the perceptions of the pregnant woman, subliminally shaping her thoughts and desires (as Hermia hoped to shape the will of her father) so that the pregnant woman "freely" chooses to behave in a particular way (by following medical advice, for example) thus avoiding express reference to the reality (enshrined in the *ratio* or the rule) that the woman had the right to refuse medical treatment but has felt compelled to forego that right.

---

25 J. Bate and E. Rasmussen, *The RSC Shakespeare: Complete Works* (London: Macmillan, 2007).

Shakespeare uses the drama of *A Midsummer Night's Dream* to test both models of behaviour control. In the court scene in the opening Act, the views of Hermia and her father are not altered by legal reasoning and argument. Hermia appears equally unmoved by the threat of legal sanctions if she continues to defy her father. Theseus's model of legal compulsion therefore appears ineffective at altering attitudes and behaviour. In the law-less realm of the forest, however, the desires and actions of Demetrius and Lysander are very effectively manipulated through the manipulation of their vision as a result of the use of Oberon's magic potion. Yet as this manipulation is covert the pair believe they are acting freely.

A Midsummer Night's Dream reaches its harmonious conclusion through this manipulation of perception, as it is the sight of the happy couples (Demetrius still subject to the effects of Oberon's potion) that convinces Theseus to "overbear [Egeus's] will" [4.1.172] and allow Hermia to marry the man of her choice, Lysander. The manipulation of perception produces harmony in a way that legal compulsion would not. In place of Oberon's magic potion, the "enforced caesarean" cases use the magic of narratives and images to alter the perceptions, and thence the conduct, of pregnant women. The subtle manipulation of their perceptions in order to bring about their willing submission to medical control is far more effective and generates far less hostility and resistance than more visible forms of compulsion.

In *A Midsummer Night's Dream*, Egeus is the villain who is trying to use the law to impose his will and in doing so, threatens the life of Hermia. In the "enforced caesarean" cases, the pregnant woman is also using the law to protect her refusal of medical treatment, and thereby also threatens the life of the foetus. Egeus justifies his right to control his daughter by saying: "As she is mine, I may dispose of her". [1.1.43]. He made her and therefore her interests must be subverted to his will. English law no longer conceives of one person having property in another. In the "enforced caesarean" cases, however, the *ratio* ignores the duality of the pregnant woman and the foetus, and denies the foetus legal standing. In effect, as the foetus is the property of/is the pregnant woman, she may "dispose" of it.

In each drama, the hero is the character that refuses to apply the legal rule: despite Egeus's "privilege" under strict Athenian law, at the conclusion of the play, Hermia's life is saved by Theseus's refusal to enforce her father's legal rights. In the same way, in the "enforced caesarean" cases, the foetus is saved by the judges' refusal to apply the legal rule that pregnant women can refuse medical treatment (justified on grounds of incompetency). In both instances, however, the failure to apply the legal rule whilst refusing to overrule it, allows a discrepancy to exist between the stated position of the law and its practical application. Hermia's father appears to be able to force his daughter to marry the man of his choice or to put her to death; the pregnant woman appears to be able to refuse medical treatment and thereby bring about the death of the foetus with impunity. However, in both cases, the operation of forces within the non-legal realm undercut this expressly stated legal position: the manipulation of sight within the forest by means of magical potion creates such happiness and harmony amongst the young couples that Theseus refrains from enforcing Egeus's "privilege"; the control of pregnant women through their antenatal medical care, as subtly facilitated by the subtext of the "enforced caesarean" cases, encourages them to refrain from insisting upon their right to refuse treatment.

Yet the certainty of the legal rule, the *ratio*, is insisted upon. Theseus tells Hermia that the "privilege" of her father can "by no means [be] [...] extenuate[d]" [1.1.122]. Yet at the end of the play, at the sight of the happy couples, he "overbears" Egeus's will with neither explanation nor excuse. In the same way, in the "enforced caesarean" cases, the judges state the legal rule that a competent pregnant woman can refuse medical treatment, whilst using the narrative of incompetency to justify the non-application of the rule in a particular instance. In each case, the veracity of the legal rule is open to doubt.

The Importance of Stories as a Means of Resistance

No stance is taken as to whether or not the covert support for the non-legal regulation of pregnant women found within the "enforced caesarean" cases is a good thing. It may be paternalism made palatable in the age of autonomy but analysis of its merits and demerits requires more space than is available here. However, within the cases, as in society at large, medical discourse and the medical profession are privileged and constructed as authoritative. Great faith is placed in them. The law also gives them great power: a medical practitioner must attend a birth, unless the woman gives birth alone or in an emergency.[26] The State also gives financial benefits to pregnant women (such as, free prescriptions during pregnancy and statutory maternity pay) but these are conditional upon written confirmation of the woman's pregnancy (a MAT B1 form signed by her doctor or midwife) thus bringing almost all

---

26 The Midwives Act 1902, as amended by the Midwives Act 1926 and s17 Nurse, Midwives and Health Visitors Act 1979, states that a qualified midwife or doctor must be in attendance at a birth unless the woman gives birth completely alone or in an emergency.

women into contact with the maternity services at some point during their pregnancy.

Once it becomes clearer, however, that the authority of medical discourse is constructed, once the links between power and knowledge are revealed, then its value becomes more open to scrutiny and challenge. If medico-legal knowledge is artificially constructed, then alternative discourses are equally valid even though they lack the imprimatur of professional power.

This is especially important as education and communication is often championed by consumer groups as the key to empowering pregnant women.[27] Yet more information is not empowering if it flows from those in power. Only by proffering genuinely alternative stories can the dominance of medical narratives be challenged.

The narrative of the pregnant woman as a threat to the foetus and the doctor as its guardian can be reimagined. We can tell stories of the bond between the woman and foetus: their mutuality rather than their separation and antagonism. Even better, stories could illustrate the complexity of the relationship: "not one but not two",[28] avoiding dualistic and reductionist narratives, and essentialised characters.

For example, in the US, in the appeal of the "enforced caesarean" case *Re AC* (1990) the depiction of the pregnant woman as intelligent, compassionate and maternal, and the medical authorities as hasty and insensitive was an important counter-balance to the usual images in such cases, unsettling the narrative of the doctor as guardian of the foetus.

However, once we expose the law's support for the covert non-legal protection of the foetus, the express legal standing of the foetus may have to be clarified and possibly reconsidered. In addition, the precarious protection afforded the foetus non-legally may fail for numerous reasons: the disciplinary power of medical care becomes problematic when medical care breaks down: when women avoid the socialising power of medicine by avoiding medical care. In the US, research has found that a large proportion of the women subjected to court-ordered caesareans was non-white and had often had little antenatal care: they had fallen through the maternity services net.[29]

---

[27] Within the UK, the chief consumer groups are AIMS (Association for Improvements in the Maternity Services: www.aims.org.uk) and the NCT (National Childbirth Trust: www.nctpregnancyandbabycare.com).

[28] An expression used by John Seymour in *Childbirth and the Law* (Oxford: Oxford University Press, 2000).

[29] V. E. B. Kolder, J. Gallagher, M. T. Parsons, "Court-Ordered Caesarean Sections" *New England Journal of Medicine* 316 (1987): 1192–1196.

It may be that an application to override a treatment refusal may come before the courts in sufficient time to allow a full consideration of the issues before the birth. Not being an emergency hearing, the woman may well be in attendance and legally represented. Not being in labour, she may be found to be competent. The express legal standing of the foetus would have to be reconsidered without the narrative of incompetency and whilst the danger to the foetus is real (the appeals in *Re MB* and *St George's* being heard after the safe delivery of the foetus).

The court would almost certainly respect the treatment refusal, however, perhaps on the basis of "giving away the basement to keep the castle". Enforced treatment could drive pregnant women away from the maternity services, fearful that their beliefs and treatment refusals would be overridden. Yet non-legal regulation through medical care protects far more foetuses than could ever be protected using the blunt instrument of legal compulsion.

Conclusion

The subtexts of the cases subtly entrench the power of the medical profession by constructing meaning and knowledge; promulgating supportive narratives and images and thereby encouraging willing submission on the part of pregnant women to medical supervision and control.

Therefore, despite or, it is submitted, in part because of the "enforced caesarean" cases, the medical profession wields a degree of *de facto* power which it does not appear to possess in law; the pregnant woman lacks the freedom she appears to be given by law and the foetus has a status not legally recognised.

Repetition of the *ratio* of these cases alone gives a false impression of the law. The *ratio* can be seen to operate at times as a confidence trick, a sleight of hand – declaiming the law's stated position whilst deflecting attention away from the often contradictory but ideologically powerful subtext of the case. It is only by reading beyond the *ratio* that the full meaning and effect of the case can be grasped and an opinion formed as to whether or not the *ratio* tells the "full story".

From a Literary Perspective

Eric S. Rabkin
University of Michigan

Science Fiction and Bioethical Knowledge

In the wake of surgical advances driven by the necessities of World War II, Bernard Wolfe extrapolated the world of *Limbo*, a sardonic, cutting dystopia. As the materials for prostheses became lighter and tougher, as joints became hyperactivated by miniature atomic motors, as their use became reliable by exquisite nerve surgery, Wolfe's future amputees literally began to outpace the uninjured. Soon there arose a class of "vol-amps", voluntary amputees seeking the power and prestige of superhuman abilities. The Olympics produced high jumpers and pole-vaulters who literally looked down on the stadium crowds. It is no wonder that the winners thought themselves Olympians in more than one sense of that word. But, as the war receded, only sponsored athletes or the wealthy could afford such advanced limb replacement, double amputees outclassing single amputees, and quadruple amputees highest of all. At that point, one somewhat old-fashioned character still wonders, should "quadros ... be considered plus-fours or minus-fours?"[1]

How should one evaluate a voluntary act meant to improve oneself? Surely a victim of accidental dismemberment is not wrong to wish to restore lost function. Then could he be wrong, if he must already rely on a prosthesis, to wish that prosthesis more capable, even more capable than his original limb? Would it be wrong for someone fortunate enough never to have suffered dismemberment to arrange to have such a limb, or limbs? What is the difference between relying on seven-league legs and on an automobile? Well, one may reply, you don't change yourself, your very body, to use the automobile. True enough, but our technology does change us, even our very bodies. *Homo sapiens* evolved the ability – except in the residual lactose intolerant population – to digest milk from other mammals only after the domestication of cattle made their milk available.[2] Ah, you may say, that may be true, but no human, then or now, knowingly changed their bodies to accommodate non-human milk.

---

[1] B. Wolfe, *Limbo* (New York: Carroll & Graff, 1952), 71.
[2] S. A. Tishkoff et al., "Convergent Adaptation of Human Lactase Persistence in Africa and Europe", *Nature Genetics* 39.1.1 (2007): 31–40.

Bioethics, the questions of right and wrong in biotechnological matters, just as in legal matters, hinges on knowing, whether or not we know and, if so, how we know. Science fiction, the literature defined by its concern for the possibilities and social implications of scientific and technological change, provides a dramatic mirror for bioethics.

As Ray Bradbury tells his story of human expansion in *The Martian Chronicles*, the Earthmen of the first three expeditions all died at the hands of the Martians. Those back home, however, did not know why the expeditions failed. When the fourth arrives, they find a depopulated planet, soon discovering desiccated corpses, killed, it turns out, by chicken pox.

> "My God, no!" the captain says, but his investigator replies, "Yes. I made tests. Chicken pox. It did things to the Martians it never did to Earth Men. Their metabolism reacted differently, I suppose. Burnt them black and dried them out to brittle flakes. But it's chicken pox, nevertheless. So [the earlier expeditions] must have got through to Mars ... God knows what happened to them. But we at least know what *they* unintentionally did to the Martians".[3]

For Bradbury to set up Mars as a second chance for essentially Midwestern American Earthmen to populate a New World without the original sins of genocide and slavery, the landscape must be cleared of our competitors without our knowledge. Bradbury's Mars is a fairyland.[4] The tests, by the way, only show that the disease agent was chicken pox. They do not show that, unlike Columbus or certain U.S. Cavalry officers in the Old West, the spacefarers transmitted their germs unintentionally. That conclusion comes not from science but from fiction, the fiction that at the time of the fourth expedition Americans are benign. Yes, the Martians died, but we bear no guilt. The ethical situation is clarified by the explicit lack of intent, an intent that cannot exist without our knowledge of Martian biology.

Of course, one could argue that Bradbury's Earthmen should have known better. Ethics, according to the *Oxford English Dictionary*, is "the science of morals; the department of study concerned with the principles of human duty". We have encountered unwitting infections of indigenes on Earth before, so we have a duty to guard against such contamination hereafter. But one could view the ethical situation differently.

"Ethos", as Aristotle uses that term in his *Rhetoric*, is the "characteristic spirit, prevalent tone of sentiment, of a people or community; the 'genius' of

---

[3] R. Bradbury, *The Martian Chronicles* (New York: Bantam Books, 1950), 50.

[4] E. S. Rabkin, "To Fairyland by Rocket: Bradbury's *The Martian Chronicles*" in *Ray Bradbury*, eds. M. H. Greenberg and J. D. Olander (New York: Taplinger, 1980), 110–126.

an institution or a system" (OED) while "ethic", which relates to "morals", comes more specifically from the Greek "ethikos", meaning "character" (OED). "Character", in turn, comes from a Greek word for "an instrument for marking or graving, for an impress, stamp, distinctive mark or nature" (OED). Let us consider what this means. When we say that person is a serious character, we don't mean that she never laughs but only that in general she is serious, that her behavior is marked by seriousness. At any given moment, she may be amused or distracted or infatuated, but in general she is serious. Her character is what marks her; it is the condition of personality to which we expect her to return. An honest character may tell a small lie to spare someone's feelings or an important lie under duress, but, unless he begins to lie habitually, the fact of a lie or two will not change our assessment that he is an honest character. In the sense that we can know him to be honest or wise or expert, Aristotle tells us that the ethical argument, the argument from character, is the most persuasive of all. It is a fundamentally conservative argument, one based on the notion that what one was, one will be. More broadly, one's ethos embodies one's ethics, one's system of morals, the behaviors that mark one. "Moral" comes from "more" which, like "ethos", in its root means "custom". The connection between personal ethics and a communal ethos is that both point to habitual behavior expressing continuing values. Ethics, like character, is fundamentally conservative. It speaks of the pull back to behaving as we know we should, as our people do.

For this reason, in extending the law through judicial argument, fictions are always created that make the current case simply a new example of an existing precedent. Legislatures may have the right to promulgate new values, to assert ethical changes, but not courts. When it comes to judgment, a fundamental conservatism, or at least the appearance of conservatism, helps validate our conclusions.

Europeans entering new territories have habitually impressed their own ways, to the extent they could, on the local populations. Bradbury's Earth Men, then, were true to character. Bradbury italicizes "*they*" in the phrase "we at least know what *they* unintentionally did to the Martians", but he slips that "unintentionally" in there, right behind "*they*", with no explicit consideration. Accidents will happen. If our messing around causes harm we didn't know about, we may count ourselves blameless.

Ethical systems always contain potentially opposing tenets. I abhor violence, but I believe in self-defense; I resist war, but I honor soldiers. The former conflict is personal, the latter communal, but both are ethical. To the extent that the ethics of the individual incarnate the ethics of the group that produced him, while ethical conflicts may be real, they are predictable and

manageable. When it comes to self-defense, I can invoke a doctrine of pro-
portionate response, no more violence than is necessary to protect myself.
When it comes to war, our society can invoke a doctrine of due process, making a collective judgment that our cause is just. But as science and technology
develop, they create options we have not encountered before. Are the
vol-amp quads glorious new forms of humanity or, as some characters in the
book argue, narcissistic masochists? We can judge them from the actual
1950s culture in which the book was written, but can we judge them from the
future world the book narrates? The fundamental problem of bioethics is
that ethics is inherently stabilizing and biotechnology is inherently destabilizing; one resists change, one promotes it.

To make bioethical judgments requires bioethical knowledge. Knowledge,
of course, like beauty, is in the eye of the beholder. It may perhaps be possible to get virtually everyone to agree that two plus two equals four, but
some people are sure the U.S. moon landing was faked, that the Nazi Holocaust is a lie, that group X is inherently more intelligent or athletic or shrewd
than group Y, and that the world owes them a living. In as much as ethics reflects both the character of individuals and that of groups, making bioethical
judgments requires examining diverse individual characters in diverse relations to groups. Science fiction runs these human thought experiments for
us. To explore the range of bioethical knowledge, I would like to consider
four characteristic situations: that of the scientist who seeks new biotechnological knowledge and that of the physician who employs it, each of whom
has bioethical agency; that of the parent, who may or may not have agency;
and that of the patient.

Victor Frankenstein, title character of Mary Shelley's famous novel, created a nameless being, popularly known by his creator's name, that has become synonymous with the horrific effects of science that has escaped the
lab. Although Victor does not truly create life, by his reanimation of charnel
parts, he makes a being – the most articulate in the book – who asks,

> And what was I? Of my creation and creator I was [at first] absolutely ignorant;
> but I knew that I possessed no money, no friends, no kind of property. I was, be-
> sides, endued with a figure hideously deformed and loathsome; I was not even of
> the same nature as man. I was more agile than they, and could subsist upon coarser
> diet; I bore the extremes of heat and cold with less injury to my frame; my stature
> far exceeded theirs. When I looked around, I saw and heard of none like me. Was
> I then a monster, a blot upon the earth, from which all men fled, and whom all
> men disowned?[5]

---

[5] M. Shelley, *Frankenstein or The Modern Prometheus* [1818], ed. M. K. Joseph (New
York: Oxford University Press, 1969), 120.

Notice that the creature's lacks – money, friends, property – are all social goods. In intellectual and physical respects, save for his superficial ugliness, the creature far surpasses all humans, a group from which Victor's biotechnology elevated the creature, and separated him. He pleads for Victor to make him a mate.

> If you consent, neither you nor any other human being shall ever see us again: I will go to the vast wilds of South America. My food is not that of man; I do not destroy the lamb and the kid to glut my appetite; acorns and berries afford me sufficient nourishment … My vices are the children of a forced solitude that I abhor; and my virtues will necessarily arise when I live in communion with an equal.[6]

The book offers nothing to suggest the creature is mendacious or mistaken, but Victor, in the midst of creating the mate, suddenly feels, or so he says, horrified by the idea that his labor would bear a lineage of monsters. He destroys the half-made female, which the male observes from outside a window. Soon the monster (from the Latin "monere", "to warn" [OED]) begins his violence against Victor's family in order to force Victor to change his ways and begin his mate-making again.

The monster's first family victim is Victor's little brother William. With his corpse is found a locket belonging to Justine, a cousin who lives with the Frankensteins as an attendant, and so Justine comes to trial. Elizabeth, the other orphan brought into the Frankenstein household and the woman destined to become Victor's wife, testifies movingly to Justine's good character, but fruitlessly. Victor, recently returned to Geneva from his solitary scientific pursuits, watches the failure of Elizabeth's testimony in court. "I believed in [Justine's] innocence", Victor recalls.

> I knew it. Could the daemon, who had (I did not for a minute doubt) murdered my brother, also in his hellish sport have betrayed the innocent to death and ignominy? I could not sustain the horror of my situation; and when I perceived that the popular voice, and the countenances of the judges, had already condemned my unhappy victim, I rushed out of the court in agony. The tortures of the accused did not equal mine; she was sustained by innocence, but the fangs of remorse tore my bosom, and would not forego their hold.[7]

Victor, son of a burgher, would have been taken quite seriously in court and, had he alerted the community to the then-unknown existence of his creature, Justine's execution could certainly have been deferred pending further investigation. But that would have required admitting guilt then, not when he does, much later and near death on a ship in the Arctic. Victor's egotism is so

---

6 Shelley, *Frankenstein or The Modern Prometheus*, 146–147.
7 Shelley, *Frankenstein or The Modern Prometheus*, 85.

great that without evidence he leaps to the correct conclusion that William's death was at the monster's hand and the incorrect conclusion that Victor's own remorse must be far worse than Justine's fear of unwarranted execution.

The power of biotechnology, like that of all science, can cut an individual from the community, narrowing his perspective, it seems, until he sees more clearly than do others both what is there and what is not. To parry the fell edge of this sharp sword requires community, the sharing of vision. The system of courts, with their judges and juries, intends just that, applying the corrective of many perspectives to the viewpoint of one. But in the case of biotechnology, those other viewpoints go uninformed. How could the judges have guessed a monster stalked the hills, unless Victor himself had informed them of the possibility by testifying to the certainty of his own unnatural acts. Had the judges had Victor's knowledge, doubtless they would have stayed execution; even without evidence, Victor should have urged that course. He mistakes his own situation, "knowing" both correctly and incorrectly, leading to a communal breach of ethics, his silence in the court, and a personal one as well, his devaluation of Justine's suffering.

Shelley has given us a warning: those who think they know how to shape life must submit in some measure to the greater knowledge of the group.

Edward Prendick, the main narrator of H. G. Wells's *The Island of Dr. Moreau*, finds himself on the title island where Moreau is assisted only by an Englishman, Montgomery, and an odd, dark servant, M'ling, in his gruesome vivisective research. Among the mysteries plaguing Prendick is Moreau's identity, but then he remembers that years earlier a reporter had somehow entered Moreau's London laboratory, found his working notes, and seen a faceless living dog. Prendick believes he recalls that

> [t]he doctor was simply howled out of the country. It may be that he deserved to be; but I still think that the tepid support of his fellow-investigators and his desertion by the great body of scientific workers was a shameful thing. Yet some of his experiments, by the journalist's account, were wantonly cruel. He might perhaps have purchased his social peace by abandoning his investigations; but he apparently preferred the latter, as most men would who have once fallen under the overmastering spell of research. He was unmarried, and had indeed nothing but his own interest to consider.[8]

Prendick, who ultimately will become an isolate himself even back in London, readily excuses Moreau's possession by scientific enthusiasm in the absence of a restraining family and the presence of an expulsive professional

---

[8] H. G. Wells, *The Island of Dr. Moreau* [1896] (New York: New American Library, 1988), 32–33.

establishment. So Moreau continues to torment animals. Without a group and its customs, ethics becomes a purely personal matter.

Despite this conceit, Moreau recognizes the practical advantages of promulgating supposedly communal ethical knowledge by the institution of law. On his island, Moreau has manufactured many Beast People, creatures he has raised to nearly human – rather than superhuman – state not from charnel parts but living animals. These animals, often in pain, always in danger of reversion, and some by nature fierce predators, pose a threat to Moreau. So, having knives and guns at his disposal, Moreau teaches them to recite, "swaying in unison and chanting:"

> "Not to go on all Fours; *that* is the Law. Are we not Men?"
> "Not to suck up Drink; *that* is the Law. Are we not Men?"
> "Not to eat Flesh nor Fish; *that* is the Law. Are we not Men?"
> "Not to claw Bark of Trees; *that* is the Law. Are we not Men?"
> "Not to chase other Men; *that* is the Law. Are we not Men?".[9]

Not to go on all fours. Of course not. Nor to suck up drink. But, come to think of it, most men eat flesh and even Jesus provided fishes to the hungry. The point, of course, is to keep the Beast People from chasing "other Men", of whom only Moreau and Montgomery are available (M'ling turns out to be an altered ape), during their period of indoctrination. In fact, men do sometimes suck up drink when no vessel is handy, and we see that near the novel's end when Prendick claims he can't quench a wounded Montgomery's thirst for want of a vessel, but not want of a hand. Montgomery – conveniently for Prendick's ascendancy on the island – dies.

On the ship from Africa where Prendick meets M'ling and his cargo of soon-to-be Beast People, the "Captain and Owner" complains loudly about the living cargo's smell and allows some manhandling. When M'ling objects, the captain shouts, "I'm the law here. I tell you – the law and the prophets",[10] foreshadowing what life under a unitary rather than group law would mean. We quickly find out it means much more than a disrespect for non-human life. "You're going overboard, I tell you", was the captain's refrain … [sic] "Law be damned! I'm the king here".[11] Moreau, like the captain on the ship, is king of his self-named island, which the outermost narrator, Charles Prendick, Edward's nephew, guesses is what on European maps is known as Noble's Island.[12] But there is nothing noble here, for in Moreau's isolation, ethics becomes merely expressions of will and law a monument to self.

---

9   Wells, *The Island of Dr. Moreau*, 59.
10 Wells, *The Island of Dr. Moreau*, 14.
11 Wells, *The Island of Dr. Moreau*, 22.
12 Wells, *The Island of Dr. Moreau*, 1.

While scientists may be praised as well as condemned for a single-minded devotion to science itself, physicians, they of the Hippocratic Oath, are supposed to function socially, and first do no harm. Given the intimate importance of knowledgeable doctors in our lives, one might expect that science fiction would represent them much more frequently than a random occupational lottery would imply. In fact, in earlier research with a representative sample of nearly eight hundred science fiction short stories,[13] I found that physicians are drastically underrepresented. To discover why that might be, I chose to look at the few medical science fiction stories in the sample that had been reprinted at least twice. I discovered two facts. First, those reprinted stories all showed a physician who either placed him- or herself firmly in the service of society, even at great personal cost, or else the physician became subject to untempered criticism and perhaps even death. Second, those stories, showing the clear dominance of the group over the physician, despite their relative rarity, were among the most highly reprinted stories in the whole genre.

In C. M. Kornbluth's "The Little Black Bag", which appeared in at least eighteen different print anthologies and at least three different television adaptations, Dr. Full, a drunken has-been, somehow stumbles on what we later learn is a highly automated medical kit accidentally sent via time travel from 2450 to 1950. Knowing his own limitations, he wants only to pawn the instruments for wine money, figuring they're just improvements since he stopped practicing, but before he can sell his find a skid-row mother manages to persuade him to attend her child. Almost magically, he heals her with this kit. He realizes he can do much good for the poor, who come to him. He sobers up, and serves, and also attracts Angie, a conniving, low-life opportunist. She becomes his attendant, and begins learning the rudimentary skills necessary to use the wondrous kit. She wants Dr. Full to build a lucrative cosmetic surgery practice, but he, once a patent pending notice on one of the instruments alerts him to their future origin, wants to donate them to the medical society for the good of humanity. He and Angie scuffle, the bag spills, and she seizes a scalpel which normally would guide itself around major blood vessels and heal minor ones in its wake, and stabs him viciously to death. Remorselessly, she gathers up the kit and her half-learned skills and sets off to make a fortune.

In her last scene, Angie argues with a possible client about the safety of removing a wattle from the fat woman's neck. The woman expresses fear,

[13] E. S. Rabkin, "The Medical Lessons of Science Fiction", *Literature and Medicine* 20.1 (2001): 13–25.

so Angie demonstrates by slipping the knife into her own neck. At that moment, a technician in the future notices a signal indicating that an instrument had been used in a murder – Dr. Full's, of course – and so deactivates the kit. Angie cuts her own throat.

> In the few minutes it took the police, summoned by the shrieking Mrs. Coleman, to arrive, the instruments had become crusted with rust and the flasks which had held vascular glue and clumps of pink, rubbery alveoli and spare gray cells and coils of receptor nerves held only black slime, and from them when opened gushed the foul gases of decomposition.[14]

Dr. Full, fallen though he may be, thinks of mankind and earns our respect; Angie, living by an egotistical sword, dies by an egotistical sword and initiates the degeneration of the very objects that made advanced biotechnology tangible. To be a physician requires that no matter how much biotechnology you have at hand, your hand act in the service of others.

Parents usually act – or believe they should – in the service of others, that is, of their children, but this social position readily produces ethical conflicts. I believe murderers should be brought to justice, but not my child. I believe that social goods should be distributed fairly, but I will seize any I can for my child. I believe in "forgive and forget", but don't expect me ever again to let you near my child. The skid-row mother in Kornbluth's tale has only our sympathy, but what should we think of the wealthy parents in Nancy Kress's *Beggars in Spain*? In that novel, it has become possible, at great cost, to modify an embryo so that the resulting child need never sleep. The ramifications of this are enormous, but the most obvious flow from the fact of work-time differential. Imagine how much more you could learn or do if you had eight more productive hours every day. Now consider how that would make your Sleeper parents feel. It is one thing to grow up to be like your parents, perhaps adopting the behaviors they modeled for you and the ethics those behaviors reflect, but what if, through the wonders of the biotechnology they gave you, you can't grow up like them? What happens to your feelings for them, and theirs for you? And how do those feelings translate into ethics?

In Wells' *The Food of the Gods*, two scientists develop Herakleophorbia, a compound that allows enormous growth. They aim to multiply the size of agricultural products to feed the world, but the compound escapes control and giant rats and wasps arise, too. The scientists feed Herakleophorbia to their children, who soon must have it or die, and who grow to forty feet. The scientists would gladly share this biotechnology, but the vast majority of or-

---

14 C. M. Kornbluth, "The Little Black Bag" [1950] in *The Science Fiction Hall of Fame, Volume 1, 1929–1964* [1970], ed. R. Silverberg (New York: Tor, 1998), 362.

dinary people don't want it. They don't want to be huge, to need new homes, to watch their children grow away from them. The Little People pass laws to constrain the Big People who at one point lob non-explosive artillery shells into a city. The Little People seek legal action for the giants' violating "the rules of war … firing shells filled with – poison". "Poison?" one of the scientists asks in astonishment. "Yes. Poison. The Food –",[15] for poison Herakleophorbia must be since it destroys a way of life.

Gift or curse? Something to seek for one's child or something to shun? If you could relinquish the need to feel connected with your child once grown, would you then provide it when young with the Food of the Gods or mortgage your future to make it Sleepless? Ethics, including bioethics, depends on what you know and how you know it, which means, in part, how you see that knowledge defining your relations to others, now and hereafter.

While a parent sometimes has the chance to reject or obtain any specific biotechnology, a child, like members of the public plagued by Herakleophorbic wasps, may have no biotechnological choice. As Kress's first generation of Sleepless grow toward adulthood, they find the social world resentful. Laws are passed against these unfairly advantaged children of the rich, just as Wells's Little People passed laws against the Big People. Leisha, one of the first Sleepless, contemplates what amounts to the unprosecutable murder of one of her friends:

> [...] the law, in its striving to be fair and treat all equally, left too much out. It was not large enough. It was not as large as the genetic and technological future which, outgrowing it, would be lawless.

> Sitting on the edge of the bed in the dark hotel room, Leisha could feel her belief in the law leave her, as if the air itself were being sucked out of the room. She was choking, falling into a vacuum of cold and dark. The law wasn't large enough. It couldn't hold Sleeper and Sleepless together after all, couldn't provide any ethical way to judge behavior, and without judgment there was nothing. Only lawlessness and the mob and the void –[16]

Leisha, here, is a patient, not an agent. As such, her allegiance to the law dissipates.

Unlike Moreau, most of us believe the law should protect those without agency, punish those who misuse agency, and foster agency wherever ethically possible. Seeking informed consent, I think, aims to shift some agency from the doctor to the patient. If I understand precisely the risks,

---

[15] H. G. Wells, *The Food of the Gods and How It Came to Earth* [1904] in *Seven Science Fiction Novels of H. G. Wells*, H. G. Wells (New York: Dover, n.d.), 796.

[16] N. Kress, *Beggars in Spain* (New York: William Morrow, 1993), 216–217.

costs, and benefits of following your advice, and I am allowed to decide my own compliance, I have ethical agency. But, in the face of modern biotechnology, is informed consent possible? In *Flowers for Algernon*, Daniel Keyes presents Charly, a sweet-tempered man of painfully limited mental capacity. Algernon, a mouse subjected to an experimental mind-enhancing drug, beats Charly at solving mazes. Charly, asked if he'd like to be smart enough to beat Algernon, agrees to become the first human subject of this study. The novel tells of Charly's growth into an intellect more than equal to that of the scientists who enhanced him, and, after Algernon reverts and dies, we feel the pathos of a fully intelligent Charles Gordon narrating his understanding of and descent into his erstwhile limits. Most critics point to the cruelty of the scientists' subjecting Charly to an inadequately tested drug, but long before his demise is known, Charly's problems begin. He learns, for example, that someone at work, whom he had thought a friend, had used Charly as a cover to steal. At once Charly finds himself betrayed, bereft, and guilty. Where there had been ignorance, there had been bliss. The biotechnological apple leads directly to Charly's fall. "Now I understand", he writes, early in his growing intelligent phase, "one of the important reasons for going to college and getting an education is to learn that the things you've believed in all your life aren't true, and that nothing is what it appears to be".[17]

Later, when his demise seems inevitable but he can still speak powerfully for himself, Charly rejects a scientist's assertion that "we've always treated you well – done everything we could for you".

> Everything [Charly replies] but treat me as a human being. You've boasted time and again that I was nothing before the experiment, and I know why. Because if I was nothing, then you were responsible for creating me, and that makes you my lord and master. [...] But what you did for me – wonderful as it is – doesn't give you the right to treat me like an experimental animal.[18]

Biotechnology has the power to change us so profoundly that it can challenge our knowledge of what it means to be human. If a raised-up Charly Gordon is human, are Moreau's Beast People, too? If Victor's monster says, "I was not even of the same nature as man", have we still an obligation to treat him as a man? Does biotechnology require that we revise our ethics, "enlarge", as Kress's Leisha says, our law? How can we do that when we can't know what we don't know and therefore even the rudimentary ethical option

---

[17] D. Keyes, *Flowers for Algernon* (New York: Bantam, 1966), 50.
[18] Keyes, *Flowers for Algernon*, 172.

of informed consent becomes impossible? The answer is both very simple and very difficult. If we cannot adjust our ethics through a change in our own customs, our behaviors, our relations with our group, our habits, we can adjust our ethics through enlarging our habits of mind. And that, of course, is the great gift of science fiction.

John Drakakis
University of Stirling

Shaping Personhood:
Problems of Subjectivity and the Self in Shakespeare's
The Taming of The Shrew and *Much Ado About Nothing*

Bio-ethics, Bio-power, Shakespeare

In a lecture that he delivered in the Collège de France in March 1976, Michel Foucault described a bio-politics that:

> will introduce mechanisms with a certain number of functions that are very different from the functions of disciplinary mechanisms. The mechanisms introduced by bio-politics include forecasts, statistical estimates, and overall measures. And their purpose is not to modify any given phenomenon as such, or to modify a given individual insofar as he is an individual, but, essentially, to intervene at the level at which these general phenomena are determined, to intervene at the level of their generality.[1]

This process exists beneath the level of "the dramatic and sombre absolute power that was the power of sovereignty, and which consisted in the power to take life", and is "the power of regularisation" that, "in contrast, consists in making live and letting die".[2] He notes the gradual disappearance of "the great public ritualisation of death" in the late 18th century, whereby "we went from one court of law to another, from a civil or public right over life and death, to a right to either eternal life or eternal damnation".[3] Foucault's concern is to locate and describe different "technologies" whereby the "body" is disciplined, in order to produce "individualising effects" that are linked to the manipulation of "the body as a source of forces that have to be rendered both useful and docile" and also whereby "life" in general, the randomness of events, and "the mass effects characteristic of a population" can be controlled. He notes that even though both technologies are concerned with the body, in one "the body is individualised as an organism endowed with capacities", while in the other "bodies are replaced by biological processes".[4]

---

[1] M. Foucault, *Society Must Be Defended* (London: Allen Lane, 2003), 246.
[2] Foucault, *Society Must Be Defended*, 247.
[3] Foucault, *Society Must Be Defended*, 247.
[4] Foucault, *Society Must be Defended*, 249.

Foucault's concern is, clearly, with those "scientific" processes that began to emerge with the Enlightenment, hence his particular emphasis on the late eighteenth century as a moment of transition, and upon the histories of particular institutions in and through which the various forms of regulation were administered. Some years later, and following the publication of *The History of Sexuality Volume 1* (1976), and commenting on the "ethical" aspects of behaviour in ancient Greece, he observed that "*ēthos* was a way of being and of behaviour. It was a mode of being for the subject, along with a certain way of acting, a way visible to others".[5] Foucault's account of what he calls "the concrete form of freedom" is, itself, part of a more general politicised practice that distinguished someone who was "free" from someone who was a "slave": "It is political in that non-slavery to others is a condition: a slave has no ethics. Freedom is thus inherently political".[6] This is, of course, a different kind of "ethics" from that dismissed by Alain Badiou in his book *Ethics: An Essay on the Understanding of Evil* (2001) in which he begins with the claim that Ethics is positioned between "conservatism and the death drive":

> [...] ethics oscillates between two complementary desires: a conservative desire, seeking global recognition for the legitimacy of the order peculiar to our 'Western' position – the interweaving of an unbridled and impassive economy with a discourse of law; and a murderous desire that promotes and shrouds, in one and the same gesture, an integral mastery of life – or again that dooms *what is* to the 'Western' mastery of death.[7]

From different perspectives both Foucault and Badiou challenge the definition of "ethics" as it emerges in, for example, the work of Emmanuel Levinas, who holds that "[P]olitics must be able in fact always to be checked and criticised starting from the ethical". What Levinas calls "this second form of sociality":

> would render justice to that secrecy which for each is his life, a secrecy which does not hold to a closure which would isolate some rigorous private domain of a closed interiority but a secrecy which holds to the responsibility for the Other.[8]

I raise these three versions of "ethics", two of which are inscribed within a recognisable "politics" and one that seeks to transcend the sphere of the

5 M. Foucault, *Ethics, Subjectivity and Truth, Essential Works of Foucault 1954–1984, Volume 1*, ed. P. Rabinow (New York: Pantheon, 1984), 286.
6 Foucault, *Ethics, Subjectivity and Truth*, 286.
7 A. Badiou, *Ethics: An Essay on the Understanding of Evil*, trans. Peter Hallward (London: Verso, 2001), 38.
8 E. Levinas, *Ethics and Infinity: Conversations with Philippe Nemo*, trans. Richard A. Cohen (Pittsburgh: Duquesne University Press, 1985), 80–81.

political as a means of approaching two Shakespearean comedies, where many of these issues surface, and in ways that might force us to modify Foucault's problematical formulation of historical periodisation at the same time as they challenge Levinas' understanding of Ethics as a "responsibility for the Other". Also raised is the question of what the representation of a "bio-power" might be if it involves a "normalising society" that arrogates to itself the capacity to exercise "sovereign power". Foucault links "bio-power" with racism, and with the sovereign capacity to kill. This capacity includes "every form of indirect murder": "political death, expulsion, rejection, and so on", in short, everything that we might bracket under the heading of marginalization.[9] "Normalising" here, of course, involves the shaping of subjectivity as something more complex than a process of self-definition. Thus, bio-power and bio-ethics, regarded as sinister forces by Badiou, are entangled for Foucault in various forms of the regulation of subjectivity that begin to be transformed during the early modern period.

Shakespeare: *The Taming of The Shrew*

We might find no better example of the shaping of person-hood within a hetero-sexist matrix than in Shakespeare's "comedy" *The Taming of The Shrew* (c. 1594). The precise connection between this play and the anonymous *The Taming of a Shrew* that was published in 1594 is uncertain. Both plays are concerned with the "shaping" of the "shrewish" Kate, described in the anonymous play as "this devilish skould" [TSB, Scene 4].[10] But in the anonymous version, unlike Shakespeare's, the framing play of the transformation of the tinker Christopher Sly is sustained through until the end, so that the process of "taming" appears to be a masculine fantasy, as much as it is an account of the process of "taming" a "devilish skould". Here Kate is tortured by being denied food, and then clearly brainwashed, by having her own sense of time distorted [TSB, Scenes 13 and 14]. When she finally succumbs to Ferando's (Petrucchio's) strategy, she participates in it when confronted with the Duke of Cestus (Old Vincentio), who, after his encounter with her, doubts his own self-hood:

---

9 Foucault, *Society Must Be Defended*, 256
10 Anon., *The Taming of a Shrew* [1594] in *Narrative and Dramatic Sources of Shakespeare, vol. 1*, ed. G. Bullough (London: Routledge and Kegan Paul, 1977), 75, hereafter indicated by the abbreviation TSB.

> What is she mad too? or is my shape transformd,
> That both of them persuade me I am a woman,
> But they are mad sure, and therefore Ile begon,
> And leave their companies for fear of harme,
> And unto *Athens* haste to seeke my son. [TSB, 100]

The Duke of Cestus experiences what Kate before him has experienced, and this incident effects a potent demonstration of the bio-power that Fernando deploys in that it causes the Duke to doubt his own gender. It is also an experiment designed to transform Kate as she occupies the position of both victim and perpetrator. Ferando does not propose to leave the Duke in doubt much longer, but he emphasises the heuristic value of the incident for Kate:

> Why so *Kate* this was friendly done of thee,
> And kindly too: why thus must we two live,
> One minde, one heart, and one content for both,
> This good old man dos thinke that we are mad,
> And glad he is I am sure, that he is gonne,
> But come sweet *Kate* for we will after him,
> And now perswade him to his shape againe. [TSB, 100]

Two other moments in the play require comment. The first is in the following scene, when the Duke discovers that the servant Phylotas has been impersonating Valerio, his son. The Duke's response is immediate: "Lay handes on them,/And send them to prison straight" [TSB, 101]. The exercise of power wakes the disguised Sly, who insists: "I say wele have no sending to prison" [TSB, 101], a moment in the play when the delusion of sovereign power momentarily gives way to a potentially subversive "otherness" that would flout the law. The second moment occurs at the end when the restored Christopher Sly wakes up and recounts his "dream" to the Tapster:

> I know now how to tame a shrew,
> I dreamt upon it all this night till now,
> And thou hast waked me out of the best dreame
> That ever I had in my life, but Ile to my
> Wife presently and tame her too
> And if she anger me. [TSB, 108]

The active sadism of this utterance contrasts with the thinly veiled eroticism of the other instances of "courtship" in the play; it is masculine bio-power red in tooth and claw. But from a very early point in the play, it is made clear that whatever danger Kate might be in, she retains some confidence in her capacity to resist. She rounds on her father asking him, "what do you meane to do with me,/To give me thus unto this brainsick man, /That in his mood cares not to murder me?" [TSB, Scene 5, 77]. But then in an aside she declares:

> But yet I will consent and marrie him,
> For I methinkes have livde too long a maid,
> And match him too, or else his manhoods good. [TSB, 77]

What we wonder about here is the extent to which the fantasy of male domination is capable of fulfilment in this instance. Or are we dealing here with a structure of power within which resistance can become a real strategy for survival *in spite of* – or as Foucault might put it, *because of* – those social mechanisms that are designed to coerce, or subjugate the feminine "other"? Is it a question of Kate's colluding with a hierarchy produced by masculine ideology merely in order to survive, or has she really internalised, through a mechanism whereby her "shrewish" behaviour is reflected in that of Fernando, the very ethical values that are designed to keep her in her place? These questions are postponed in *The Taming of a Shrew* by the return of Sly to a male identity within which sadistic domination is nothing more than a dream, and one that is, at that, remote from ever being fulfilled.

Shakespeare's play, however, is a little more subtle (though not by much). The Christopher Sly frame is quickly dispensed with, making the ending more problematical in that Kate's capitulation is balanced by the new-found "shrewishness" of the married Bianca. What in the earlier play is offered as a biblical justification for masculine supremacy: "Then to his image did he make a man,/Olde *Adam* and from his side asleepe,/A rib was taken, of which the Lord did make,/The woe of man so termd by *Adam* then,/Woman for that by her came sinne to us" [TSB, 107], becomes in Shakespeare's play an explicit statement about the fundamentally *political* nature of subjection:

> Thy husband is thy lord, thy life, thy keeper,
> Thy head, thy sovereign; one that cares for thee,
> And for thy maintenance; commits his body
> To painful labour both by sea and land,
> To watch the night in storms, the day in cold,
> Whilst thou liest warm at home, secure and safe;
> And craves no other tribute at thy hands
> But love, fair looks, and true obedience;
> Too little payment for so great a debt.
> Such duty as the subject owes the prince
> Even such a woman oweth to her husband. (5.2.147–57)

The woman who is "froward, peevish, sullen, sour" is nothing more than "a foul contending rebel,/And graceless traitor to her loving lord". (5.2.158–61). In a later play, *The Merchant of Venice* (1597), Portia invokes the same hierarchy as she fulfils the conditions of her father's "will". Speaking of herself in the third person, she performs an act of allegiance to Bassanio similar to that which Kate performs before Petrucchio: "her gentle spirit/ Commits itself to yours to be directed,/As from her lord, her governor, her king" (3.2.163–65). Clearly, in these plays the "self" (particularly the female "self") is not the result of autonomous human agency, nor is it part of an abstract appeal to some form of higher "good". The "good" outlined (perhaps ironically) by Shakespeare's Kate, and enthusiastically embraced by Portia, is *politically* contingent, where "happiness" and "the good life" support a masculine ideology. Departures from fixed hierarchically defined identities derive, in large part, from *male* fantasies of power in these plays, but whose structure we are now able to read dialectically. And, of course, the exclusively masculine institution *par excellence* where these fantasies can take on a certain degree of mobility is in, and through, theatrical performance.

Much Ado About Nothing

The anonymous *The Taming of a Shrew* deals gingerly with the issue of female chastity. Kate admits that she has "livde too long a maid" [TSB, 77], and Ferando's servant Sanders alludes impertinently to her sexual power to cuckold her husband: "I hope sheele make you one of the head men of the parish shortly" [TSB, 78]. But in both cases the woman's identity is eradicated. This is exactly the point that William Blackstone makes in the relevant sections of his *Commentaries on the Laws of England* (1769) when he observes:

> By marriage the husband and wife are one person in law: that is, the very being or legal existence of the woman is suspended during the marriage, or at least incorporated and consolidated into that of the husband: under whose wing, protection and *cover*, she performs every thing; and is therefore called in our law-french a *feme-covert, foemina viro co-operta*; is said to be *covert-baron*, or under the protection and influence of her husband, her *baron*. or lord; and her condition during her marriage is called *coverture*.[11]

And this is a central issue in a later Shakespearean comedy, *Much Ado About Nothing* (1598).

The opening scene-heading of the 1598 quarto of *Much Ado About Nothing* includes a reference to one "*Innogen his* [Leonato's] *wife*" and in a later scene-heading at Act 2 scene 1, "*his wife*". Editors, of whatever gender, have come

---

[11] W. Blackstone, *Commentaries on the Laws of England* (Oxford: Clarendon Press, 1769), 4 vols., I. 422.

to regard this as an error, in what are assumed to have been Shakespeare's foul papers, and have expunged Innogen from the text altogether. For example, A. R. Humphreys claimed that her presence in the quarto reflected "an unrealised intention"[12] while the most recent Arden 3 editor of the play, Claire McEachern, regarded her as a "ghost character", who, she claims, Shakespeare found "more powerful in her absence than her presence" and so "she has been retired from the fray".[13] Of course there is another way of viewing the ubiquitous silent presence of Innogen, the married woman whose silence performs the very legal condition of her status, and whose subjectivity is the *terminus ad quem* of the play's females; but what the play discloses about the care of the female self also impinges upon its definitions of masculine subjectivity. Indeed, at one level, and in terms of a teleology of comic form, the play appears to confirm what Alain Badiou, in his critique of the ethics of modern subjectivity calls "the absence of any project, of any emancipatory politics, or any genuinely collective cause".[14] But of course, we know that in the process of distancing ideology – a project of which art is perfectly capable – whatever absences we may *read* into the text disclose the conditions of their own production. Far from heralding an end of ideology, and "a return of ethics", or a resignation to the manifest teleology of the text as a heuristic process designed to prescribe "care of the self", such a move "signifies", according to Badiou, "an espousal of the twistings and turnings of necessity, and an extraordinary impoverishment of the active, militant value of principles".[15] Thus, the presence of the silent Innogen at crucial points – and she should appear in all of the play's "public" scenes, particularly at the abortive wedding ceremony of Claudio and Hero at Act 4 scene 1, and in the final scene of the play – is both an affirmation of how patriarchy shapes female subjectivity, *and* a feminine rebuke to its methods of caring for the self. In this way, the play's own conservatism is held up for scrutiny.

Nowhere is this process more clearly revealed than in the fates of Beatrice and Hero, two females who occupy extreme positions, and yet who are respectively "refashioned" through the same mechanism of deceit that drives one plot centripetally, and the other centrifugally. Whatever the antagonism between Beatrice and Benedick, if "the world must be peopled" (2.3.233–4), then some form of sexual encounter between them must take place, and it can

---

12 W. Shakespeare, *Much Ado About Nothing* [1598], ed. A. R. Humphreys (London: Arden, 1981), 77.
13 W. Shakespeare, *Much Ado About Nothing*, ed. C. McEachern (London: Arden, 2006), 139–140.
14 Badiou, *Ethics*, 31.
15 Badiou, *Ethics*, 32.

only do so *legitimately* within the institution of marriage. Hero's "plot" to transform the "disdainful" Beatrice from one whose "spirits are as coy and wild/As
haggards of the rock" (3.1.35–6) and who is "so self-endered" (3.1.56), into
an obedient, "selfless" and passive wife, is effected by means of a dangerous
deception. Her plan involves approaching Benedick and maligning Beatrice:

> And truly I'll devise some honest slanders
> To stain my cousin with: one doth not know
> How much an ill word may empoison liking. (3.1.84–86)

Of course, Hero has already just managed to escape being the victim of
slander, and when this perverse logic is applied to her case later, it will have
devastating effects. In his chapter on *The Pharmakon* in his essay "Plato's
Pharmacy", Derrida aims to locate the constitutive *difference* that resides at
the heart of a term that can mean both "remedy" and "poison", and that is,
after all, a problem for language and for translation. Commenting on Plato's
treatment of writing in the *Phaedrus* Derrida notes:

> Writing is no more valuable, says Plato, as a remedy than as a poison. Even before
> Thamus has let fall his pejorative sentence, the remedy is disturbing in itself. One
> must indeed be aware of the fact that Plato is suspicious of the *pharmakon* in gen-
> eral, even in the case of drugs used exclusively for therapeutic ends, even when
> they are wielded with good intentions, and even when they are as such effective.
> There is no such thing as a harmless remedy. The *pharmakon* can never be simply
> beneficial.[16]

From the play's title, through its unstable language of courtship, to the linguistic infelicities of the Watch and Dogberry's mangling of grammar, there
is something dangerous in a strategy that will produce both benefit *and* harm.
The benefit is that Hero's "honest slanders" will serve to correct Benedick's
and Beatrice's stated independence, and overhaul what is, in part, their ethical objections to each other, thereby neutralising their hostility and bringing
their behaviour into line with a dominant ideology. But, of course, Hero's
reduction of language to what Derrida might call "one of its simple elements
by interpreting it"[17] will prove her own undoing, with the consequence that
she herself is subjected to the indignity of rejection at the altar, and to having
her own body "read" and interpreted as though it were a text susceptible to
translation. Claudio gives Hero back to her father as "but the sign and semblance of her honour" (4.1.32), and Leonato's response is to "interpret" his
daughter in accordance with a homosocial ethic:

---

[16] J. Derrida, *Dissemination*, trans. Barbara Johnson (Chicago and London: University
of Chicago Press, 1981), 99.
[17] Derrida, *Dissemination*, 99.

> Why, doth not every earthly thing
> Cry shame upon her? Could she here deny
> The story that is printed in her blood?
> Do not live, Hero, do not ope thine eyes;
> For did I think thou wouldst not quickly die,
> Thought I thy spirits were stronger than thy shames,
> Myself would on the rearward of reproaches
> Strike at thy life. (4.1.120–27)

But Hero is a multiple text, as the "reading" of the Friar demonstrates, and the very symptoms that persuade her father of her guilt, can be adduced as evidence of her innocence:

> I have mark'd
> A thousand blushing apparitions
> To start in her face, a thousand innocent shames
> In angel whiteness beat away those blushes,
> And in her eyes there hath appear'd a fire
> To burn the errors that these princes hold
> Against her maiden truth; call me a fool;
> Trust not my reading nor my observations,
> Which with experimental seal doth warrant
> The tenor of my book; trust not my age,
> My reverence, calling, nor divinity,
> If this sweet lady lie not guiltless here
> Under some biting error. (4.1.158–170)

The presence of a silent Innogen in this scene would certainly sharpen a sense of what is at stake in the "reading" of the silent Hero, just as her appearance at the end would serve as a comment on the silent Hero, "given away" by her father, and the silencing of Beatrice. Of course, Hero is never placed in the position of needing to "know" herself, and she therefore does not need to have access to any of the technologies of the "self" to which Foucault refers.[18] She is what she is, a pawn in a patriarchal marriage game who is as much a victim of the duplicity of language as she is a perpetrator of "honest slanders". The case is different for Beatrice since, as "Lady Disdain" (1.1.109), and "My Lady Tongue" (2.1.258), she simply refuses to enter into dialogue with Benedick and *vice versa*. From Benedick's point of view the transformation of Claudio from soldier to lover involves entry into a bewildering minefield of linguistic representation:

---

[18] See M. Foucault, "Technologies of The Self" in *Essential Works of Foucault 1954–1984, vol. 1: Ethics, Subjectivity and Truth*, ed. Paul Rabinow (New York: Pantheon, 1994), 230–238.

> I have known when he would have walked ten mile
> afoot to see a good armour, and now will he lie ten
> nights awake carving the fashion of a new doublet.
> He was wont to speak plain, and to the purpose, like
> an honest man and a soldier, and now he is turned
> orthography – his words are a very fantastical banquet,
> just so many strange dishes. May I be so converted
> and see with these eyes? I cannot tell; I think not.
> (2. 3. 15–23)

Of course, like Beatrice, Benedick is drawn, by a piece of social engineering, to what he openly resists; he *does* "turn orthography", just as Beatrice is "persuaded" to submit in silence to him. Dangerously volatile though the discourse of "love" may be, it is finally "interpreted", as it were by consensus, within the institution of marriage. In this play an ethic of gender "freedom" is articulated negatively as a form of self-absorption that "society" cannot tolerate. In this respect the play "names" what we might call a simulacrum that heads off "other possible kinds of truth processes" in its attempts to manage what Alain Badiou calls "the amorous event". It is an understanding of the social formation that brings with it "error and violence"[19] issues that we would not normally expect to find in Shakespeare's comic universe.

The reconstitution of Beatrice and Benedick as subjects within the discourse of patriarchy is effected with greater subtlety than Petrucchio's shaping of Kate's "self" in *The Taming of The Shrew*, although even in the earlier play it is difficult to evaluate the extent to which her capitulation to his brainwashing techniques is ironical. In *Much Ado About Nothing* the reconstitution of Benedick as "married man" takes place alongside that of Beatrice as prospective wife. But what is also surprising is the masculine subjectivity that Beatrice constructs for Benedick. Following the metaphorical "death" of Hero at the altar, both are bewildered by their proclamations of their love for each other. In a moment of romantic extravagance Benedick exhorts Beatrice to "Come, bid me do anything for thee" (4.1.287). Her response is an immediate, but surprising appeal to a distinctly masculine chivalric code: "Kill Claudio" (4.1.288). Subsequently when she encounters Benedick she calls him to account, and his reluctance to carry out the deed provokes a withdrawal of her affection; her desire to know "what hath passed between you and Claudio" is met with "Only foul words – and thereupon I will kiss thee" (5. 2. 47–8). Her resistance is a reflection upon Benedick's *lack* of masculinity, arising from his refusal to act according to a prescribed subjectivity:

---

[19] Badiou, *Ethics*, 77.

> Foul words is but foul wind, and foul wind is but
> foul breath, and foul breath is noisome; therefore
> I will depart unkissed. (5. 2. 49–51)

But this is more than a casual matter, since by refusing to play his masculine part, Benedick releases Beatrice's destructive alter-ego that proceeds to upbraid her lover for his refusal to convert words into action. There is something wrong with the ethics of a discourse that *avoids* action, and Benedick almost half realises this, just as Beatrice half realises the shortcomings of the very chivalric code that she persists in invoking. Benedick's strategy is to accuse her of exploiting the ambivalence of language:

> Thou hast frighted the word out of his right sense,
> so forcible is thy wit. But I must tell thee plainly,
> Claudio undergoes my challenge, and either I must
> shortly hear from him, or I will subscribe him a
> coward. And I pray thee now tell me, for which of
> my bad parts did you first fall in love with me?
> (5. 2. 52–57)

Both parties then proceed by their denials to deconstruct the very discourse in whose net both their subjectivities are now inscribed. Beatrice has fallen in love with *all* of Benedick's "bad parts", "which maintained so politic a state of evil that they will not admit any good part to intermingle with them" (5. 2. 58–60), while he "suffers" love for her because like the unattainable mistress of established sonnet sequences, she maintains a decorum of keeping aloof. Here both invoke in their denials the very formal language of love and its prescriptions for their social behaviour towards which they gravitate. And sure enough, both are caught in the act of writing poetry that brings them to exactly the position at the end that resembles the situation of Claudio and Hero at the beginning of the play.

On the next, and final, occasion upon which Beatrice threatens a return to her old ways, she is silenced – and we can see why Innogen's presence on the stage in Act 5 scene 4 might make that silence "speak". In both Q and F the line "Peace I will stop your mouth" is given to the patriarch Leonato, and it was Theobald who initiated a lasting editorial tradition that re-ascribed the line to Benedick. This emendation misses completely the point of the line: that the patriarchal care of the female self involves an authoritative silencing, that effectively exposes a bio-power in which ethics and ideology are aligned with each other. These are the values that Claudio, Benedick and Beatrice are required to internalise in order for society to function productively.

But there is one obstacle that remains a threat: the bastard brother of Don Pedro, Don John. While efforts are made to curtail the act of interpretation (that, at its most comically difficult, involves Dogberry and the Watch, who are in possession of a sequence of "truths" that escape any formal grammar), there is a danger that the "cure" can never be permanent. The play begins with the ending of a war against the "outlaw" brother Don John, whose bastardy stands as a flagrant threat to the social order and to the authority of Messina. As a being wholly devoid of selfhood, "the son of nobody, and sometimes called *filius nullius*, sometimes *filius populi*" as Blackstone describes the figure of the bastard,[20] he represents a serious threat to what we might call a Renaissance bio-ethics and the patriarchal power that sustains it.

The "evil" in *Much Ado About Nothing* reduces sex to an anarchistic and animalistic level. His "sadness" is "without limit" (1.2.4), he is "trusted with a muzzle and enfranchised with a clog" (1. 2. 30–31) and if, as he says, "I had my mouth I would bite" (1. 2. 32–3). His is an evil that functions as "an unruly effect of the power of truth", to use Alain Badiou's phrase,[21] articulated as a radical evil, that is the consequence of a consensual identification. As the alter ego of Don Pedro, Don John fulfils the function of a linguistic *supplement*, a "mouth" whose negative anti-social potential must be "muzzled" and contained, formally outlawed. The non-subject, Don John, the *filius nullius*, who is also at the same time the demonised vitalism of the *filius populi*, is the binary opposite against which the play's gendered subjects measure their own institutionally sustained subjectivities. In the play, the bastard, who is the overdetermined "other" of gendered subjectivity, and the means whereby the regulation of sexual practice achieves its ideological justification, counterbalances the figure of the "silent woman". As the play's title implies, both are different facets of the "nothing" that serves to define an entire ideological ethos. Don John may be again "contained" at the end of *Much Ado about Nothing*, but for the two pairs of lovers who are finally united, Benedick and Beatrice's "persevering ethic" places a strain upon ideology, whereas the sexual jealousy that has earlier "killed" Hero remains as a poisonous supplement, capable of surfacing whenever Don John evades his captors and becomes "unmuzzled", releasing an extreme social violence. In this play, the "nothing" that is at the centre of the play is "virginity", the immaterial object that structures male dominance and female subjection. The containment of sexual energy, and the essentially *political* threat that it poses to community, is there in both *The Taming of The Shrew* and *Much Ado about Nothing*. In the earlier play subjectivity and self-hood are achieved my means of a mirroring effect, forcing the shrewish woman to recognise her own imaginary identification,

---

[20] Blackstone, *Commentaries on the English Laws, vol. 4*, 459.
[21] Badiou, *Ethics*, 61.

but at the same time rendering fragile the male fantasy of total control. In the later play, faults are distributed across the gender divide, but it is Don John, the bastard who represents the displaced, and potentially anarchic evil whose poison threatens an orderly social dance. The teleological thrust of both plays, their coming to rest in a conservative solution, is undermined by the forces that each action unleashes. The energies that are shown to be the "unruly effect(s) of the power of truth" infect the very society that would seek to regulate them. And in spite of aesthetic closure, the representation of a conservative deployment of a bio-ethics, and a bio-power that seek to sustain each other, gendered subjectivity is shown to be inscribed along the faultline that the supplementary forces of linguistic instability uncover. Perhaps the last word should remain with Dogberry, the possessor of the truth, the champion of hierarchy, and the arch de-constructor of grammatical form, whose very quasi-legal subjectivity enshrines the contradictions that this comedy glimpses as it moves to its formal conclusion:

> Write down that they hope they serve God: and write 'God' first, for God defend but God should go before such villains. (4. 2. 17–19)

Patrizia Nerozzi Bellman
IULM University Milan

On the Sciences of Man in Eighteenth-Century English Literature and Art: Anatomizing the Self

> *Is it with blood that we think, or with the air and fire that is in us? Or is it none of these, but the brain that supplies our sense of hearing, and sight and smell, and from these that memory and opinion arise, and from memory and opinion, when established, that knowledge comes?*[1]

This essay arose from both my interest in the relationship between man and machine in eighteenth-century art and literature and from the interdisciplinary research project on law and literature in which I have been involved over the past few years. The complex, multifaceted, somewhat magmatic and elusive discipline which we now call "bioethics" provides here the privileged field of discussion on which the limited scope, albeit couched under the rather ambitious title of my paper, relies for its meaning.

The starting point is practically a must. The idea of the Self which indelibly marks the transition to modernity takes on consistency in the eighteenth century when an intersecting of disciplines – literature, law, art, medicine and philosophy – cooperate in shaping the so-called "science of man" in a way which was to remain unique in Western intellectual history. It is at this seminal crossroads, where the great cultural myth of "modern individualism", together with that of the "machine", happened to be born that many of the roots which still influence the contemporary debate on bioethics are constantly renewed and updated in the processes and products of invention. The construction of the first automata, while adding spectacular evidence to the debate on the statute of the individual in relation to the machine, gave concrete form to the dilemma affecting the right of the artist/scientist to arrange mechanical and organic components to create artificial life, under the sign, howsoever ambiguous, of the human being. In a time of "expanding horizons",[2] the debate

---

[1] From Plato in *Phaedo.* I owe this quotation to S. Zeki's seminal essay "Neural Concept Formation & Art. Dante, Michelangelo, Wagner", *Journal of Consciousness Studies* 9. 3 (2002): 53–73.

[2] M. Berger, *Real and Imagined Worlds. The Novel and Social Science* (Cambridge, Mass., and London: Harvard University Press, 1977), 13.

moved from scientific experiments to new literary genres, extending the same ontological nature to the latter, under the mask of fiction.

The realistic novel emerged as the genre which best epitomized the problem of identifying the precise moment at which the existence of a character actually begins, thus legitimizing the author's poetics based on the conventional, but not-so-obvious, identification of fiction with reality. Framing the modern ethical consciousness becomes the acclaimed purpose of the production and diffusion of new ideological perspectives, promoting sound moral and religious values, through narratives which highlight the connection between religion – whether natural or revealed – and the welfare of man in society.[3] A few pertinent, although desultory, fictional examples may prove functional to illustrate this issue. In Daniel Defoe's documentary descriptions of reality, overfurnished with useful information and moral principles, the theme of abortion and infanticide is largely present, significantly interwoven with the narrator's reflections on the right to exist of his characters. In Laurence Sterne's *Tristram Shandy* the author's attempt at dissecting the "nature" of man, seen as an individual confronted with the snares of time and place, remains the focus of narrative interest, whereas the novel's aesthetics, the meaning of writing itself,[4] is based not only on the search for the biological and cognitive origins of life *ab ovo*, but also on the legal statute of the "persona" endowed with rights from birth: the homunculus "consists, as we do, of skin, hair, fat, flesh, veins, arteries, ligaments, nerves, cartilages, bones, marrow, brains, glands, genitals, humours, and articulations".[5] In symmetrical correspondence with John Locke's "first ideas" which are not innate but derive from sensations already perceived by the foetus when still in the mother's womb,[6] Sterne's narrator discusses the question, at the same time theological, legal and aesthetic, of the need to christen children before birth. In Samuel Richardson's *Clarissa*, an early *Bildungsroman* about the process of knowing oneself and the nature of individual liberty, the heroine lets herself die after being raped, as she feels that the integrity of her being, as a physical, psychic and juridical person, has been destroyed.

---

3   See P. Nerozzi Bellman, "Equity on Trial: Judicial Cases in the Novels of Richardson and Fielding" in *Practising Equity, Addressing Law. Equity in Law and Literature*, ed. D. Carpi (Heidelberg: Winter, 2008), 295–316.

4   See P. Carbone, *La lanterna magica di Tristram Shandy. Visualità e informazione, ordine ed entropia, paradossi e trompe l'oeil nel romanzo di Laurence Sterne* (Verona: Ombre Corte, 2009).

5   L. Sterne, *The Life and Opinions of Tristram Shandy Gentleman* [1759–1767] (London: Penguin, 1985),Vol. I, Chapter II, hereafter indicated by the abbreviation: *TS.*

6   John Locke, "Of Perception", in *Essay Concerning Human Understanding*, Book II, Chapter IX.

Re-writing,[7] or re-inventing the Self, is one of the great utopias of the century. But what exactly does the word "self" mean? A short excursus into the
history of the word "self" can reveal how its meaning and use respond to the
cultural impulses which mark the transition to modernity. The first evidence
of "self-" dates back from the tenth century, when its use overlapped with
the other personal pronouns "myself", "yourself" and "herself". Until the
fifteenth century the number of words composed with "self" is extremely
low: besides "self-will" and its cognates, "self-willing" and "self-willed", we
find "self-heal" (a botanical variety) and "self-same". At the beginning of the
sixteenth century the number of English words compounded with the prefix
"self" increased with astonishing speed. Initially, these had a strongly
negative connotation in philosophical or theological manuscripts, expressing
the human distance from God or the isolation of man unable to check or
counteract the power of evil forces. This condition of spiritual isolation per-
meates *Paradise Lost* where Milton insistently associates the compounds of
"self" with the images of Satan and sin. In the eighteenth century the
number of words linked to divine, supernatural or religious elements plum-
meted, while scientific terminology increased and became generally more
widespread as the end of the century drew near. Significantly, the prefix
"self" denoted the mechanical capacity of a machine to perform automatic
functions: "self-acting", "self-charging", "self-winding", "self-moving", etc.
Since man himself was viewed primarily as a social being and society was
seen to play a prominent role, the majority of newly coined words belonged
to the concepts of gentility, morality and reputation. "Self-government",
"self-stability", "self-possessing" must be pursued and achieved through an
escalation of values, from virtue through gentility and morality to reputation.

"Self is not determined by Identity or Diversity of Substance, which it
cannot be sure of, but only by identity of consciousness". In the second
edition of his *Essay Concerning Human Understanding* (1694), Locke defined the
relationship between self, persona and consciousness that is the foundation
of the modern conception of the individual, where morality is placed on
a rational basis.[8] Human nature is uniform "from China to Peru" Samuel
Johnson, the age's foremost intellectual, will add a few decades later: "The
interests and passions, the virtues and vices of mankind, have been diversi-

---

[7] Cfr. C. Taylor, *Sources of the Self. The Making of the Modern Identity* (Cambridge: Press
    Syndicate of the University of Cambridge, 1989) and R. Porter, *Rewriting the Self*
    (London: Routledge, 1977).
[8] J. H. Tufts, *The Individual and his Relation to Society as Reflected in British Ethics* (New
    York: Augustus M. Kelley-Publishers, 1970), 49.

fied in different times, only by unessential and casual varieties" (*Adventurer,* 95) as "such is the constitution of the world, that much of life must be spent in the same manner by the wise and the ignorant, the exalted and the low. Men, however distinguished by external accidents or intrinsick qualities, have all the same wants, the same pains and ... the same pleasures" (*Idler,* 51).[9] The word "constitution" suggests an immediate analogy with human anatomy; since all men and women share the same anatomical characteristics, their differences can only be "casual" and "inessential".[10] Essential human nature is considered independent of historical times and circumstances, permanent and uniform, even if man knows himself as a single, unique individual, conditioned by that sense of self, "ce sentiment individual de notre existence [...] ce sentiment du *moi,* si unique, si simple".[11] However, "the more we know of the human Machine, the more simple it appears" claims reassuringly Robert James in the preface to his *Medicinal Dictionary* (1743–1745).[12]

The eighteenth century does not herald in any epoch-making changes or discoveries, nor is its *Weltanschauung* similar to the disquiet and uneasiness generated by the "new astronomy" during the Renaissance with the diffusion of "Copernicanism". The dualistic construction of the corporeal and incorporeal cosmos was inherited from the seventeenth century:

> Eighteenth century man was the beneficiary of seventeenth-century scientific discovery; as such, he was also the first modern man deeply to question the role of science and the effects of technology on his normal everyday life.[13]

What is new and modern is the perception of scientific thought; this is the result of a popularization of science geared to capture the imagination and not exempt from darker sides which already prelude the dangers of current uncontrolled scientific communication. Medical learning was becoming popularized and even the use of Latin, important in enhancing the prestige and aloofness of the profession, was gradually being abandoned. The *Literary Magazine* (1756–1757), though predominantly a political journal, in-

---

9   Quoted in J. Wiltshire, *Samuel Johnson in the Medical World. The Doctor and the Patient* (Cambridge: Cambridge University Press, 1991).

10   Cfr. Wiltshire, *Samuel Johnson in the Medical World, The Doctor and the Patient,* 207.

11   P. L. Gérard, *Le Comte de Valmont, ou Les égarements de la raison* [1776], in John McManners, *Death and the Enlightenment. Changing Attitudes to Death among Christians and Unbelievers in Eighteenth-century France* (Oxford and New York: Oxford University Press, 1985), i.552, 120.

12   Quoted in Wiltshire, *Samuel Johnson in the Medical World. The Doctor and the Patient,* 80.

13   G. S. Rousseau, "Science" in *The Eighteenth Century,* ed. P. Rogers (London: Methuen & Co Ltd., 1978), 154.

tended to extend its scope, as may be seen in Johnson's introductory preface "To the Public", to scientific information and "physiological discoveries":

> Our regard will not be confined to Books; it will extend to all the productions of Science. Any new calculation, a commodious instrument, the discovery of any property in nature, or any new method of bringing known properties into use or view, shall be diligently treasured up wherever found … We hope to find means of extending and perpetuating physiological discoveries, and with regard to this Article, and all others, intreat the assistance of curious and candid correspondents.[14]

Compiling medical handbooks such as William Buchan's *Domestic Medicine* (1769) would enable "the reader to defend himself against 'ignorant pretenders' and upstart empirics".[15]

Eighteenth-century "science" can be considered as a unified whole, pervading and interconnecting all other societal activities and needs – politics, economics, philosophy and the fine arts. Mathematics enjoyed a unique position as the science to be applied to all the other developing disciplines, but medicine was considered to rival with mathematics in its new discoveries. What else are "questions of science" if not "questions about man"? as David Hume states, epitomizing the spirit of the age. It was a time of men and groups, eager not only to found and attend scientific societies but also to organise private meetings in which they would mull over scientific experiments, such as those attended by the intellectuals, manufacturers and artists of the Birmingham "Lunar Society".[16]

The connection between the visible surface and invisible depth of an organism became a recurrent question both in scientific circles and during "drawing room" experiments like the ones painted by Joseph Wright of Derby which, as in the case of "Experiment with an Air Pump" (1768), show

---

[14] Quoted in Wiltshire, *Samuel Johnson in the Medical World, The Doctor and the Patient*, 111.

[15] Wiltshire, *Samuel Johnson in the Medical World, The Doctor and the Patient*, 91.

[16] The Lunar Society of Birmingham was a dinner club and informal learned society of prominent Midlands industrialists, scientists, natural philosophers, artists and intellectuals who met together regularly between 1765 and 1813. The name of the society arose because the group met each month during the full moon when the extra light would make the journey home easier and safer. The members of the Lunar Society were all prominent in British society. Amongst those who regularly attended the meetings were Matthew Boulton, Erasmus Darwin (grandfather of Charles), Josiah Wedgwood, James Watt and James Keir. Less regular attendees and correspondents of the Society included Sir Richard Arkwright, James Wyatt, John Smeaton, Thomas Jefferson and even Benjamin Franklin. See J. Uglow, *The Lunar Men. Five Friends Whose Curiosity Changed the World* (London: Chatto & Windus, 2003).

how the new sciences can sustain or destroy the breath of life with a single gesture. The visualization of knowledge[17] is considered central to the process of demonstration and optical illustration is deemed a necessity.[18] Anatomical dissection is believed to lead to scientific discoveries as well as to new aesthetic experiences:

> William Hunter's insistence on teaching anatomy through direct student dissection of cadavers helped to transform the medico-scientific standing of the surgical profession. His involvement with the plastic and figurative arts performed a similar function, elevating the figure of physician as guardian of culture and presaging the complex dialectic of arts and sciences so integral to romanticism.[19]

However, the reverse side of the century's pragmatic optimism is exhibited in William Hogarth's "The Reward of Cruelty", the last of "The Four Stages of Cruelty" (1751). The macabre setting shows the highwayman Tom Nero after his hanging at Tyburn; his autopsy is being carried out in the Cutlerian theatre near Newgate Prison which was used by the Company of Surgeons between 1745 and 1751. As was the practice with a number of executed criminals, his body is being dissected for the study of anatomy and is shown as reduced, under the surgeons' pitiless hands, to the utmost grossness of its bones and entrails' physicality. The scene, rich in symbolic references and gruesome allusions – not exempt from a whiff of cannibalism – presents a satirical, ghastly parody of a good death, as in a medieval allegory of the triumph of death. Tom's hideous grimace could suggest that he is being represented as still alive, a realistic depiction of a particular fear that many criminals had that they would survive the hanging and be conscious at their dissection.[20]

Anatomical waxes offered both scholar and layman less traumatic access to what could only be previously imagined, the innermost parts of the human body. Throughout the eighteenth century and up to the beginning of the nineteenth, Italy is significantly the most renowned centre of production in Europe for anatomical sculptures, which solved, though only partially, the Church's age-old problem of cadavers and their dissection. European universities were well-equipped with polychrome wax models of entire figures or of separate parts and individual organs of the human body. Conversely,

---

17 See B. M. Stafford, *Body Criticism. Imagining the Unseen in Enlightenment Art and Medicine* (Cambridge, Mass., and London: The MIT Press, 1991), 350.

18 See T. Chico, "Microscopy and Eighteenth-Century Narrative", *Mosaic: a Journal for the Interdisciplinary Study of Literature* 39.6, II (2006).

19 W. F. Bynum and R. Porter eds., *William Hunter and the Eighteenth-Century Medical World* (Cambridge: Cambridge University Press, 2002), 2.

20 See P. Linebaugh, Hay, Doug, and Thompson eds., *Albion's Fatal Tree: Crime and Society in Eighteenth-Century England* (London: Pantheon Press, 1975), 65–118.

the numerous collections in Europe, such as the one in Paris, which opened as early as 1711, were a very popular attraction:

> The somewhat grand-guignolesque spectacle of the hidden gears and mechanisms of the human body out of which both death and life mischievously peeped, could not fail to attract the curiosity of the urbanized masses often beset by the concrete reality and imaginary fears of by now epidemic and sinful diseases like syphilis.[21]

The spectacularity of science is already on its way while brand new questions regarding the constitution of personhood are coming to the fore.

The problem of making the invisible visible, of finding visual strategies for imagining the unseen, became critical in both the fine arts and natural sciences, taking on the form of a major modern epistemological, artistic and scientific quest. During the course of the eighteenth century visual strategies and theories were put forward for "imagining the unseen".[22] The scientist as well as the writer and the painter felt compelled to look inside and explore "the internal fabric" to discover the relationship between exteriority and interiority, public conduct and private pathos, the social and the private self:

> How painful soever this inward search or enquiry may appear, it becomes, in some measure, requisite to those, who would describe with success the obvious and outward appearances of life and manners. The anatomist presents to the eye the most hideous and disagreeable objects; but his science is useful to the painter in delineating even a Venus or an Helen. While the latter employs all the richest colours of his art, and gives his figures the most grateful and engaging airs; he must still carry his attention to the inward structure of the human body, the position of the muscles, the fabric of the bones, and the use and figure of every part or organ.[23]

The new scientific instruments, such as the new optical devices, provided new vistas onto an "other", hidden, world, transforming the way of perceiving the human body. The so-called "Microscopical Eyes" which artificially improved the sense of sight, could now allow the viewer to look inside an organism. Robert Hooke (1635–1701), founding member of the Royal Society and author of *Micrographia* (1667), had already claimed that the microscope enabled viewers to widen "the narrowness and wandering of our Senses" by focusing on details that could not ordinarily be seen – and therefore perceived or understood. According to Hooke, vision leads to the imagination

---

21 P. Bellasi, "Signore e signori, il corpo!" in *Corpo. Automi. Robot tra Arte, Scienza e Tecnologia*, eds. B. Corà and P. Bellasi (Milano: Mazzotta, 2009), 18.
22 Stafford, *Body Criticism. Imagining the Unseen in Enlightenment Art and Medicine.*
23 D. Hume, *Philosophical Essays concerning Human Understanding* [1748], quoted in "Introduction" to *The Analysis of Beauty* [1753], ed. Ronald Paulson (New Haven and London: Yale University Press, 1997).

and the mind's eye, as Joseph Addison was later to develop in his essays on the "Pleasures of the Imagination" in the *Spectator* (1712) where he praises "the discoveries they have made by glasses":

> Our Sight is the most perfect and the most delightful of all our Senses. It fills the Mind with the largest Variety of Ideas, converses with its Objects at the greatest Distance, and continues the longest in Action without being tired or satiated with its proper Enjoyments.[24]

Examining the minute world was seen to amplify our senses and, by extension, our reason and understanding, leading to a new awareness of man. Since true knowledge derives from empirical observation, microscopists conceptualized the minute particular as both a product of visual observation and subject to theoretical considerations. However, the particular must be inspected by the natural philosopher's discriminating eye since only he can make sense of it. In his Cutlerian Lecture to the Royal Society (1691–1692), Hooke lamented the fact that this visual machine was developing into a spectacular toy. Only the Dutch naturalist Anton van Leeuwenhoek (1632–1723) still used microscopes for science

> besides whom, I hear of none that make any other Use of that Instrument, but for Diversion and Pastime, and that by reason it is become a portable instrument, and easy to be carried in one's Pocket.[25]

The infatuation with optical instruments, typical of the period, shows the interdependence of science, fine arts and technology. Aesthetics and biology, neurology and research into visual perception, physiology and philosophy cooperate in the foundation of a new cultural approach. What particularly worried Hooke was the true understanding of the relationship between minute particulars and the acquisition of general knowledge by fitting fragmented details into whole organisms. By mid-eighteenth century, microscopes were seen to "furnish us as it were with a new Sense, unfold the amazing Operations of Nature, and present us with Wonders unthought of by former Ages" as Henry Baker wrote in the preface to his *Microscope Made Easy* (1742), probably the most successful and popular handbook of the times.[26] The microscope's influence, while bringing new convictions into the world of ethics and aesthetics, introduced new themes to literature. The human body, traditionally conceived as the organizing, coherent structure of

---

[24] Addison, *Spectator* (1712), 411.

[25] R. Hooke, "Discourse", quoted in Chico, "Microscopy and Eighteenth-Century Narrative", 261.

[26] See Hooke, "Discourse", quoted in Chico, "Microscopy and Eighteenth-Century Narrative".

all structures, could now be minimized or magnified. It is difficult to believe Swift had never looked through either telescopes or microscopes:

> Swift satirized the new science with characteristic vehemence, but *Gulliver's Travels* would not be what it is had Swift not looked through a microscope – perhaps the one he bought for Stella – and felt the fascination and repulsion of grossly magnified nature.[27]

In a century fascinated by measures and models, it was the model of the body as an integral whole which was finally falling apart[28] through dissecting and recombining. Life was shown to be made up of "minute particulars" which the new novel's realistic engagement was due to reproduce. In *Tristram Shandy* Sterne's intense engagement with the possibilities of his medium voices a persistent anxiety about capturing and relating the mass of "minute particulars" to the general system of narration. The expression "minute particulars" transmigrates to the domain of literature from Richardson's novels onwards, and can still be found later when Jane Austen narrates with "minute particulars" the descriptive, fragmented details of ordinary daily experience in *Emma*'s story.

It is the microscopical anatomist's task to penetrate beyond the superficial level and probe into the multiple layers of subjectivity, sometimes only to discover what was already expected. In one of the *Spectator*'s satirical essays, the "beau's head" when "opened" seemed like any other head "but upon applying our glasses to it, we made a very odd discovery, namely, that what we looked upon as brains, were not such in reality, but an heap of strange materials wound up in that shape and texture, and packed together with wonderful art in the several cavities of the skull". Moreover the pineal gland is found "encompassed with a kind of horny substance, cut into a thousand little faces or mirrors, which were imperceptible to the naked eye".[29] Anatomizing the "Coquette's Heart" shows the connection between the heart and the eye but not with the brain. It must. however, be remembered that women's intellectual engagement could no longer be overlooked: the "learned lady", the "scientific girl", the "philosophical girl" are already part of the *Spectator*'s readers, ready to appreciate the English translation of *Il Newtonianismo per le dame*,[30] be constant readers of Eliza Haywood's *Female Spectator*

---

[27] M. Nicolson, *Science and Imagination* (Ithaca, New York: Great Seal Books, Cornell University Press, 1956).

[28] Stafford, *Body Criticism. Imagining the Unseen in Enlightenment Art and Medicine*, 350.

[29] Addison, *Spectator*, 281.

[30] The English translation by E. Carter of Francesco Algarotti, *Il Newtonianismo per le dame: ovvero, dialoghi sopra la luce e i colori* (Napoli, 1737) appeared as *Sir Isaac Newton's Philosophy Explain'd* (London, 1739).

(1744–1746) and even to be given the fictional role of intellectual as in *The Bassett-Table* (1761), a novel by Mrs. Susannah Centlivre.[31]

Not unlike the anatomist who dissects a cadaver, the painter is involved "in laying open the internal constitution of Man" in order to explore the relationship between the visible surface and the invisible interior, between the social self and the inside of his body, the face and the mind. The paintings and engravings of William Hogarth illustrate the desire to trace the origin of form through an optical trip inside the physical world of "objects", guided by his "pleasure of pursuit" and his theory of perception:

> [...] let every object under our consideration be imagined to have its inward contents scoop'd out so nicely, as to have nothing of it left but a thin shell, exactly corresponding both in its inner and outer surface, to the shape of the object itself [...] the imagination will naturally enter into the vacant space within this shell, and there at once, as from a centre, view the whole form within, and mark the opposite corresponding parts so strongly, as to retain the idea of the whole, and make us masters of the meaning of every view of the object, as we walk round it, and view it from without.[32]

Students of the visual and medical arts become more and more mutually interdependent[33] in exploring the functions and workings of the human body.

In his lessons on anatomy at Oxford's Christ Church, Thomas Willis (1621–1675) had already experimented a new way of dissecting the brain and, for the first time, located the functions of specific nerves in localized areas of the brain.[34] Nerves are found to be ruled by the brain which operates as a sort of heart. Sensibility can be described as the effect of "the nerves in the heart" (*Spectator*, 281) and produces a physiological communicative effect: tears flooding the eyes of sentimental heroines and men of feeling[35] make public what is inwardly felt. The interplay of body and mind cannot be interrupted or overlooked: "Now, I (being very thin) think differently" [TS, Vol. VII Chapter XIII].

The major eighteenth-century discovery of the nervous system changed the physicians' attitude to the interaction between body and mind.

> [Medical Theorists] believed it their business as physicians to set forth the circumstances, extent and means of control of the interaction between body and mind. They believed also, in keeping with an unbroken tradition originating in ancient

---

31 See Nicolson, *Science and Imagination*, 181–187.
32 William Hogarth, *The Analysis of Beauty* (1753), 21.
33 See Stafford, *Body Criticism, Imagining the Unseen in Enlightenment Art and Medicine*.
34 See G. S. Rousseau, "Science", 191.
35 See Mackenzie, *The Man of Feeling* (1771).

Greek medicine, that the mind could be influenced corporeally, that is, by means of drugs, diet, climate and other factors acting primarily on the body.[36]

Dr. William Battie, the author of *A Treatise on Madness* (1758) and founder of St. Luke's Hospital for lunatics, the first house for the treatment of the mentally ill, saw madness as a physiological rather than a psychological condition. Dr Cheyne, the author of *Essay on Health and Long Life* (1724) and *The English Malady, or a Treatise of Nervous Diseases of all Kinds*, printed by Samuel Richardson in 1733, described the interaction of the vascular and nervous systems with the brain. The diseases of the mind, including "melancholy", were referred to the diseases of the body, conceived as a hydraulic machine:[37] "The same sanguinary or serious obstructions are capable in any other nervous part of the body of exciting false ideas as well as in the brain".[38] The rational part had to be defended: so that the defect was attributed to the corporeal part. With systematic empiricism the source of insanity is proved to be "not Satan but the soma".[39] The human being can be envisaged as mechanized into a clock: "For man is as frail a piece of machinery, and, by irregularity, is as subject to be disordered as a clock" as Mr. B ostentatiously affirms in Samuel Richardson's *Pamela*.[40] And in *Tristram Shandy*:

> A man's body and his mind, with the utmost reverence to both I speak it, are exactly like a jerkin, and a jerkin's lining; – rumple the one – you rumple the other. [TS, Vol. III, Chapter IV]
>
> Though man is of all others the most curious vehicle, said my father, yet at the same time 'tis of so slight a frame and so totteringly put together, that the sudden jerks and hard jostlings it unavoidably meets with in this rugged journey would overset and tear it to pieces a dozen times a day – was it not, brother *Toby*, that there is a secret spring within us – Which spring, said my uncle *Toby*, I take it to be Religion. [TS, Vol. IV, Chapter VIII]

---

[36] L. J. Rather, *Mind and Body in 18th-Century Medicine* (London: Wellcome Historical Medical Library, 1972). On this point see A. Cattaneo, "Dr. Cheyne and Richardson: Epistolary Friendship and Scientific Advice" in *Science and Imagination in XVIIIth-Century British Culture*, ed. S. Rossi (Milano: Edizioni Unicopli, 1987), 113–132 from which the above quotation is taken.

[37] "Ever since 1690, the date of publication of the *Principia*, iatromechanists who sought to quantify the sciences of the body – anatomy, physiology, morphology, medicine – had been proliferating" in G. S. Rousseau, "Science", 166.

[38] W. Battie, *A Treatise on Madness* [1758] (New York: Milford House Inc., 1969), 49. Quoted in J. Wiltshire, *Samuel Johnson in the Medical World, The Doctor and the Patient*, 168.

[39] R. Porter, *Flesh in The Age of Reason* (London and Harmondsworth: Allen Lane, an imprint of Penguin Books, 2003), 308.

[40] Samuel Richardson, *Pamela or, Virtue Rewarded* [1740] (London and Harmondsworth: Penguin Books, 1981), 394.

From a jerkin and its lining to a vehicle, from fashion to mechanics, the stage is set with a contemporary frame. In a predominantly literary culture the frequent mechanical similes used with reference to the body clearly reflect the cultural impact provoked by the new wonders of mechanical improvements. The satire on mechanical determinism pervades Swift's works from *A Tale of a Tub* to *A Modest Proposal*; in the fantastic realm of *Gulliver's Travels* Gulliver, who is a physician, becomes not insignificantly the victim of a sort of retaliation. His body is the permanent target of mechanical cruelty, tortured as he is from the very beginning by "these People [...] most excellent mathematicians, and arrived to a great Perfection in Mechanicks by the Countenance and Encouragement of the Emperor, who is a renowned Patron of Learning":

> [...] in an Instant I felt above an Hundred Arrows discharged on my left Hand, which pricked me like so many Needles [...] Sometime they determined to starve me, or at least to shoot me in the Face and Hands with poisoned Arrows, which would soon dispatch me: But again they considered, that the Stench of so large a Carcase might produce a Plague in the Metropolis, and probably spread through the whole Kingdom.[41]

When the culture of technology articulates our existence, men are transformed into sense-machines. The dangers of their proposals and solutions foreshadow darker skies obscured by the related and perverse human tendency to transform the complex and the subtle into the oversimplifying mode of reasoning.

Generally speaking, the eighteenth century was an age of projects, academies and institutions in an increasingly professionalized world marked by the tendency to replace questions about the world with questions about the mind, what exists, with how that existence is known.[42]

> The mappings of disease revealed a prominent aspect of Enlightenment medicine, in that it was patient-oriented [...] Doctors relied on patients' accounts of their own feelings and symptoms to make their diagnoses [...] it was a time of impressive medical entrepreneurialism. Health mattered and people were prepared to pay for it. This meant that ambitious (or devious) healers of all stripes could seek to carve out their niche in the medical market place. Telling the difference between the "quacks" and the "regulars" was not always easy, since many so-called quacks also generally operated within the cultural cosmology of medicine, and "regulars" might advertise their therapies, use secret remedies, and cultivate no-

---

41 J. Swift, *Gulliver's Travels* [1726] (New York and London: W. W. Norton & Company, 1970), 6 and 15. See P. Fussel, *The Rhetorical World of Augustan Humanism, Ethics and Imagery from Swift to Burke* (Oxford: Clarendon Press, 1963).
42 T. Davies, *Humanism* (London and New York: Routledge, 1997), 120.

toriety as a means of attracting attention, and thereby patients [...] The second striking characteristic of Enlightenment medicine was its busy optimism.[43]

In the history of man the borders between nature and culture have been subject to a continuous shifting. The so-called modern "liberal" tradition of the Western world sees the body as belonging to the individual and considers the individual as something greater, superior to the body in which s/he is implanted. In the English tradition, the idea of improving the body, enhancing physical and intellectual abilities, dates back from Francis Bacon's *The New Atlantis* (1627) where he spoke about prolonging life by postponing old age, healing diseases considered incurable, mitigating pain, changing character, height and physical traits, transforming a body into another one, establishing new species and producing new foods. But it was the intellectual context of eighteenth-century pragmatic optimism and ideals of public virtue which imagined medical care being implemented on a grand scale, for both the rich as well as the poor.[44]

Man is to be considered primarily a moral actor as Johnson stresses in the *Life of Milton:* "We are perpetually moralists, but we are geometricians only by chance".[45] His well known expression "Physick of the mind" derives from a traditional metaphor for the analogy of moral philosophy with medicine, first introduced by Plato in the *Phaedrus*. As Werner Jaeger remarks, the applicability of the analogy "rests on the fact that both the art of the physician and that of the ethical philosopher always deal with individual situations and with practical action".[46] Moreover, the experimental artist and the clinical physician "were both devoted to a discriminating observation of signs and symptoms, to a contextualized recognition of pattern, to an informed and refined sensory judgment of appearances or looks".[47] All the senses were held to depend on "the precise workings of the nerves, their intricate morphology and histological arrangement, and their anatomic function".[48] The mind was seen as incarnated in the body: "[Our minds] are wrapt up here in a dark covering of uncrystalized flesh and blood". [TS, Vol. I, Chapter XXIII]

---

[43] W. Bynum, *The History of Medicine. A Very Short Introduction* (Oxford: Oxford University Press, 2008), 40–41.

[44] See Bynum, *The History of Medicine. A Very Short Introduction*, 40–41.

[45] See Fussell, *The Rhetorical World of Augustan Humanism. Ethics and Imagery from Swift to Burke*, 7.

[46] Quoted in Wiltshire, *Samuel Johnson in the Medical World, The Doctor and the Patient*, 146.

[47] Stafford, *Body Criticism. Imagining the Unseen in Enlightenment Art and Medicine*, 39.

[48] G. S. Rousseau, "Science", 192.

God remained the Great Watchmaker of the mechanists, as well as of the deists. The human body could be decomposed and visualized as a system of parts, but if it were not the anatomist's investigations which could "disprove the soul's existence or describe its destiny",[49] where exactly was the soul to be located? Not the answer, but at least a clue is given in Sterne's portrait by Joshua Reynolds. At the beginning of March 1760, after publishing the first two volumes of *Tristram Shandy*, Laurence Sterne reached London where he was enthusiastically received by fashionable society. It is in this period that he sat for Reynolds' portrait. In the painting a sheaf of papers lies on the table and visible on one page we see the title of his novel and on another, "J. Reynolds/pinx 1760". Sterne is dressed in his dark ecclesiastical robes. The viewer's eyes are drawn in succession to the writer's head, to his eyes, to his half smile which has something of the mysterious in it. The gesture traced by his hand is the most luminous part of the painting. As has been noted, Sterne points upwards in a trajectory that passes through his head; it was thought that the index finger seen to be pressing against the temple, was taken from the figure of John the Baptist in a painting attributed to Leonardo da Vinci's school (after c. 1509, Oxford, Ashmolean Museum).[50] The Baptist points towards heaven, Sterne points upwards through his own head. Does this gesture encapsulate a shandean reference to the true origin of his inspiration? Is the image of the world in our mind?[51] Or does it echo Tristram's father's statement that he has discovered the exact location of the soul in the "medulla oblongata"? And just where may the "spirits" be located? According to Harvey's work on the circulation of the blood, the heart must be regarded as a pump, not as a fireplace. In *Tristram Shandy*, trying to find a visual device to show the way death works when this occurs due to heart failure, Sterne has recourse to punctuation:

> – The blood and spirits of Le Fever, which were waxing cold and slow within him, and were retreating to their last citadel, the heart, – rallied back, – the film forsook his eyes for a moment, – he looked up wishfully in my uncle Toby's face, – then cast a look upon his boy, – and that ligament, fine as it was, – was never broken. – Nature instantly ebb'd again, – the film returned to its place, – the pulse fluttered – stopp'd – went on – throb'd – stopp'd again – moved – stopp'd – shall I go on? – No. [TS, Vol. VI, Chapter X]

---

49 McManners, *Death and the Enlightenment, Changing Attitudes to Death among Christians and Unbelievers in Eighteenth-century France*, 150.

50 See M. Postle, *Joshua Reynolds e l'invenzione della celebrità* (Ferrara: Ferrara Arte Editore, 2005), 144.

51 The phrase alludes to the title of V. Braitenberg, *L'immagine del mondo nella testa* (Milano: Adelphi Edizioni, 2008).

Hooke had demonstrated that in man's body supposedly inexplicable functions were in fact performed by "small machines". No less perplexing appeared the marvels operated in factories by the new mechanical looms by
which "Tapestry or flowred Stuffs are woven".[52] The comparison of the
human body with new mechanical knowledge is dominated by the pervasive
presence of the clock, not only as a seductive metaphor but as a more and
more sophisticated instrumentation dedicated to measuring the centrality of
man in time and space. One of the grand mechanical utopias[53] of the Enlightenment, the construction of the "homme-machine"[54] offered the concrete possibility of applying a mechanistic conception to a living organism,
of grasping the link between life and death in a unique celebration of the art
and science of the human body.[55] The automaton is not an eighteenth century invention. Its history may be traced back to Talos, the servant of Minos,
depicted in an Attic vase of 400 B.C.:

> The first references to automata are found in mythology and there are no refer-
> ences to historical texts to prove whether the mythological automata were an ex-
> ercise in theoretical mechanics or whether they could be practically made. The
> word *automaton* is Homeric and is found in the *Iliad*.[56]

From Hephaestus's mechanical maids who could even speak to the golden
tripods and the wheels that could move on their own in the *Iliad* to the ships
that could navigate without men and the gold and silver dogs that guarded
the palace of Alcinoos also made by Hephaestus in the *Odissey*. However, it
was in the mid years of the century that the automaton definitely left the
boundaries of mythological and literary interpretation to play a crucial role in
efforts to understand the nature of existence. The "enlightened"[57] transpar-
ency of the Jaquet-Droz three automata, "The Scribe", "The Draughtsman"
and "The Lady Musician", showed the wonders of intellectual performances
in a "cognitive dramatization":

---

[52] In Stafford, *Body Criticism. Imagining the Unseen in Enlightenment Art and Medicine*,
350.

[53] Stafford, *Body Criticism. Imagining the Unseen in Enlightenment Art and Medicine*, 343.

[54] J. O. de La Mettrie, *L'Homme machine* (1747).

[55] See the title of M. Kemp and M. Wallace, *Spectacular Bodies. The Art and Science of the
Human Body from Leonardo to Now* (London and Los Angeles: Hayward Gallery
Publishing, 2000).

[56] V. Vassilopoulou, "Gli automi e la tecnologia dell'antica Grecia" in *Corpo. Automi.
Robot tra Arte, Scienza e Tecnologia*, 61.

[57] S. Schaffer, "Enlightened Automata" in *The Sciences in Enlightened Europe*, eds.
W. Clark, J. Golimski and S. Schaffer (Chicago and London: University of Chicago
Press, 1999).

In the triad of automata created in the first half of the 1770s by the watchmakers of La Chaux-de-Fonds the external movement was represented by the actions of writing, drawing and playing music and the internal one resulting from the mechanism.[58]

The confluence of organic and inorganic was exemplified in the automata with physiological needs such as de Vaucanson's famous duck.

As a creature with a complex, many-faceted identity, the eighteenth century automaton finds his multiform origin at the intersection of art, philosophy, jurisprudence and science, a "symbolic product in which the great philosophical dichotomies are reflected: the relationship between humanity and machines, between spirit and body, between God and humankind".[59] As father to a vast progeny of hybrid creatures, from robots to cyborgs, he will find his disturbing companion a few decades later when galvanism gives Mary Shelley the inspiration for *Frankenstein; or, the Modern Prometheus* (1818). Doctor Frankenstein brings to life his monster-automaton, born from anatomical parts of different cadavers, by means of an electric discharge. Leaving the Enlightenment, the automaton and his descendants are confronted with a twofold destiny. Wearing the mask of tightrope walkers, apes, jugglers, songbirds, magicians … they are turned into a source of entertainment, exhibited in fairs and markets as fascinating toys for privileged children and grown ups. But as aesthetic artefacts uniting the artificial with the living, they herald in the alliance between art and constructive or aesthetic surgery.

Nowadays it is widely recognized that "the new research on the nature of ethics is located at the interface of neuroscience, evolutionary biology, molecular biology, political science, anthropology, psychology, and ethology"[60] but the pressing need for an ethic of culture is often disregarded as an obsolete humanistic obsession with keeping man as "the central focus" (Denis Diderot). In this short survey made of memories and suggestions, I have tried to uncover some of the roots which lie behind the contemporary debate on bioethics, at the confluence of science, ethics and fine arts in the eighteenth century.

---

[58] R. Baldi, "Quando la meccanica si fa meraviglia: il *Disegnatore* di Henri-Louis Jaquet-Droz" in *Corpo. Automi. Robot tra Arte, Scienza e Tecnologia*, 77.

[59] M. G. Losano, "Le alterne vicende delle macchine calcolanti e semoventi" in *Corpo. Automi. Robot tra Arte, Scienza e Tecnologia*, 368.

[60] P. Smith Churchland, "Moral decision-making and the brain" in *Neuroethics. Defining the Issues in Theory, Practice, and Policy*, ed. J. Illes (Oxford: Oxford University Press, 2005), 3.

Daniela Carpi
University of Verona

The Beyond: Science and Law in *The Island of Doctor Moreau* by H. G. Wells

> Dr. Moreau's Monkey Man seemed a monstrous fantasy at the time, but the questions Wells raised about the ethics of creating chimeras have a new relevance today.[1]

> [Literature] can contribute to a richer understanding and deeper appreciation of our humanity, necessary for facing the challenges confronting us in a biotechnical age.[2]

The Island of Doctor Moreau by H. G. Wells[3] anticipates many of the legal problems that the latest scientific discoveries are posing man in our century. There is a close fit between Wells' specific criticisms of chimeras[4] and the concerns that trouble medical ethicists today.

What is mainly at stake is an ever-changing concept of "persona" that is extended by the new cloning experiments and by the recourse to organ transplantation. Wells' novel actually speaks of vivisection, but the connection with genetic experiments is very strong. Such experiments, being extreme, undermine human beings' uniqueness by suggesting the possibility of a serialization of beings. A new law is necessary to keep these new beings within society, or a new concept of society is required so as to include them.

The perspective from which I intend to analyze the novel is the persona/human being dichotomy: these terms in fact do not necessarily converge.

---

[1] J. Clayton, "Victorian Chimeras, or, What Literature Can Contribute to Genetics Policy Today", *New Literary History* 38 (2007): 569–591, 569.

[2] L. Kass, "Being Human: an Introduction" in *Being Human: Readings from the President's Council on Bioethics* (New York: W.W. Norton & Co., 2004), xx.

[3] H. G. Wells, *The Island of Doctor Moreau* [1896] (New York and London: Penguin Books, 2005). (All quotations are taken from this edition: further references in the text will be abbreviated as IDM, followed by the page number.)

[4] Chimeras: interspecies mixtures.

Persona/Human Being

What are the legal limits of the technical-scientific manipulation of human and non-human life? To what extent is it licit to use the new possibilities of artificial intervention on life, made possible by the recent developments in science and technology? The intuitive coincidence between person and human being is questioned by the bioethical and bio-juridical debate, in whose spheres the concept of person is revived.

In bioethics and biolaw, precisely because of the unprecedented solicitations deriving from scientists and technologists' need for experimentation, a reversal of the traditional notion of person is taking place. Following this conception, not every human being is a person and not every person is a human being: as a consequence, it would be licit to act (even experimentally, in a non-therapeutic way, and thus suppressively) on some human beings (particularly in the initial, final and marginal parts of their existence), while it would be illicit to act on some non-human beings (animals and robots). This reversal, which is often not explicit, can generate momentous confusions.[5]

The concept of person is undoubtedly a problematic issue in bioethics and biolaw, and it indeed requires a solution. A critical analysis of the recent debate within the fields of bioethics and biolaw reveals a common element (despite the specificity and peculiarity of the semantic value and of the practical uses of the term within the different trends of thought): on the one hand, the unanimous acknowledgement of the pragmatic relevance of the concept (as a criterion for establishing, at the emergence of actual problems due to technical-scientific progress, what is licit and illicit with reference to artificial intervention upon human and non-human life, by identifying who is a person and who is not), and, on the other hand, the recognition of the theoretic ambiguity surrounding the use of the term.

The notion of person has been theorized by Western philosophy precisely in order to characterize the human being in an appropriate way and validate its axiological-normative centrality, the moral and juridical subjectivity. While admitting that the category of person is not essential for the construction of a moral and juridical theory of the human being, it is nonetheless important to underline that this concept belongs to our cultural tradition and can help us thematize the objective task of protecting and respecting the individual. However we realize, when we start analyzing the term, that it does not have a universal meaning: heterogeneous disciplines indicate different

---

5 See L. Palazzani, *Il concetto di persona tra bioetica e diritto* (Torino: Giappichelli, 1996).

concepts and different subjects with the term "persona" that bring about discrepancies and discriminations.

History has demonstrated that there can be a separation between the notions of persona and human being: let us consider for instance the problems of slavery, Jewish persecutions, racism. In these cases the living being is denied the status of legal persona.

But let us go back in time to see, very briefly, how the notion of persona has developed across time. In classical Greece "persona" was "prosopon", "mask": in fact it was used for actors and for their action of impersonation. *Prosopon* was the mask that was to create the actor's fictitious personality; it stood for the actor's intrinsic characteristic of changing personality and face. The mask in this case allowed the actor to exist as a new person.

With Cicero the position the individual occupied in society was added to the idea of mask, of role.[6] So far, however, "persona" is not meant in what intrinsically and spiritually denotes an individual in its essence. This connotation comes about with Christianity: a real philosophy of person is rooted in Christianity. In classical times the Greek-Roman culture considered the concept of person only depending from a being's social and racial status, not as a value in itself; whereas Christianity adds an absolute value and dignity to the notion of person: the "persona" is the mirror of God. It is the broadening into the Stoic and Christian idea of one who has moral value.

With Boetius the concept of persona is secularized: "persona est rationalis naturae individua substantia".[7] Boetius is the first to apply the concept of persona to the human context in order to define it as an individual substance endowed with rationality. What Boetius first and Thomas of Aquinas subsequently offer us is an ontological concept of persona.

Cartesio marks a schism with a well balanced concept of persona through his distinction between "res cogitans" and "res extensa". If Boetius and Thomas of Aquinas had reached an ontological definition of persona as spiritual absolute essence, with Cartesio the persona is reduced to a pure act of self-knowledge: "cogito ergo sum".

With John Locke and David Hume new elements are inserted in the definition: persona is a thinking and intelligent being, capable of subtle reasoning and endowed with psychological self-awareness. Reason is the essential

---

[6] B. Mondin, *Storia della metafisica* (Bologna: Edizioni Studio Domenicano, 1998); E. Mounier, *Il personalismo*, eds. G. Campanini and E. Pesenti (Roma: AVE, 2004); G. Lauriola, "La persona: storia di un concetto" in *Trattato di bioetica*, ed. F. Bellino (Bari: Levante Editori, 1992); D. De Grazia, *Human Identity and Bioethics* (Cambridge and New York: Cambridge University Press, 2005).
[7] S. Boetius, *De duabus naturis et una persona Christi*, ch. 3.

element of knowledge. The mind does not possess any innate ideas, but is a *tabula rasa*. Knowledge is reached only through experience. Locke's empiricism brings about the dissolution of the ontological essence of persona.

> [...] [s]elf is not determined by identity or diversity of substance, which it cannot be sure of, but only by identity of consciousness. [...] Person, as I take it, is the name for this self. [...] It is a forensic term, appropriating actions and their merit; and so belongs only to intelligent agents, capable of a law, and happiness, and misery. This personality extends itself beyond present existence to what is past, only by consciousness, – whereby it becomes concerned and accountable; owns and imputes to itself past actions, just upon the same ground and for the same reason as it does the present.[8]

> The modern concept defines persons as beings with the capacity for certain complex forms of consciousness, such as rationality or self-awareness over time. Here is John Locke's classic formulation: "A thinking intelligent being, that has reason and reflection, and can consider itself, as itself, the same thinking thing, in different times and places". This concept is closely associated with, and arguably includes, the idea of someone who has moral status [...] and perhaps also moral responsibilities.[9]

On one hand Kant stresses the predominance of thought in the concept of persona, on the other hand he recovers the ethical side of the concept, by also adding the juridical element. Persona is one who has a value in itself, a dignity, rights and duties and is a goal and not a means.[10] In *Metaphysics of Morals* Kant defines persona as the subject whose actions can be charged, thus stressing the juridical side of his concept. Persona in Kant is at the basis of principles concerning public law, which are the external laws that make the harmonious coexistence of many individual freedoms possible.

At this point we may summarize the conceptions so far examined by saying that human beings are certainly beings with the capacity for complex forms of consciousness, for they are psychologically complex, highly social, linguistically competent and richly self-aware.[11] But must a person be human? Perhaps some nonhumans display equally sophisticated forms of consciousness.

Literature has always dealt with the analysis of the moral and psychological characteristics of human beings. In the many cases dealt with by literature

---

[8] J. Locke, *An Essay Concerning Human Understanding*, Book II, Chapter XXVII, "Of Identity and Diversity", 23 and 26. Available at: http://oregonstate.Edu/instruct/phl302/texts/locke/locke1/Book2c.html#Chapter%20XXVII.

[9] DeGrazia, *Human Identity and Bioethics*, 3. J. Locke, *Essay Concerning Human Understanding*, Ch. 27, sect. 9.

[10] Palazzani, *Il concetto di persona tra bioetica e biodiritto, passim*.

[11] DeGrazia, *Human Identity and Bioethics*, 3.

(*E.T, The Planet of the Apes, 2001 Space Odyssey*, the many characters in *Star Trek*, *The Island of Doctor Moreau, Blade Runner*, etc., just to mention a few) the concept of personhood seems to extend beyond humanity. Can the idea of moral status serve the function of distinguishing between human and non-human beings?

Mankind, which has lived for centuries under the protection of natural laws, suddenly discovers new surroundings that reveal unforeseen areas of existence where "normality" loses its well acquired connotation and where the laws of nature must be readapted.

The Island of Doctor Moreau: Is It Science?

Science appears in the novel in its utter violence: such violence is stressed by the terror and groans of pain that emanate from the poor animals under vivisection. Moreau's experiments when he was living in England were called "The Moreau Horrors". He had transposed his experiments on blood transfusion into a pamphlet which had been defined "gruesome" and he himself was considered to be "wantonly cruel". He is considered to have "fallen under the overmastering spell of research" where the terms clearly recall the Faustian devilish pact. Research is considered to be risky and morally dangerous, so much so that "conscience has turned against the methods of research" [IDM, 34] and Moreau has been deserted by "the great body of scientific workers". The results of his experiments are "crippled and distorted men" [IDM, 35] thus underlining the utter inhumanity of his stubborn will. The violence of such scientific attempts becomes evident in chapter VIII "The crying of the puma", as the narration is interspersed with the poor brute's cries of pain, while Moreau proceeds relentlessly on his scientific path.

Medicine meant as a Hippocratic science implies a dialogue with the patient, a mutual understanding between physician and patient; while in the bioethical experiments anticipated by *The Island of Doctor Moreau* the pathology is instilled in the patients and not just cured. The being is victimized by it and such horrific practices forerun the monstrous actions of Nazi soldiers' prison camps. So much so that bioethics rises precisely to correct the monstrous medical experiments that took place in the Nazi camps: medicine must not forget its Hippocratic roots.

What Are These New Beings?

The problem that concerns me is whether these new beings that are being created are human beings, *personae.*

The ambiguity of the term "man" comes to the forefront even before Prendick is openly told about the experiments carried out by Moreau because in his descriptions of what he sees the characteristics pertaining to man and beast are often intermingled. In the chapter "The thing in the forest" we read:

> It bowed its head to the water and began to drink. Then I saw it was a man, going on all fours like a beast. [...] I could hear the suck of the water at his lips as he drank. [IDM, 40]

The beast-like attitude of these apparent men always puzzles Prendick, who defines them as "brutes", "grotesque half-bestial creatures" and "crippled and distorted men". The verbs the text uses for these beings' actions are very often animalistic terms: "suck", "scuttle", "growl", "go on all fours" etc. Even before knowing what takes place on the island we are given key words for the correct interpretation of the mystery. But are the results of Moreau's experiments real personae? Considering the fact that personhood may even extend beyond humanity, we may certainly assert that all the beings on the island are "persons" with rights and therefore need protecting. The most obvious examples of this assertion are the cries of the puma.

> The emotional appeals of these yells grew upon me steadily, grew at last to such an exquisite expression of suffering that I could stand it in that confined room no longer. [...] It was as if all the pain in the world had found a voice. [...] It is when suffering finds a voice and sets our nerves quivering that this pity comes troubling us. [IDM, 38]

The strong insistence on pathos stresses the fact that animals also have rights that are enforced through the law: any civilized legal system imposes some duties concerning animals.

But what are the traits of real human beings? Philosophers throughout the centuries have stressed the fact that to be a human being one must first of all possess the physical characteristics of a man. This step is reached by Moreau in some way, even if his creatures remain something in between: fur does not totally disappear, they may present pointed ears, faces that are more similar to muzzles, a peculiar way of looking ("a peculiar furtive manner, quite unlike the frank stare of your unsophisticated savage") [IDM, 33] or an ungainly gait. But more than this, to be considered a human being one must have a conscience and introspective awareness and some sort of psychological continuity. If we centre our analysis on the group of new beings that live in the forest in a sort of community, we may say that they represent the closest we can get in the novel to the concept of legal persona. They have attained a deep sense of what constitutes a man and what constitutes an animal and they ground this knowledge on a strong behavioral distinction. Such behavior is strictly codified by a list of rules that are constantly being repeated to them by one of the group whose function is exactly that of saying the rules out aloud (the so-called Sayer-of-the-law). The constant repetition of the law marks their process of evolution: the more ingrained the rules become, the more akin to human beings they become. Such process of "personification" moves through a long list of prohibitions towards an enforced sense of guilt and of total adhesion and obedience of the rules: the penalty hinted at in case of the violation of any rule insists on the physical pain that would stem from the violation itself.

> His is the House of Pain.
> His is the Hand that makes
> His is the Hand that wounds
> His is the Hand that heals. [IDM, 59]

The creatures' awareness of the distinction between man and beast, their awareness of the necessity of the law, their capacity to realize when they have committed a crime according to the rules, the fear deriving from the awareness of the awaiting punishment, all suggest an introspective capacity. They appear to be beings concerned with their well-being and responsible for their actions: in their minds present and past moments are connected by memory. In fact the law for them is particularly threatening because they remember what happened to those who did not conform to the rules. If memory and anticipation are constitutive of our identity, Moreau's creations try to imprint the rules into their minds so as to constantly remember them, and they anticipate the future action of punishment in case they break them. The law in this case serves the function of making them into human beings and creating their identity in psychological continuity. The literal application of the law and the awareness of the consequences stemming from its violation give rise to a sort of superimposed morality making them persons in the sense of moral agents. So much so that when the punishment does not immediately follow the breaking of the law they end in mistrust.

What Law For These New Beings?

If the aim of law is not that of commanding, but to order a society whose characteristics must be respected, only by respecting these characteristics will the law not be considered as violence and will society actually be ordered. To establish order means respecting society's complexity.[12] In *The Island of Doctor Moreau* the law is enforced through violence, both as rules and punishment.

The new biological experiments must go hand in hand with new legal rules that should keep the new beings under the restraint imposed by Doctor Moreau. Law is invested with a particularly authoritarian hue: it is an imposition and not the reflex of a shared way of life, thus originating a violent reaction by the "excluded". We are faced with a flexible borderline between law and non-law, by the necessity of new rules and their refusal, by the necessity of a new organization of society and the correct assertion of one's natural and cultural identity. The new persona that emerges from Moreau's laboratory should be protected by new rules that recognize their diversity, while the only law we perceive in the text is an instrument of the new scientific power on the island. In fact the contemporary new bio-juridical rules are based on a dialogical procedure that takes the various political, religious, scientific and cultural aspects of society into account. These procedures require a large consensus to avoid creating a law based on juridical absolutism.

> The Master of the House of Pain will come again. Woe be to him who breaks the Law! [IDM, 120]

> [...] all of them swaying in unison and chanting: Not to go on all-Fours; that is the Law. Are we not Men? Not to suck up Drink; that is the Law. Are we not Men? Not to eat Flesh or Fish; that is the Law. Are we not Men? Not to claw Bark of Trees; that is the Law. Are we not Men? Not to chase other Men; that is the Law. Are we not Men? [IDM, 59]

This list of strong prohibitions pronounced in a sort of religious chant has the function of stressing the biological changes violently imposed on these creatures.

This novel anticipates a deep problem that pervades our epoch with its rapid scientific changes. The law must adapt itself to the new social surroundings but the challenges that science and technology impose on the legal system are such that the law may feel expropriated and cannot keep up with the ever-changing situation. This is what happens on Moreau's island, where the concepts of human dignity and human rights are annihilated for

---

12 P. Borsellino, *Bioetica tra autonomia e diritto* (Milano: Zadig, 1999), 208.

the sake of keeping these new biological creatures at bay. Such creatures are not considered human from a legal point of view and the only law they come to know is the law of duties and not that of rights.

One of the fundamental rights an individual possesses is that of controlling his/her own private sphere: human beings create a common space, a community and this is the case also with the new creatures Moreau has given form to. They have found a common dwelling, where they share food and shelter, besides creating a sort of common protection. What is violated in their case is the right to own a proper genotype that is unrepeatable and clearly distinguished, while in the case of these creatures the genetic program precedes their entering a moral/physical community. In this way their self is degraded even before their very existence. Their rights are violated *a priori*.

The text presents two concepts of society at clash: a natural world dominated by a sense of causality and by norms admitting no exceptions, and a civilized world, a social world ruled by conceptions behind which the interests of individuals (here Dr. Moreau's) who, upon reaching power, present them as norms, are hidden. The Beast Folk live in an in-between world where their natural animal instincts (the rules of nature) still survive, but are stifled in favor of the norms imposed by a superior power.

> Before they had been beasts, their instincts fitly adapted to their surroundings, and happy as living things may be. Now they stumbled in the shackles of humanity, lived in a fear that never died, fretted by a law they could not understand. [IDM, 95]

What creates these beings as *personae* is only the law ("The Law held them back") [IDM, 94]. The law's effectiveness is particularly evident here: "ars imitatur naturam". Normative artifice takes the place of natural law.

In this novel, the law takes on sinister connotations, insofar as it appears as an attempt to extend the control of the political power (represented by Moreau) on life. Foucault reminded us how for centuries one of sovereign power's privileges was the right of disposing of life and death. On the island, Moreau claims such power for himself. The fact that this novel seems to reaffirm the law's authoritarian dimension and claim that rights depend on a limitless power leads to a weakening and negation of both freedom and law. In other words, the new beings created by Moreau are not aware of the law as a link among men (a fundamental aspect of natural law), but see it only as an authoritarian power, and this has destructive consequences for the shattering of the principle of humanity.

In this sense, the novel anticipates the crisis of Western democracies and their ethical-political foundations: the law is at the core of a crisis of reason, a loss of meaning which is the consequence of the reduction of the law's content to that of the norm, in the marginalization of ethics and knowledge as elements of the autonomous organization of social life.

The exasperation in the production of laws, which become substitutes for the natural awareness of basic rules of human behavior, shows that such beings only exist insofar as they are created and kept together by laws, thereby revealing the exasperation of the law's pervasive power and self-referentiality.

The consequent sense of panlegalism appears as the substitute for a real awareness of being human. The men-beasts are men only insofar as they are involved in the laws' performance: they are created by the law. In this sense, the law does not simply codify/regulate social life, but it creates it. This represents a superimposition that is not helpful for defining these new beings as *personae*.

Conclusion

This novel foreruns the problems concerning the body that characterize our electronic era. Who does the body belong to? To the person concerned, to the social circle, to nature or to the physician that has given form to it? A body made of parts assembled together presents us with the hypothesis of "l'homme machine", a one dimensional person that is the object of an implacable coercion.

Nowadays the body's physicality is at the centre of juridical and social negotiations. Dating as far back as the *Habeas Corpus* of the *Magna Charta* in 1215 the body, formed by nature, has fallen under the human law and keeps coming up as a problem whenever the juridical system feels the need of controlling either the souls' or the State's salvation.

Such experiments anticipate the following problem: is whatever is scientifically and technologically possible also ethically valid? Bioethics in fact stems from the epistemological swerve between techno-scientific innovations and ethical-legal possibilities. A hermeneutical mediation must be found between the freedom of experimentation that is necessary for science and respect for human values. Bioethics is an inter-discipline which involves biology and medicine, philosophy and theology, psychology, sociology and anthropology, economics, law and politics: all these disciplines in fact converge towards bioethical knowledge.

Yvonne Bezrucka
University of Verona

Bio-Ethics *Avant la Lettre*: Ninenteenth-Century Instances in Post-Darwinian Literature

The attention to people as ontological beings and the necessity to redefine words like "human being", "identity", "nature", has received a new boost after the publication of Charles Darwin's *Origin of the Species* (1859). The semantic reconceptualization was fostered by a shared feeling, amongst intellectuals, that, in Charles Kingsley's words, "the great fairy Science, who is likely to be queen of all the fairies for many a year to come",[1] would indeed triumph as a mere rational orientation, a scientism without proper ethical mitigation. Darwin's work, with its emphasis both on progressive and retrograde evolution, forced to a revision of the canonical view of man as a once and for all created being and of his/her place in the universe. The humanities responded to the new scientific challenges at a symbolical level with a reconfiguration and recovery of the utopian-or-distopian genre in its topical development into science-fiction. Science fiction once embodied, gave vent and processed via transmutation, as it were, the common fears of the people which were the result of the appearance, on the one hand, of the new scientific hypotheses and, on the other hand, of the pseudosciences to which these in due course gave rise.

The necessity for an investigation of the possibilities science could offer to man was deeply fostered by the debate that arose in English scientific circles in connection with the opposing views of two important institutions: *The Ethnological Society* (established in 1843)[2] and *The Anthropological Society* (founded in 1863) which disputed strongly in the wake of the evolutionary vision proposed by Darwin on opposed standpoints: that of monogenesis

---

[1] C. Kingsley, *The Water Babies*, ed. R. Kelly (Peterborough, Ont., and New York: Broadview Edition, 2008), see also A. Tennyson who also speaks of "the fairy tales of science, and the long result of Time" in his poem, *Locksley Hall*, v. 12, ed. C. W. Eliot, *English Poetry III: from Tennyson to Whitman* [1909–1914], The Harvard Classics, vol.42 (New York: P. F. Collier and Son, 2001).

[2] Members of the Society, who supported Darwin strongly, were: Thomas Henry Huxley, Augustus Pitt Rivers, Edward Tylor, Henry Christy, John Lubbock, and A. W. Franks.

(single origin of man) and that of polygenesis (multiple origin of man). The difference between the two is not irrelevant in that from the second a new strong case for the classification of the varieties of man based on racial differences and discriminations was drawn by James Hunt, a disciple of Robert Knox (1791–1862), and founder of *The Anthropological Society*. Hunt used the polygenetic theory in order to conjure up the racial standpoint that the white races had to be located in hierarchical evolutionary terms on the top of the *scala naturae*. The monogenist theory was upheld by James Cowles Prichard (1786–1848). The two societies were later united in *The Anthropological Institute* (1871), to be later transformed, in 1907, into *The Royal Anthropological Institute*.[3] The poligenists upholding biological differences justified an innate origin of differences between peoples rather than considering them acquired cultural differences and as a result of an adaptation to the cultural environment, deliberately confusing nature with nurture. The two societies, as has been proven, hand in hand, acted by "identifying issues of race and class as questions of heredity and environment".[4] This of course has to be linked to the deconstruction of the creationist hypothesis due to Darwin's revolution, a not secondary consequence of which was the emancipation of ethics from religion, and it is exactly this issue which sets ethics and its themes strongly into the foreground because new secular values had to be discussed in order to safeguard those domains where religion had till then exercised its monopoly. After the Enlightenment this was the strongest affirmation of secularism and its values. It was once more demonstrated that people needed not to be believers to behave according to common ethical principles, and that being an atheist or an agnostic did not necessarily mean that one had no ethical principles of reference. One needed not to be a religious person to behave fairly. Atheists had ethic principles as believers. Morality and ethics were not the private enclosures of religion, they could rather be emancipated from it. Darwin himself hypothesized that the moral faculty was the result of living in groups and in itself a faculty subject to development as all human other faculties and organs:

> In the distant future I see open fields for far more important researches. Psychology will be based on a new foundation, that of the necessary acquirement of each mental power and capacity by gradation.[5]

---

3 See D. Lorimer, "Theoretical Racism in Late-Victorian Anthropology, 1870–1900", *Victorian Studies* 31. 3 (1988): 405–430.

4 Lorimer, "Theoretical Racism in Late-Victorian Anthropology, 1870–1900", 430.

5 Online Variorum of Darwin's *Origin of Species*: fourth British edition (1866): 576, available at http://darwin-online.org.uk/Variorum/1866/1866-576-c-1869.html.

Ethics could thus become an evolutionary issue, a culture-specific problem, as it was for Darwin, strictly and only a cultural matter of a specific environment.

Another strong emphasis that forced to a revision of one's standpoints on personhood, identity, and ethnicities stemmed from that series of pseudo-sciences that came into being in the circles around the two anthropological societies. In particular, in the 1880s, under Francis Galton's patronage of *The Anthropological Society*, the eugenic movement was given a strong hearing and emphasis. Francis Galton, half cousin of Charles Darwin, had been a disciple of Johann Caspar Spurzheim,[6] in his turn a follower of the phrenologists Franz Joseph Gall (1758–1828) and Pieter Camper (1722–89). We must not forget that a hierarchical, physical anthropological typologisation of peoples had already been presented by J. F. Blumenbach's (1752–1840) taxonomical human subdivision. In his reading he insisted on the fact that: "The Caucasian must, on every physiological principle, be considered the primary or intermediate of the[se] five principal Races".[7] Gall, for his part, devised an anthropo*metry* – making use of the tyranny of numbers on which science always relies – a detailed evaluative method that made use of the mathematical measurements of the skull (and of its bumps data) which came of ideological use to polygenists to promote the issue of a biological, i.e. innate, wrongly "natural" rather than a cultural explanation for differences between peoples.[8] The theory gained popularity just by filtering results through a quantitative and apparently strictly rational and scientific epistemogical means, in reality confusing tool with theory. The term "phrenology" (from the Gr. *phrén* = "mind"[9] + *logos* = "science,

---

[6] Johann Caspar Spurzheim published his works, due to a misunderstanding with his master, separately: cfr. *Phrenology, or the Doctrine of the Mind, and of the Relations between Its Manifestations and the Body* [1825], where he says that he uses the term phrenology to define the special faculties of the mind, and his *Phrenology, or the Doctrine of the Mental Phenomena*. See P. S. Noel and E. T. Carlson, "Origini del termine Frenologia" in *Frenologia, Fisiognomica e Psicologia delle Differenze Individuali in Franz Joseph Gall. Antecedenti storici e sviluppi disciplinari*, eds. G. P. Lombardo and M. Duichin (Torino: Bollati Boringhieri, 1997), 44–45.

[7] J. F. Blumenbach, *A Manual of the Elements of Natural History* (London: W. Simpson & R. Marshall, 1825), 37, which I take from J. W. Griffith, *Joseph Conrad and the Anthropological Dilemma: Bewildered Traveller* (Oxford: Oxford UP, 1995), 79.

[8] The cultural explanation is present in the work by E. B. Tylor, *Primitive culture: researches into the development of mythology, philosophy, religion, art, and custom*, 2 vols. (London: John Murray, 1871).

[9] The Greek word *phrén* has always been associated to the mind and its feelings. Translated literally with mind it also defined the heart, the precordium and the diaphragm, all those organs which were usually thought to be the seat of the mind and its spiritual faculties, cfr. P. S. Noel and E. T. Carlson, "Origini del termine Frenologia", 44.

ratio"), with which this pseudoscience is still known today, has to be attributed rigorously to Benjamin Rush, an American psychologist who used the term in his unpublished manuscript *Lectures upon the Mind* (1805) and in his later publication *Sixteen Introductory Lectures*, in 1811,[10] which was finally used by Spurzheim (the actual divulger of this pseudo-science) to define Gall's cranioscopy. The new anthropo*metry* (the actual reification of people to numbers) was spread in England by Spurzheim and his follower, the Scottish phrenologist George Combe. Galton became one of its champions. Phrenology was by now a popular fashion: Charlotte Brontë had her head read, and refers to the pseudo-science in her novel *Villette*,[11] but not only, for also in previous works she often refers to organs and to their supposed characteristics to describe character.

Besides phrenology, another well-known instance of a pseudoscience, craniology, used as a means of a body politics application, is that devised by the Veronese doctor Cesare Lombroso, Max Nordau's master, who, in his work *La mente criminale*, taxonomises the, according to him, "family characteristics" of criminals by following recurrent somatic and phrenotypical features. The most famous literary instance of a phrenologist doctor appears in Joseph Conrad's *Heart of Darkness*, where the Belgian doctor measures Marlowe's head because he thinks there is madness in "those who want to go out there [Congo]".[12] Craniology, the comparative study of races based on skulls' measurement has to be seen as the direct derivation of the invalid innatist theory of character and intelligence as it had been devised by Johann Kasper Lavater's (1741–1801) in *Physiognomische Fragmente* (1775–1778),[13] in

---

10   The term was later used by Thomas Forster who, in 1815, wrote a study on phrenology entitled *Sketch on the New Anatomy and Phisiology of the Brain and Nervous System of Drs. Gall and Spurzheim*.

11   As Sally Shuttleworth has demonstrated, "Brontë's fiction is permeated by the language and assumptions of phrenology", and she makes use of phrenological elements not only in her major novels. Besides having her own skull examined in 1851, in one of the short stories of the Angria cycle dated July 21, 1838, "The Duke of Zamorna", Charlotte attributes to her character Jane Moore an hyperdeveloped secretive organ, which being related to instinct could manifest itself as reservedness, but also shrewdness, dissimulation and hypocrisy: cfr. S. Shuttleworth, *Charlotte Brontë and Victorian Psychology* (Cambridge: Cambridge U.P., 1996), 64–67.

12   J. Conrad, *Heart of Darkness* [1902](London: Penguin, 1995), 15.

13   J. C. Lavater, *Physiognomische Fragmente zur Beförderung der Menschenkenntnis und Menschenliebe* [1775–1778] (Zürich: Orell Füssli, 1968–1969) first translated into English in 1792 as *Essays on Physiognomy*. Lavater's work defined features according to "national characteristics" without hesitation attributing small eyes to Italians, light-coloured, wrinkled eyes to Germans, open, steadfast eyes to the English

itself a recovery of the medieval theory of the humours. A contemporary of Gall, Lavater translated in his work body appearances into a semiological code by linking external traits to supposed internal, innate characteristics. Unfortunately, through this work invalid scientific categories came to be strongly tied to apparently neutral aesthetic categories. That is, body features became, in Lavater's study, apparently only "aesthetic" categories, but in reality they were used to back up the supposed superiority and validity of a dictatorial aesthetics of the beautiful, which took the, in itself regional, "Greek" ethnic type as the standard prototype for "universal" beauty. Exactly in the same way, and according to the same premises, in Gall's case, somatic features became racial ones, used to back up the supposed superiority of the white ethnic type, a somatics at that point, as seen, inevitably connected with European (western) aesthetics.

Indeed, Lavater's work provided racists with an imaginary but ideological taxonomy that claimed to read moral worth in external appearance, a pseudo-science widely used in fiction, but with an enormous and telling increase during the nineteenth-century rather than in eighteenth-century fiction.[14] By the characteristics of the nose, mouth, forehead, eyebrows, chin and so forth, one could, it was affirmed, infer character. Skin colour as a difference-marker has an even longer history because of its direct visibility, but in the nineteenth-century other characteristics became relevant and were used as outward signs to highlight character within, i.e. to permit a direct reading of supposed *innate* biological differences. Profiling, stereotyping became a rule. Phrenotypical marks (the set of observable traits of an organism) became a strong discriminating code, made of signs and signifiers.

The fact is that, combined with the Whig interpretation of history, the myth of a continuous teleological progress, that co-opted Darwin's misread thesis of a supposed teleological and progressive myth of evolution, the phrenotypical reading was used to back up a colonial and imperialistic body-

---

and dull eyes to the Swiss. Treatises on physiognomics are present since antiquity: cfr. R. Giampiera, *Pseudo-Aristotele, Fisiognomica, e Anonimo latino. Il trattato di fisiognomica* (Milano: Rizzoli BUR, 1993) and Polemone di Laodicea, *De Physiognomonia Liber* (II A.D., summarized by Adamanzio il Sofista, *Physiognomica* in the 4th c. A.D.) in Greek, later collected in the Latin work, *De Physiognomonia Liber*, by an anonymous author. See also G. della Porta, *De humana Physiognomia* (1586).

[14] For the relation of physiognomy to the nineteenth-century novel see G. Tytler's study, *Physiognomy in the European Novel. Faces and Fortunes* (Princeton N.J.: Princeton UP, 1982), and my analyses of one of Conan Doyle's short stories and one by Bram Stoker in Yvonne Bezrucka, *Oggetti e collezioni nella letteratura inglese dell'Ottocento* (Trento: A.R.E.S., 2004), 147–164.

politics based on the supposed biological, read "innate", backwardness of those "other" peoples, who were not interpreted as simply having and using "different" cultural codes, but who were accordingly taxonomized as being "primitive", "inferior", "savage" and "barbarian". Peoples and bodies, in short, who needed to be "civilised".

These ideological projections found their alleged factual evidence in physiognomy, phrenology, pathognomy which provided the pseudo-scientific contribution to the playing down and "belittlement" of what simply by being "other", becomes "ugly" and, as in Lavater's case, how a sign is read as a stigmata of "evil" as happens later in Nordau and Lombroso.[15] Postcolonial literatures have done their best to dismantle such ridiculous aesthetic presumptions through their new "dissenting" body-politics aesthetics.[16] Stevenson, a lover of the Pacific who in *Jekyll and Hide* demonstrates that an outward respectable and normal body does not necessarily correspond to beauty within, did the same. Knowledge became embodied and readable. The difference with former theories of race (based mainly on skin colour) is the application of a taken for granted scientific epistemology: i.e. mathematical measurement applied to an invalid content. A case in which the tool is used to guarantee the validity of the theory.

As an example of the conflation of the aesthetic and the pseudoscientific strand, a popular nineteenth-century work on physiognomy was Henry Frith's study *How to Read Character in Faces, Features and Forms. A Guide to the General Outlines of Physiognomy*, which appeared in 1891.[17] The work shows a marked preference for the Greek type and the fair complexion that although generally connected with weakness, becomes the main characteristic of the "intellectual" type which shows "mental superiority" so that "the weaker

---

[15] Physiognomy thus contributed, needless to say, to the climax of the British imperial enterprise which paraded and self-fashioned itself as bringing "light" to those places of the earth where barbarism reigned. If this is scandalizing, let me remind you that Plato, on his side, thought of and promoted beauty as an "implicit" guarantee of goodness. This essentialism, I think, forces us to highlight whenever the old aesthetics of the beautiful, which would render us complicit with the ideology it hides, appears. The discriminatory body politics needs to be deconstructed, over and over again, in all its Prometeo-like historical reappearances.

[16] Cfr. Y. Bezrucka, "Albert Wendt's 'Flying-Fox in a Freedom Tree': Contagious Infection and 'Regional' Dissenting Bodies", in ed. A. Righetti, *Theory And Practice Of The Short Story: Australia, New Zealand, The South Pacific* (Verona: Valdonega, 2006), 245–263.

[17] H. Frith, *How to Read Character in Faces, Features and Forms. A Guide to the General Outlines of Physiognomy* (London: Ward, Lock and Co., 1891).

ones in intellect become the servants"[18] concluding, in disquieting social
Darwinistic, i.e. teleological terms, that:

> the White Man [...] excels all other species [...] fitted for changes of climate, and
> possessing also intellect, the white races naturally rule. It is Nature's law. The fit-
> test survive.[19]

The ambiguity of a too direct homology between fittest and best defines the
distance from Darwinism to Spencerism. The short quote, with its uncom-
plicated use of a Darwinian discoursivity, gives us an idea of the political
application of social Darwinism which sets races along an evolutionary and
evaluative progressive teleological ladder. The Whig arrow-of-time set
provided then to set peoples on a Western hierarchy passing from what the
West considers to be primitive to a supposed height of civilization and evolution defined according to occidentalist standards (without taking into any
account the western and boasted utilitarian principle of the happiness of
"others").

At that point, in the field of science T. H. Huxley, himself an initial sup-
porter of phrenology, was filled with indignation at the misreading of the
Darwinian hypothesis set off by the comparative anatomy of peoples, that
imperial England had espoused and applied on a hierarchical basis and
politics, and came to reject pseudo-science as a valid method.

Eugenics,[20] the term created by Galton, was a sort of direct consequence
of these various strands: indeed, the new science could transform "spontaneous" selection – indeed it is better not to use the word "natural" for the
difficulty one finds in giving the adjective a qualifying definition – into "artificial" selection. By promoting a policy of guided mating (as in the breeding
of animals) which would produce a eu-genic race (derived from the Greek
εὖ, "good" or "well") that is to say a good (better) race, result of selective
breeding. The Victorians felt that eugenics' possibilities were in their hands.
Something similar happened to us in February 1997 when the cell of an adult
mammal was used to grow another genetically identical clone, which forced

---

[18] Frith, *How to Read Character in Faces, Features and Forms*, 14.

[19] Frith, *How to Read Character in Faces, Features and Forms*, 15.

[20] George Mosse has shown the considerable success with which the "Archiv für
Rassen- und Gesellschaftsbiologie" (*Journal for Racial and Social Biology*), founded
in 1904, propagated Galton's and Pearson's ideas in D. Stone, *Breeding Superman:
Nietzsche, Race and Eugenics in Edwardian and Interwar Britain* (Liverpool: Liverpool
UP, 2002), 132. See also G.L. Mosse, *Toward the Final Solution: A History of European
Racism* (Madison: Wisconsin UP, 1985, 2nd ed.) and his *Nationalism and Sexuality:
Middle-Class Morality and Sexual Norms in Modern Europe* (Madison: Wisconsin UP,
1985).

us to consider that "Dolly" might have been a human being, and that organs could be created only for body replacement, which makes of our bodies a sort of mechanical machine whose obsolete parts can be substituted. For us at the other end of the issue stands also the fact that life can be prolonged beyond the spontaneous human span, because machines can again intervene in lengthening life beyond the body's organic clock. Bioethics has to discuss how far science can go and law has to enter, in order to make sure that people's will and the single person's will is safeguarded.

Important to note is the fact that eugenists did not believe in the influence of the environment for the evolution of species, on the contrary they supposed biological criteria for the fit and the unfit and therefore insisted on a "system of legal restraint over the reproduction of dysgenic strains", whereas in Darwin species were "not an essentialist entity but a historical population".[21] Even for Spencer in reality it had been so; indeed, he, like Darwin, "retained in his biology a strong environmental emphasis, insisting in particular that acquired characters could be inherited" but to "the eugenist [...] this was [...] an old fashioned biology inappropriate to modern genetic science".[22] In particular this view had been strongly emphasized in the work of August Weissman.[23] As a consequence for these people there existed a correlation between pauperism, unemployment and genetic inferiority. Eugenics became soon allied with a certain kind of politics, as R. J. Halliday points out:

> As political doctrine, the choices were simple: to limit the birth rate of the "unfit" by means of eugenics, or to dismantle welfare administration and "grand-motherly" legislation in favour of the free labour market and the voluntary beneficence of individuals.[24]

As such:

> Social Darwinism is defined as that discourse arguing for eugenic population control; an argument requiring a complete commitment to an exclusively genetic or hereditary explanation of man's evolution. In practice, the discourse was carefully aimed at two specific and definable social groups – the native urban proletariat and the alien immigrant. [...] practitioners were characteristically concerned with public morals, temperance, sex-education, medical statistics, and elementary hygiene. They were grouped, in particular, around the Eugenics Education Society,

---

21 See R. J. Halliday, "Social Darwinism: A Definition", *Victorian Studies*, 14.4 (Indiana University Press, June 1971): 389–405, 399.
22 Halliday, "Social Darwinism: A Definition", 399.
23 A. Weissman, *The Germ-Plasm: a Theory of Heredity*, ed. R. Robbins (London: Electronic Scholarly Publishing, 1893). See also his earlier work, "Essays on Heredity and Kindred Biological Problems" (Oxford: Clarendon, 1888).
24 Weissman, *The Germ-Plasm*, 394.

the National Eugenic Laboratory at the University of London, the B.M.A.'s section on Medical Sociology, and the National Council of Public Morals – the forerunner, of course, of the British Board of Film Censors.[25]

It is a whole body of institutions that, as R. J. Halliday has pointed out, crystallized its work into commissions and laws:

> the Immigration Reform Association, the National Birth Rate Commission, the Interdepartmental Committee on Physical Deterioration, the committees on [sweated] labour and Alien Immigration, the Aliens Act itself (1905) [and the Assistance given to German research into race-hygiene]. One might also include, on the recommendation of the practitioners themselves, the Voluntary Sterilization Bill and the Prevention of Crimes Act.[26]

All this is done "In the name of science".[27] Dan Stone affirms: "In the English literature on eugenics there existed for some forty years before the Holocaust a notion – the 'lethal chamber' – which can be differentiated from the Nazi gas chambers 'only' in the fact that the English versions never went into operation".[28] All these facts could be read in historical sociological terms in relation to what the population of England probably perceived as a sort of alien invasion when 150,000 people arrived in Britain.

---

[25] Weissman, *The Germ-Plasm*, 401–402.

[26] Weissman, *The Germ-Plasm*, 402: in note 19 he pointed out the following: "The Immigration Reform Association was founded in 1903 and continued the investigations of the Sweated Committees into the "evils" of foreign immigration. The Association sponsored an Alien Immigration Bill and pressed for the repatriation of aliens convicted of crimes. The Eugenics Education Society was formed in 1907 and was successful in pressing for a National Birth Rate Commission and for a eugenics section within the British Medical Association".

[27] Cfr. EMBO Reports, 2.10 (2001): 871: "It was scientists who interpreted racial differences as the justification to murder … It is the responsibility of today's scientists to prevent this from happening again". See also Paul Lombardo, "Eugenics Laws Restricting Immigration", essay in the Eugenics Archive, available online at http://www.eugenicsarchive.org/html/eugenics/essay9text.html, and his "Eugenic Laws Against Race-Mixing", http://www.eugenicsarchive.org/html/eugenics/essay7text.html. See also S. Jay Gould, *The Mismeasure of Man* (New York: Norton, 1981).

[28] Stone, *Breeding Superman: Nietzsche, Race and Eugenics in Edwardian and Interwar Britain*, 124. He also writes: "It is true that the Nazi path to the gas chamber was a twisted one, a path which (once the actual murder process started) began with the face-to-face shootings of the Einsatzgruppen (the mobile killing squads which accompanied the Wehrmacht into the Soviet Union), 'progressed' through the gas-vans of Serbia and Chelmno, and then into the carbon monoxide gas chambers of the Operation Reinhard death camps (Belzec, Sobibor and Treblinka), based on those used in the 'euthanasia' programme, before ending with the most technologically sophisticated version in Auschwitz, the zyklon B gas chamber": 125.

In the 19th century, Tsarist Russia was home to about five million Jews, at the time the largest Jewish community in the world obliged to live in the Pale of Settlement, on the Polish-Russian borders, in conditions of great poverty and, were subjected to religious persecution. About half left, mostly for the United States, but many – went to England. This reached its peak in the late 1890s, with "tens of thousands of Jews […] mostly poor, semi-skilled and unskilled" settling in the East End of London.[29]

Due to the restriction that eugenists were planning, it is therefore no surprise that the decade of the 1880s witnessed an incredible concentration of works on ethics. To mention the main ones: W. E. Lecky, *A History of European Morals* (1869), Herbert Spencer, *Data of Ethics* (1879), Leslie Stephens, *The Science of Ethics* (1882), T. H. Green, *Prolegomena to Ethics* (1883), Henry Maudsley, *Body and Will* (1883), W. R. Sorley, *The Ethics of Naturalism* (1885), Henry Sedgwick, *Outline of the History of Ethics* (1886), J. G. Romane, *Mental Evolution in Man, the Origin of Human Faculty* (1888), Henry Drummond, *The Ascent of Man* (1894), T. H. Huxley, *Evolution and Ethics* (1894), and Paul Topinard, *Science and Faith* (1899).

Ethics, which had at that point been fixed on the utilitarian principles of the happiness of the greatest number, and which in its turn was vying with the duty based ethics of Christian and Kantian origin, had to re-elaborate its principles in order to take into account the disenchantment Darwin, once more, had brought about, since the Enlightenment. It was felt that a needed salvage of humanism, which scientism and pseudo sciences were failing to apply, was needed. The safeguard of the principle of life, the bio-essence of the "objects" under consideration, the people, became of extreme relevance. This new bio-orientation took the form of an ethics of care.

In *The Data of Ethics* (1879), Herbert Spencer defined his need to write a work on ethics as the result of a time-specific issue:

> Now that moral injunctions are losing the authority given by their supposed sacred origin, the secularization of morals is becoming imperative.[30]

The fact indeed was that another authority was gaining extreme power: science. Spencer saw Natural Selection in the terms of the Survival of the Fittest (seen as the best) issue, whereas Darwin had been careful to speak of the fittest as the fitter to survive within a given, relative environment which presupposed also, as it happened with the cirripede,[31] a parasite. Darwin

---

[29] Cfr. http://www.museumoffamilyhistory.com/20c-evans-gordon.htm, February 23, 2010.
[30] H. Spencer, *The Data of Ethics* [1879], (Preface 3), http://fair-use.org/herbert-spencer/the-data-of-ethics/preface, last visit May 4, 2009.
[31] Cfr. C. Darwin, *The Lepadidae; or, Pedunculated Cirripedes* (London: Palaeontographical Society, 1851).

took pains to study for eight years a possible retrograde development and a
loss of one's complexity (exactly what happens to Dr Jekyll when, dreaming
of an evolution, he degenerates into the atavistic Hyde). This point is of primary importance, in that it marks the distance of Darwin from Lamarck's
theory of heredity (progressive tendency to perfection, the theory espoused
by eugenists and initially by Darwin himself[32]) and Darwin's consciousness
of indefinite developments, of random changes, but always based on effective causes. But for Spencer and the social Darwinists there was an innatistic
core that produced the correspondence of the weakest with the worst.

Thomas Henry Huxley responded to eugenists and to Social Darwinists
by attacking Spencer in his Romanes Lecture: *Evolution and Ethics* (1893). He
denied clearly that the Fittest were necessarily the Best by stating clearly that
"what is 'fittest' depends only upon the condition".[33] He also championed
and pleaded for an ethics of evolution:

> Social progress means a checking of the cosmic process at every step and the sub-
> stitution for it of another, which may be called the ethical process: the end of
> which is not the survival of those who may happen to be the fittest in respect of
> the whole of the conditions which they obtain, but of those who are ethically the
> best. [...] Laws and moral precepts are directed to the end of curbing the cosmic
> process and reminding the individual of his duty to the community, to the protec-
> tion and influence of which he owes, if not existence itself, at least the life of
> something better than a brutal savage [...] Let us understand once and for all, that
> the ethical progress of society depends, not on imitating the coming process, still
> less in running away from it, but in combating it.[34]

This only could, according to Huxley: "repudiate[d] the gladiatorial theory of
existence".[35]

Two novels respond to the cultural upheaval Darwinism had created and
to the fears that the English people were experiencing at the end of the century. The one novel which sets a direct critique at the nightmares eugenists
were proposing is *The Island of Dr. Moreau*.[36] The novel presents many Darwinian legacies.

---

[32] Cfr. C. Darwin, *The Origin of Species* [1859] (Oxford: OUP, 1996).
[33] T. H. Huxley, "Evolution and Ethics" in *The Fin de Siècle: A Reader in Cultural History,
      c. 1880–1900*, eds. Sally Ledger and Roger Luckhurst (Oxford: OUP, 2000), 238.
[34] T. H. Huxley, "Evolution and Ethics", 238.
[35] T. H. Huxley, "Evolution and Ethics", 238.
[36] H. G. Wells, *The Project Gutenberg EBook of The Island of Doctor Moreau* [1896] [EBook
      #159], October 14, 2004 and first posted in August, 1994, created by Judith Boss,
      of Omaha, Nebraska, from the Garden City Publishing Company. Minor correc-
      tions made by Andrew Sly in October 2004, hereafter quoted parenthetically as
      DM and line no.

Moreau is a eugenist: in fact he collects embryos with the aim of showing that species have a common origin, and tries his experiments on animals in order to make them into human beings: an implicit artificial quickening of a natural process. He has been driven out of England for the sufferings he has inflicted on animals (he probably infringed England's *Cruelty to Animals Act* (1835[37]), and therefore he considers himself a victim of his "overmastering spell of research". In trying to civilise animals ("They were animals, humanised animals, – triumphs of vivisection") [DM, 2543], Dr Moreau, a parody of Galton, is just setting into practice what eugenists were advocating. Galton himself in his "Eugenics: Its Definition, Scope and Aims" (1904) had affirmed that:

> We must […] leave morals as far as possible out of the discussion, not entangling ourselves with the almost hopeless difficulties they raise.[38]

Once done: "The race as a whole […] should be better fitted to fulfil our *imperial* opportunities". The program wanted to quicken what eugenists in acceptance of Nature's own finality considered to be the inevitable "improvement of our stock": "What Nature does blindly, slowly, and ruthlessly, man may do providently, quickly, and kindly". That this is far from being a kind process is what, in *The Island of Doctor Moreau*, "The Cry of the Puma" testifies, emitting: "sharp, hoarse cry of animal pain" [DM, l. 1242].

Darwin was very careful in his work *The Expression of the Emotions in Man and Animals* (1872) where he pointed out that the expressions of emotions by man and animals confirmed their common origin and supported "the specific or subspecific unity of the several races".[39] Animals are therefore a gradation towards men, but they also confirm Darwin's main thesis that possible retrograde development (degeneration) was possible,[40] actually it is

---

[37] Founded as the Society for the Prevention of Cruelty to Animals (SPCA) in 1824, it adopted its current name after being granted royal status by Queen Victoria in 1840.

[38] F. Galton, "Eugenics: Its Definition, Scope and Aims", *The American Journal of Sociology*, X.1 (June 1904): 1–6.

[39] C. Darwin, *The Expression of the Emotions in Man and Animals* (London: John Murray, 1872), 367.

[40] Cfr. C. Darwin, *The Origin of Species*, 395–96: "the manner in which all organic beings are grouped, shows that the greater number of species of each genus, and all the species of many genera, have left no descendants, but have become utterly extinct. We can so far take a prophetic glance into futurity as to foretell that it will be the common and widely-spread species, belonging to the larger and dominant groups, which will ultimately prevail and procreate new and dominant species". Perfection will, therefore, be measured according to the relative criteria present in

exactly what in moral terms happens to Moreau who still shows human features but reverts and degenerates "morally" into an animal for the brutality he exerts.

If Moreau's loyalties are clearly Galtonian, Prendrick represents the threat of a correct interpretation of Darwin's thought, through the attack that Huxley had set to the non-ethic position of the eugenists:

> "Montgomery says you are an educated man, Mr. Prendick; says you know something of science. May I ask what that signifies?"
>
> I told him I had spent some years at the Royal College of Science, and had done some researches in biology under Huxley. He raised his eyebrows slightly at that.
>
> "That alters the case a little, Mr. Prendick", he said, with a trifle more respect in his manner. "As it happens, we are biologists here. This is a biological station – of a sort". [DM, ll. 949–959]

The last addition makes things clear. Indeed having been trained under Huxley, Prendick speaks of them not as beasts but as the Beast People: Dog-man, Leopard-man, and a Monkey-man, who was: "for ever jabbering at me, jabbering the most arrant nonsense" and whose peculiarity was that "he had a fantastic trick of coining new words". Not to talk of the silvery-hairy-man (the Sayer of the Law) and of M'ling (man-link), "a satyr-like creature of ape and goat"[41] a sort of "missing link" between animal and man, a creature that Spencer had hypothesized as the intermediate species between animal and man, and which journals like *Punch* were, in the meantime, identifying with the Irish.[42]

---

a *habitus*: "And as natural selection works solely by and for the good of each being, all corporeal and mental endowments will tend to progress towards perfection". Indeed, all this being subject to the main laws of variability and selection, it will lead to: "a Ratio of Increase so high as to lead to a Struggle for Life, and as a consequence to Natural Selection, entailing Divergence of Character and the Extinction of less-improved forms".

41 "There were three Swine-men and a Swine-woman, a mare-rhinoceros-creature, and several other females whose sources I did not ascertain. There were several wolf-creatures, a bear-bull, and a Saint-Bernard-man. I have already described the Ape-man, and there was a particularly hateful (and evil-smelling) old woman made of vixen and bear", H. G. Wells, DM, ll, 3018–3022.

42 See *Punch* 1848, where the Irish occupy a position just above the Africans, and *Punch* 1862's satire: "The Missing Link": "A gulf certainly, does appear to yawn between the Gorilla and the Negro. The woods and wilds of Africa do not exhibit an example of any intermediate animal. But in this, as in many other cases, philosophers go vainly searching abroad for that which they could readily find if they sought for it at home. A creature manifestly between the Gorilla and the Negro is to be met with in some of the lowest districts of London and Liverpool by adventurous explorers. It comes from Ireland, whence it has contrived to migrate; it be-

The other novel I find relevant in presenting issues that deal with typifications, in which race discourses have been made common, is *Dracula*.[43] The novel is a good instance of the literary response to the taboo of miscegenation and hybridization. The imperial gothic novel *Dracula* textualizes eugenists' fears of being invaded by "alien" immigrants (the "oriental" people from the colonies and "others" in general) representing, in contrast with Stevenson's novel *Dr Jekyll and Mr. Hyde,* the menace from without. The Count, the "reverse colonizer", as both S. Arata[44] and H. L. Malchow[45] have shown, represents the fear of the "other" who menaces a "reverse colonization" of the West by the seduction of his "sexual exuberance" on women, thus hybridizing the "purity" of the western race. It is not by chance that Mina's patronymic, Dracula's first victim, is Westenra, a homophone of westerner, typifying her as his victimized antagonist. He is represented as a virus settling itself in the organism of the English community, and is therefore killed by the pack of Westerners, "The Crew of Light", who take revenge on him, and also of the obscurantism (occultism and superstition) that are usually projected, as a counterpart of western rationality, on the East in general.[46] The fact that the novel is concentrated on blood – the blood of a body representing a political body – makes it clear that we are here dealing with a symbol of race (the blood that Dracula sucks from his victims' bodies and which he infects with his selfsame vampiric need), and earth, a symbol of geopolitics (the native earth Count Dracula has to take with him to England). The novel's racial undertone (in the same terms that will later be used by the Nazi propaganda of *Blut und Boden* ideology) speaks out clearly that the novel fuels a racist behaviour in the mind of those who perceive themselves as being under the siege of an alien invasion.

These two novels then in a paradigmatic way speak of the fears the Victorians were experiencing towards the *fin de siècle* when the intelligentsia felt that science, together with the pseudo-sciences it gave rise to, had failed men

---

longs in fact to a tribe of Irish savages: the lowest species of Irish Yahoo. When conversing with its kind it talks a sort of gibberish. It is, moreover, a climbing animal, and may sometimes be seen ascending a ladder laden with a hod of bricks", 18.

43   B. Stoker, *Dracula* [1897] (London: Penguin, 1992).

44   S. Arata, *Fictions of Loss in the Victorian Fin de Siècle* (Cambridge: Cambridge UP, 1996).

45   H. L. Malchow, *Gothic Images of Race in Nineteenth-Century Britain* (Stanford: Stanford UP, 1996).

46   Cfr. E. Said, *Orientalism* [1978] (New York: Penguin, 1995), 5, where it is clearly stated that "such locales, regions, geographical sectors, as 'Orient' and 'Occident' are man-made".

and were threatening a transformation of science into scientism, as the following dialogue from *The Island of Dr. Moreau* confirms:

> "You see, I went on with this research just the way it led me. That is the only way I ever heard of true research going. I asked a question, devised some method of obtaining an answer, and got a fresh question. Was this possible or that possible? You cannot imagine what this means to an investigator, what an intellectual passion grows upon him! You cannot imagine the strange, colourless delight of these intellectual desires! The thing before you is no longer an animal, a fellow-creature, but a problem! Sympathetic pain, – all I know of it I remember as a thing I used to suffer from years ago. I wanted – it was the one thing I wanted – to find out the extreme limit of plasticity in a living shape".
>
> "But", said I, "the thing is an abomination –"
>
> "To this day I have never troubled about the ethics of the matter", he continued. "The study of Nature makes a man at last as remorseless as Nature".
> [DM, ll, 2709–2724]

Huxley's *avant la lettre* bio-ethic answers are thus topical of his age, but they are also a reminder of the accountability we have of ours. Law and ethics, as he says, but let us add, provocatively, art also, provide good vehicles of thought to reflect on in order to remain in control of those changes in our culture-specific environments to which we need, as intellectuals, to oppose cultural resistance, to not risk falling into acquiescent adaptation.

Silvia Monti
University of Pavia

Rhetoric, Lexicography and Bioethics in Shelly Jackson's Hypertext *Patchwork Girl*

The issues contemporary bioethics deals with prove to be strictly related to the forms of biogenetic manipulation of the body made possible by the advances in biology, medicine and genetic engineering. Indeed, the biotechnological revolution has led to the blurring of boundaries between natural and artificial, between man and machine, raising debates and moral dilemmas that, if considered from an interdisciplinary perspective, find their origins in the literary field and, in particular, in the Frankenstein myth.

As a matter of fact, bioethical issues seem to reopen discussions already to be found in literature and, viceversa, the complexity of some fictional events seem to represent the testing ground for present-day bioethical theories. In American post-modern science fiction, in particular, we often witness the death of the human subject as a unitary and coherent being, as well as the emergence of a new conception of human life and the appearance of the so called "posthuman subject" or "cyborg", an entity whose boundaries undergo continuous (de)construction and reconstruction.[1] Indeed, the posthuman appears to be a revisionary conception of the "human" category, a coupling of the human and the technological, in which it is "no longer possible to distinguish meaningfully between the biological organism and the informational circuits in which it is enmeshed".[2] In the posthuman there are no essential differences or absolute demarcations between bodily existence and computer simulation, cybernetic mechanism and biological organism,[3] and the cyborg itself can be invoked in relation to engineering and biomedical advances that produce new genetic hybrids;[4] if Mary Shelley's creature was a collage of organs, the so-called transgenic monsters created by modern biotechnologies can be seen as a collage of genes.

---

[1] Cfr. K. Hayles, *How We Became Posthuman. Virtual Bodies in Cybernetics, Literature and Informatics* (Chicago: University of Chicago Press, 1999).

[2] K. Hayles, "Virtual Bodies and Flickering Signifiers", *JSTOR* 66.10 (1993): 69–91.

[3] D. Tofts *Prefiguring Cyberculture. An Intellectual History* (Cambridge: The MIT Press, 2003), 3.

[4] Cfr. T. Armstrong, *Modernism, Technology and the Body* (Cambridge: Cambridge University Press, 1998).

Starting from these assumptions, this paper sets out to explore Shelley Jackson's hypertext fiction *Patchwork Girl, or a Modern Monster*[5] (1995) as a sort of linguistic and cultural prism, diffracting aspects of a digital re-mediation and reproduction of Mary Shelley's *Frankenstein* (1818). Whereas in *Frankenstein* the female monster is denied life as she is destroyed by Victor, in *Patchwork Girl* Shelley Jackson reawakens the female monster by having her patched together by Mary Shelley – a character created by Jackson rather than an author who herself writes – with a double operation: of "sewing" together body parts belonging to deceased people and of "writing" the hypertext. Indeed, as we will see, this patching is specifically identified with the characteristically feminine work of sewing and quilting, thus turning reading into a sort of needlework and the computer itself into a reproductive technology of monstrous bodies and artificial identities.[6]

From a general point of view, both *Frankenstein* and *Patchwork Girl* look at the motives lying behind artificial creation, illustrating the potential of science to manipulate life forms, something previously reserved to nature and chance: it is "artificially" created and not naturally born life that is challenging.[7] In this sense, *Frankenstein* can be considered as an early example of the ambivalent attitude of both enthusiasm and anxiety surrounding bio-technologies and genetic experimentation which are redefining the parameters of human identity: biology now comes to be the enactment of man's estrangement from nature and from his own body in a context where technology is allotted the place of the monster-maker and biomedical scientists seem to be urged by the same ambition that led Doctor Frankenstein to create his monster;[8] all this according to the contemporary vision of the human body as a mere biological machine whose parts are interchangeable, a vision whose origins can be traced back to classical anatomy treating the body as a set of interlinking yet separable parts and the organs as in-themselves technologies that nevertheless work together to form a living body.[9] It is from this perspective that Frankensteinian science can be considered very close to

---

[5] *Patchwork Girl*, a hypertext novel built using Storyspace (a software program specifically designed for literary hypertext and available from Eastgate Systems), is issued in CD-ROM format and requires installation on the hard-drive of the computer before it can be accessed.

[6] Cfr. P. Carbone, *Patchwork Theory. Dalla letteratura postmoderna all'ipertesto* (Milano: Mimesis Eterotopie, 2001).

[7] Cfr. D. Carpi, ed., *Letteratura e scienza* (Bologna: Re Enzo, 2003).

[8] Cfr. N. Spinrad, *Science Fiction in the Real World* (Carbondale: Southern Illinois UP, 1990).

[9] Tofts, *Prefiguring Cyberculture. An Intellectual History*, 30.

field surgery and contemporary genetic engineering techniques and that both Frankenstein's monster and Patchwork Girl come to represent the "open-ended nature of the body's becoming in the techno-scientific age",[10] threatening of altering the very core of what makes the human. It is through this "new biology" that postmodernism expresses the sense of loss, of inadequacy and frustration suffered by men, posing ethical questions similar to those raised by *Frankenstein* in the 19th century: what does it mean to be made and not to be born? What does it mean to become a cyborg? How are such becomings related to identity and subjectivity?

Trying to answer these questions and taking into consideration aspects related to contemporary reconfigurations of bodies, *Patchwork Girl* and its bioethical implications will be analysed from a linguistic perspective, highlighting the function that rhetorical figures such as metaphors, synecdoches and similes fulfil of expressing the hybridity of Patchwork Girl's existence.

Shelley Jackson herself points out the presence of many rhetorical figures hidden behind the text's linguistic surface that add further meaning to the narrative; such rhetorical figures are often linguistically represented as "dreams" or "demons" as we read in the lexia [it thinks]: "Language thinks. When we have business with language, we are possessed by its dreams and demons, we grow intimate with monsters".

The most recurrent rhetorical figures to be found in *Patchwork Girl* are metaphors. Patchwork Girl herself is seen as a sort of living metaphor as she states in [metaphor me]:

> I am a mixed metaphor. Metaphor, meaning something like "bearing across", is itself a fine metaphor for my condition. Every part of me is linked to other territories alien to it but equally mine.

This illustrates the essence of metaphors themselves: as metaphors connect seemingly unrelated subjects and create a linguistic relation between concepts, expressing the unfamiliar in terms of the familiar,[11] Patchwork Girl can be seen as a reference to other persons: indeed, the parts that form her artificial body preserve the personalities of their former owners and are significantly defined as "borrowed parts, annexed territories", thus showing that Patchwork Girl's multiple identity is her inner essence, as she herself recognizes: "I cannot be reduced, my metaphors are not tautologies [...]. The metaphorical principle is my true skeleton".

---

[10] Tofts, *Prefiguring Cyberculture. An Intellectual History*, 9.

[11] Z. Kövecses, *Metaphor and Emotion: Language, Culture, and Body in Human Feeling* (Cambridge: Cambridge University Press, 2000).

The generative metaphor to be found in *Patchwork Girl* focuses on a parallel between her body and the hypertext. Why does Shelley Jackson use such metaphor? According to the ancient studies of rhetoric, a metaphor replaces a totality with another totality and is composed by vehicle and tenor: in *Patchwork Girl* we could say that the vehicle is Patchwork Girl's body and the tenor is the hypertext, or the hypertextuality. What does hold them together? Why is it possible to use the body as a metaphor for the hypertext? Probably because both the body and the hypertext have a peculiar structural aspect: as the body is an entity that displays the specific features of the material dimension, of the signifier, the hypertext reveals the specific, unconventional structure and contents of the new type of non-linear, narrative text of the 21st century. They are thus liable to be related to each other from a metaphorical point of view considering metaphors as comprehensive figures, inclusive and synthetic, rather than analytic, and fulfilling a unifying function. Shelley Jackson thus takes the whole "textual body" into consideration as a first step and then uses other rhetorical figures, such as synecdoches and metonymies, which have a fragmenting function and work by replacing the whole with the part, to illustrate the single aspects of the bodily text/hypertext connection, clearly seen in the lexia [typographical]:

> The comparison between a literary composition and the fitting together of the human body from various members stemmed from ancient rhetoric. *Membrum* or "limb" also signified "clause".

As the human body is composed of different parts and organs, a text is composed of different clauses and sentences and a hypertext – overcoming the old linear constraints of written text – is composed of different lexias. In *Patchwork Girl* Shelley Jackson thus disembodies the dead people from which she takes the different body parts in order to create a new body of text: the body of Patchwork Girl as well as the hypertext.

Shelley Jackson describes the process of sewing Patchwork Girl's body together, and therefore of writing the hypertext, opposing a process of artificial creation to natural birth, thus considering artificial life as both parallel and competitor to biological life;[12] this can be seen in the lexia [graveyard] → [sewn]:

> I had sewn her, stitching deep into the night by candlelight, until the tiny black stitches wavered into script and I began to feel that I was writing, that this creature I was assembling was a brash attempt to achieve with artificial means the unity of life-form.

---

[12] Kövecses, *Metaphor and Emotion: Language, Culture, and Body in Human Feeling.*

This lexia is strictly linked with another one, significantly entitled [writing] where Mary Shelley states:

> I had made her, writing deep into the night by candlelight, until the tiny black letters blurred into stitches and I began to feel that I was sewing a great quilt.

Shelley Jackson uses the quilt metaphor here to refer to the body also to plunge the reader into the realms of postmodern discourses of fragmentation, dispersal and the vexed questions of identity and of natural or artificial origin.[13] Indeed, Jackson's female monster constitutes a hybrid assemblage of divisionary parts as her body is created combining the "natural" element of nature with the "unnatural" characteristic of technology. Similarly, in our techno-scientific age, bodies are often made more powerful by biomedical technologies as well as physically enhanced by cosmetic surgery procedures and therefore, in a sense, they are artificially "recreated" and transformed into hybrid and, sometimes, unnatural forms of humanness.

The act itself of sewing can be metaphorically interpreted as an act of surgery and the connection between surgery and writing is made explicit when Shelley Jackson says: "Surgery was the art of restoring and binding disjointed parts": in this sense, as Patchwork Girl is stitched together and patched up with pieces of dead bodies, Jackson's hypertext stitches together different portions of material from all types of sources which are randomly displayed in the hypertextual realm where Patchwork Girl is "buried", as she herself explains in the lexia [hercut 4] → [graveyard]: "I am buried here. You can resurrect me but only piecemeal. If you want to see the whole, you will have to sew me together yourself". The reading of Patchwork Girl therefore appeals to our demiurgic power and turns readers into a sort of doctor Frankenstein putting together the different pieces of the textual corpus, "resurrecting" it and creating their own reading.

It is interesting to observe that the parallel between surgery, sewing and writing is represented from both a linguistic and a visual point of view. As a matter of fact, readers of *Patchwork Girl* encounter in the first instance an image of a woman [her], reminiscent of a diagram of the homunculus, arms outstretched and palms displayed.[14] Linked to [her], the opening graphic, is "phrenology", a graphic that further performs the metaphoric overlay of body and text. Showing a massive head in profile, "phrenology" displays the

---

[13] D. Harter, *Bodies in Pieces. Fantastic Narrative & the Poetics of Fragment* (Stanford: Stanford University Press, 1996).

[14] A. M. van Baren, *Stitch and Split. Feminist Alternatives to Frankensteinian Myths in Shelley Jackson's* Patchwork Girl, available at http://igitur-archive.library.uu.nl/student-theses/2007-1008-200415/MAThesisAMvanBaren.pdf.

brain partitioned by lines into a quilt of women's names and enigmatic phrases. When the reader clicks on the names, she/he is taken to lexias telling the stories of women from whose parts Patchwork Girl was assembled; clicking on the phrases takes the reader to some other lexias that meditate on the nature of "her" fragmented subjectivity. These textual blocks are thus entered through a bodily image, implying once again that the hypertextual form echoes the monster's body and foregrounds the fragmented identity of the postmodern and post-human hybrid it represents.[15] Indeed, the peculiar structure of this disassembled body emphasizes the many "patches" that make up each person's identity, the array of distinct personalities that produce what is called the self, a self that in the case of Patchwork Girl is "dispersed", as the girl points out in [dispersed]: "My real skeleton is made of scars. […] What holds me together is what marks my dispersal. I am most myself in the gaps between my parts".

Addressing such interrelations between the hypertextual body and the human body in *Patchwork Girl*, we can draw a parallel with contemporary biomedical practices where, in the lexia [body of text] → [mixed up] → [crazy], the "great quilt" that is Patchwork Girl's body is defined as a "crazy quilt": "Crazy quilts, unlike their geometrical counterparts, were constructed without planning or plan". The expression "Crazy quilts" can be seen as a metaphorical reference to bodies manipulated to such an extent by genetic and biomedical experimentation that they become "crazy", that is to say either monstrous or anyhow far from the conventional aesthetic parameters human bodies conform to; "their geometrical counterparts" is a metaphor for normal and well-proportioned bodies, whereas the fact that "crazy quilts […] were constructed without planning or plan" could be interpreted as a metaphor for biomedical experiments that are often performed without respect for human nature and go against the principles of natural reproduction.[16]

The connection between body and text is further pointed out in [bodies too] when body parts are metaphorically seen as letters of the alphabet that should follow a specific order to build semantically acceptable sentences; similarly, body parts should be properly distributed to avoid the risk of a monstrous appearance; in [bodies too] we read:

---

[15] S. Jackson, *Stitch Bitch: the Patchwork Girl*, available at: http://media-in-transition. mit.edu/articles/jackson.html.

[16] Cfr. M. Cozzoli, "Il dibattito in bioetica. Rivoluzione biotecnologia e domanda bioetica", *Medicina e Morale*, 6, 2002.

> [...] our infinitely various forms are composed from a limited number of similar elements, a kind of alphabet, and we have guidelines as to which arrangements are acceptable, are valid words, legible sentences and which are typographical or grammatical errors: "monsters".

Here again "valid words" and "legible sentences" are metaphors for well-proportioned and finely adjusted bodies, whereas "typographical or grammatical errors" could be seen as metaphors for physical anomalies as well as for unsuccessful techniques of genetic manipulation, often leading to the appearance of monstrous creatures and thus inevitably raising ethical questions.[17]

It is interesting to notice that Shelley Jackson also compares the creation of a body to a work of art, thus implying a metaphor for the artificial creation of perfect bodies, as we can see in [advice]: "The artist must create an integrated corpus out of detachable elements. Natural defects and imperfections need to be remedied as soon as possible". "The artist" can be interpreted as a metaphor for a surgeon who should remedy illnesses and physical imperfections through biomedical interventions, organ transplantation (thus implanting "detachable elements", that is to say donor's organs, into the recipients' bodies) and cosmetic surgeries in order to improve one's physical aspect.

If the body is compared to a written text, human life is metaphorically compared to a piece of writing, as we can see in [all written] "You could say that all bodies are written bodies, all lives pieces of writing". Furthermore, in the lexia [lives] human lives are seen as novels, whose end can come unexpected and surprise the readers as some events in our own life might come as a surprise overtaking us:

> We live in the expectation of traditional narrative progression; we read the first chapters and begin already to figure out whether our lives are romantic comedy or high tragedy, mystery or adventure; [...] most of us do our best to adhere to the conventions of our chosen genre and a kind of vertigo besets us when we witness plot developments that had no foreshadowing in the previous chapters. We protest bad writing.

It is in particular on the sentence "We protest bad writing" that we could focus our attention as it could be seen as a metaphor for the unsatisfactory conditions contemporary men often suffer as well as for risky and abortive biogenetic experiments.

The dangers implied in the manipulation of the body are further clearly expressed in the lexia "universal", where Shelley Jackson uses a metaphor prefiguring a future full of monsters: "You will all be part of me. You already

---

[17] Cfr. F. D'Agostino, *Il corpo de-formato. Nuovi percorsi dell'identità personale* (Milano: Giuffrè Editore, 2002).

are; your bodies are already claimed by future generations, auctioned off piecemeal to the authors of future monsters". The "authors of future monsters" can be metaphorically considered as doctors and biomedical scientists, carrying out experiments to create the monsters of the future, that is to say potential human clones whose genes already live in the body and cells of the living. The fact that also in *Frankenstein* the creator of the monster is an anatomist makes perfect sense, given the role of medicine in building a "proper" body.[18]

With regard to monstrous bodies, it is interesting to notice that the lexeme "monster" and its possible derivations – "monstrosity", "monstrous" – are recurrently used throughout the hypertext with different metaphorical connotations.

First of all, Patchwork Girl's monstrosity appears to be essentially due to her fragmentary and hybrid nature, as it happened for Frankenstein's monster that is described as follows:

> His yellow skin scarcely covered the work of muscles and arteries beneath; his hair was of a lustrous black, and flowing; his teeth of pearly whiteness; but these luxuriances only formed a more horrid contrast with his watery eyes, that seemed almost the same colour as the dun white sockets in which they were set, his shrivelled complexion and straight black lips.[19]

Similarly, the fact that Patchwork Girl's hybrid nature prevents her from finding a place among human beings is expressed when, in [I am], she points out:

> I am tall, and broad-shouldered enough that many take me for a man; others think me a transsexual and examine my jaws and hands for outsized bones, my throat for the tell-tale Adam's Apple. My black hair falls down my back but does not make me girlish. Women and men alike mistake my gender […] I am never settled. I belong nowhere.

Patchwork Girl is therefore perfectly aware that her identity is "something in-between", both female and male: "I was a half-man, a half-woman […] Whatever I was, I was not easy in my skin". She is considered a monster because of the inability of her character to be classified within a traditional gendered, taxonomic hierarchy; indeed, the cyborg rejects these boundaries, it is a post-gender creature that exists outside the traditional conceptions of male and female. Furthermore, in the lexia [why hideous?], Patchwork Girl states:

---

18 P. Youngquist, *Monstrosities. Bodies and British Romanticism* (Minneapolis: University of Minnesota Press, 2003).

19 M. Shelley, *Frankenstein, or The Modern Prometheus* [1816–1818] (New York: Bobbs-Merrill, 1974), 57.

I've learned to wonder: why am I "hideous"? They tell me each of my parts is beautiful and I know that all are strong. Every part of me is human and proportional to the whole. Yet I am a monster – because I am multiple, and because I am mixed, mestizo, mongrel […] Identities seem contradictory, partial and strategic. There is not even such a state as "being" female, or "being" monster, or "being" angel. We find ourselves to be cyborgs, hybrids, mosaics, chimeras.

In this sense, Patchwork Girl radically alters the traditional view of the subject as an individual with a unique personality and a definite identity: for the female monster it is mere common sense to say that multiple subjectivities inhabit her body, as the different creatures whose parts she is made of make her appear as an assemblage rather than as a unified self.[20] This could also be metaphorically interpreted as a reference to those people having had organ transplantation, as, in a sense, the transplanted organs belonging to deceased donors survive and have an autonomous life within the bodies of the recipients. But in real life transplants involve risks. Indeed, one of the major obstacles to transplanting any tissue from one person to another is the body's natural rejection response: our immune system, protecting our bodies from foreign invaders such as bacteria and viruses, views other people's organs as unwelcome intrusions that must be attacked. Even though Patchwork Girl is "re-constructed" from different body parts belonged to dead men and women and it is this multiplicity that forms the essence of her personality, thus excluding any risk of rejection, such risk is all the same taken into consideration within the hypertext, when in the lexia [mutinies] Patchwork Girl implies the possibility that body parts could go out of control and rebel against their owner:

> And should I fear this mixed company? A body part turning against its host is an oft-told tale […] A movie (*Body Parts*) in which the survivor of a car crash who receives a donor's arm wakes to find himself strangling his wife, single-handedly.

A proper transplant is performed within the hypertext, when Mary and her creature, who have become lovers and grown physically intimate with each other's bodies, decide to swap patches of skin, as we can see in [journal] → [female trouble]: "I have a crazy wish! I wish that I had cut off a part of me … and given it to be a part of her. I would live on in her and she would know me as I know myself". It is here clear the parallelism with contemporary medicine, as the sentence "I would live on in her" can be referred to the fact that, as already mentioned, dead people "resurrect" in the bodies of those who survive thanks to the organ transplantation they have had.

---

[20] Van Baren, *Stitch and Split. Feminist Alternatives to Frankensteinian Myths in Shelley Jackson's* Patchwork Girl, 18.

Patchwork Girl's body itself can be seen as a "resurrected" body and it is assimilated to a body that, aspiring to perfection, has cosmetic surgery: such connection is expressed in [body of text] → [resurrection]:

> The human, more than human resurrected body is a body restored to wholeness and perfection, even to a perfection it never achieved in its original state. [...] What is the age of the resurrected body? Is perfection at twenty years old, at thirty, at forty? Plastic surgeons must ask themselves the same question.

This ethical question could be interpreted considering the use and abuse of cosmetic surgery in contemporary society; indeed, many debates have been recently raised concerning the dangers implied in cosmetic surgery (especially for young girls), as well as in extreme manipulations of one's own body in an ever-growing quest for physical perfection.

Besides metaphors, there are other rhetorical figures such as synecdoches and similes that play a crucial role in defining Patchwork Girl's essence as a hybrid and in debating ethical issues regarding her "unnatural" body.

Patchwork Girl itemises her body parts functioning as synecdoches both of the persons they once belonged to and of her own body: these parts are represented in the hypertext as if they were real persons, refusing a proper integration in the girl's body, as we can see in [story] → [séance] → [lives & livers]:

> It seemed to me that each of my parts brought with it a trace of the whole person who was once attached to it. There was a crowd, a whole gaggle of persons, competing for the space occupied by my one limited body.

Patchwork Girl clearly recognizes her multiple identity in [body of text] → [double agent] when she states:

> I am muscular and convincing because I am whole; I am devious and an escape artist because I am broken. [...] It's me, this one: Jennifer – Bronwyn – Elisabeth – Roderick – Kate – Alise – Germain – Aphrodite and all the others, who can take any corners [...] I pay homage to Jennifer and the others and I wonder if I can detect their diverse personalities in my multiple parts.

An interesting aspect is that all Patchwork Girl's body parts and internal organs are assigned the human qualities and the behavioural attitudes of their former owners: the tongue of Susannah, a very talkative woman; the ears of Flora, which overhear secrets; the "large-nostrilled and discerning" nose of Geneva, "a woman of otherwise normal proportions" [nose]; the stomach of Bella, a glutton; the "shy" toe of Constance, "the nun with the beautiful toe" [names]; the strong heart of Agatha "who knocked dead her daughter's betrayer with a soup bone and grew a very fertile crop of kale and cabbages from her kitchen garden the next spring" [names]; the twitching finger of Dominique, a pickpocket; the "twitching", "jumping" and "joggling" legs of Jane, a nanny whose leg "wants to go places [as] has had enough of wait-ing" [left leg]; the "firm and spherical" eyes of Tituba, who loved to read, thus allowing Patchwork Girl to "peruse with equal clarity the finest of print and the faint script of smoke from a distant chimney" [eyeballs]; the ca-pacious lungs of mountain-bred Thomasina.

The fact that each body part functions as a synecdoche for the whole body is constantly emphasized also when, in America, Patchwork Girl visits Madame Q, a spiritualist, and sees ghostly hands hovering in the room with-out a body attached to them, something that she doesn't consider as particu-larly strange or unusual, as we read in [solid ghosts]:

> Ghostly hands hovering above the table didn't seem so far from reality to me, whose own hands often seemed severed from my body, and which I could easily imagine living their own lives entirely separate from my purposes. Stirring a cake; brandishing a soup bone [...] caressing people I'd never known.

Such attention on body parts also refers to the Lacanian fantasy of the *corps morcelé*, the body in bits and pieces,[21] which informs the whole hypertext and is strictly related to the contemporary biomedical vision of the body as an assemblage of parts. In the section [bio] Shelley Jackson, led by her interest in "facts concerning modern biology's understanding of the multiple nature of the living organism" [bad dreams], carries out a medical and scientific study of the protagonist's body parts, and, drawing on the contemporary dis-courses of biology and technoscience, she states that:

> the body as seen by the new biology is chimerical. The animal cell is seen to be a hybrid of bacterial species. Like that many-headed beast [the chimera], the micro-beast of the animal cells combines into one entity, bacteria that were originally freely living, self sufficient and metabolically distinct.

Within this chimerical vision of the body, bacteria, which were originally sep-arated, combine in units and the human cell appears to be a hybrid of the bacterial species, an amalgam of parts. This concept is further expressed in the section [mosaic girls] where the chromosomes of the female body are compared to mosaic tiles:

> Because of her Barr bodies a female mammal is a genetic mosaic, made up of two kinds of cells: those with active X#1 and those with X#2, just as a mosaic might be made up of blue and green tiles.

Furthermore, in [mixo], a sort of medical treatise on the functioning of the human cells, Shelley Jackson uses a scientific and medical language to state

---

21 Cfr. also S. Monti, *Le vicissitudini della corporeità. Anima e anatomia nella narrativa inglese e americana dell'Ottocento* (Milano: Arcipelago Edizioni, 2006).

that unity is just an illusion as the real self is made up of millions of separate parts:

> Take *Mixotricha Paradoxa*, an autonomous nucleated cell and a good friend of mine. It looks like one organism on first glance, but in fact it's made of several different very cooperative cells.

In this sense, generally speaking, human beings themselves could be considered as hybrids, patchworks of different parts from a strictly biological point of view, according to a fragmentation principle to which the entire human genre should conform: the hybridity of a "patchwork existence" would therefore reveal itself to be the norm rather than the exception. In this sense, it is also necessary to say that biomedical practices like organ transplants surgery and transgenic engineering (where bits of the DNA code of one organism may be spliced into that of other quite different organisms) generate models of selves in constant intimate interchange with others, they create living organisms whose every cell bears the mark of technoscientific intervention.[22]

Following this philosophy, the hypertext not only normalises the subject-as-assemblage but also presents the subject-as-unity as a grotesque impossibility.[23] And Patchwork Girl's body can indeed be seen according to the peculiar aesthetics of the grotesque that flourished during the late Middle Ages and that let the body appear as disproportionate and chaotic. From this perspective, it is interesting to notice that the grotesque body manifested itself not only in late medieval and early modern medicine, but also in penal practices of the same period; for example, in the first chapter of *Discipline and Punish* (1975) Michel Foucault describes how the bodies of criminals were boiled in pitch and oil, butchered and skinned, torn apart and chopped into pieces.[24] And as the legal punishment and the medical dissection of human bodies were spectacular events performed in market squares and in the anatomical theatres of late medieval and renaissance towns, similarly, we could say that in contemporary society the hypertextualization of the body in *Patchwork Girl* is the spectacular representation of the hybrid, post-human body peculiar to our techno-scientific age.

The other rhetorical figures often to be found in *Patchwork Girl* and fulfilling a very important function with regard to the representation of Patch-

---

[22] Tofts, *Prefiguring Cyberculture. An Intellectual History*, 89.

[23] Van Baren, *Stitch and Split. Feminist Alternatives to Frankensteinian Myths in Shelley Jackson's* Patchwork Girl, 21.

[24] H. Zwart, "Medicine, symbolization and the 'real' body. Lacan's understanding of medical science", *Journal of Medicine, Healthcare and Philosophy* 1.2 (1998): 107–117; available at http://www.filosofie.science.ru.nl/research/dl1.pdf.

work Girl's artificial nature are similes. Indeed, Patchwork Girl's body is often compared to natural elements as well as to animals, following Patchwork Girl's desire to be assimilated to what is natural rather than to what is artificial. Her desire to mix with natural elements is clearly expressed when her hand is defined as a "heavy fruit" in [story] → [falling apart] → [more partings]:

> My hand dropped off in a supermarket, where it sounded like a heavy fruit falling [...] so the produce person gave me a stern look from across the avocados until I picked it up and plopped it in my basket, between the mushrooms and the cabbage.

The similes comparing her body parts to natural elements pervade the whole hypertext, and are also used in the section [falling apart] where we witness a dissolution of her physical dimension, representing a sort of "apotheosis of fragmentation". Here she imagines herself to be in a jungle, full of body organs rather than of vegetation: the heart, the liver, the intestines, have an autonomous life and are compared, through similes, to natural elements as they hover in the air:

> The hearts are alive … intestines pile up like tires at the ankles of legs become trees … Ovaries hang like kumquats from delicate vines … [...] my palms and fingertips will drift down like aged leaves. [...] My uterus will rise up from the mulchy ruin of my abdomen like a mushroom's fine dome [...] My teeth will fill my mouth like loose pebbles.

Patchwork Girl's desire to recover from her unnatural condition is so deeply felt that she would like her body to be subject to medical experimentation in order to gain a normal physical appearance. In [appearances] she uses a simile comparing her body to a flowerbed, her organs to flower bulbs and doctors to gardeners and she says:

> I wanted to be a victim, to be toiled over, as innocent of device as a flowerbed, by doctors in their gardening gloves, plucking out unruly sprouts and coddling the well organs like the firm bulbs of tulips.

Drawing a parallel with contemporary medical practices, this could be metaphorically related to the surgery doctors perform to either cure ill organs and body parts or simply improve one's physical appearance.

From this perspective, surgery itself is seen as a natural process. Indeed, when the transplant of a patch of skin between Mary and the monster is carried out in [story] → [severance] → [join] it is compared, by means of a simile, to the process of preparing dough for pastries, thus appearing as something natural and, in a sense, even pleasant:

> I held her leg steady as she unblinking scored a circle the size of a farthing in the skin of her calf, then from the perimeter of the circle toward the center slid the blade under the topmost layers of skin, lifting it. I could see the dark metal through her fair skin. "Like detaching a round of pastry dough from a table top", she said, lifting the bloody scrap on the tip of the blade and holding it out toward me. I wiped the piece of skin off the blade onto a bit of cotton and set the sharp edge of the knife against the knotty scar that crosses my thigh to meet my groin. We have decided that as my skin did not, strictly speaking, belong to me, the nearest thing to a bit of my flesh would be this scar ... I sliced off a disc of scar tissue the same size as the bit that lay on the pink twist of cotton, and slid it off to the point of the knife onto the raw spot on her leg; she took the knife and laid her piece on me.

The sort of obsession Patchwork Girl develops towards nature is also represented by the recurrent presence of similes related to animals, as when, in [trunk], she says:

> My trunk belonged to a dancer, Angela [...] her movements were fervent. I shake my hindquarters like a little dog and arch my back like a cock crowing over his chickens

or when in [more partings] she states:

> When I bathed, I sat in the steaming fragrance bathwater amidst the warm nudging bodies of my vagrant parts. They seemed companionable, they seemed to have personalities of a rudimentary sort, like small agreeable dogs.

Also blood cells are compared to animals as we can see in [blood]: "The letters come alive like tiny antelopes and run in packs and patterns"; here "the letters" are a metaphor for the cells moistened by blood that are alive and sound and run fast within the body as antelopes run fast in the wild. And even when her dissolution takes place in [story] → [falling apart] → [diaspora] she says: "my torso fell like a cat, turning": in this case she assigns her body a graceful agility that is more peculiar to animals than to human beings, as if she felt more akin to the animal world than to the human world from which she is excluded.

From a general point of view, all similes seem to be used to make the description more vivid, to add colour to the characterization and to convey the sense of Patchwork Girl's strangeness: Shelley Jackson creates freaks in her hypertext as genetic engineering often creates monsters in genetics laboratories.

Concluding Remarks

Patchwork Girl incorporates new ways of thinking about the body on a level that is both fictional and real, dealing with the dilemmas raised by recent developments in technology and science and rhetorically charging them with the connotative bid of figures such as metaphors, synecdoches and similes. Through this cunning rhetoricity it succeeds in conveying the sense of hybridity and fragmentation overlaying body and text as well as the sense of loss of the traditional notion of what is the "human", as the new techno-scientific modes of embodiment, entangled between human and technical, clearly change the material terms in which the human takes place.[25] In this sense, the female monster of *Patchwork Girl*, with her torments and insecurities, with her longing for a place in the society in which she lives, represents the man of our time, lacking in self-confidence and searching for an identity that is never well defined and always fragmented, just as Patchwork Girl's own body.

Indeed what both *Frankenstein* and *Patchwork Girl* show, in their being fictional devices of thinking about what the human being could potentially become in the future, is the uneasiness with which man lives in contemporary society.[26] As 18th century science considered the body as a mere mechanism with interchangeable parts, similarly contemporary medicine seems to dehumanize the physical dimension, considering it as a complex collection of interacting parts liable to be replaced and manipulated often without respect for the body seen as a whole.

However, as the biological body cannot be reduced to a mechanical organism made up of dismountable parts, there emerges the need to further control the new modes of artificial manipulation of man to protect one's genetic identity. And in this sense bioethics aims to free man from alienation, trying to preserve only at a fictional level new forms of technocracy through the advent of the bionic man as a natural evolution of the biological man, always struggling in coming to terms with his identity.[27]

Such a struggle is nowadays clearly visible through bodily modifications that alter the human body primarily for aesthetic reasons rather than for medical ones, in an attempt to artificially beautify the natural form of the body often leading to charges of disfigurement; similarly, the attempt to ar-

---

[25] C. Waldby, *The Visible Human Project: Informatic Bodies and Posthuman Medicine* (London and New York: Routledge, 2000).

[26] Cfr. also L. Guerra, *Il mito nell'opera di Mary Shelley* (Pavia: CLU, 1995); A. K. Mellor, *Mary Shelley. Her Life, her Fiction, her Monsters* (New York and London: Methuen, 1989).

[27] Cfr. also L. Sfez, *Il sogno biotecnologico* (Milano: Mondadori, 2002).

tificially create life in both *Frankenstein* and *Patchwork Girl* led to monstrous and ethically unacceptable outcomes that the technological progress now threatens to make real as a symbol of the dangers of science without conscience, of the utopian quest for physical perfection as well as of bioethics gone awry. Indeed, whereas hyperfiction emphasizes the fragmentary, the decentredeness of body and text, at the core of contemporaneity there seems to be, parallel with the numerous figurations of the postmodernist body, a search for a unified identity. The fantasy of cloning a human being brings this dream to the fore; the Patchwork Girl as a clone focuses these tendencies, illustrating, also from a strictly linguistic point of view, the contradictions between her body as a textual collage and the hopeless yearning for a unified body with a singular origin.

Paola Carbone
IULM University Milan

One *Monstrous Ogre* and One *Patchwork Girl*: Two Nameless Beings

It is widely acknowledged that poetry explores the hues of feelings, aspirations, myths which colour the human essence of a person. Nevertheless, philosophy is said to investigate similar horizons. One difference, among others, between poetry and philosophy is that the former concentrates on particulars and individuality, while the latter focuses generalizations. Philosophy seeks general names; literature proper, unique names.[1] Roland Barthes expressed this clearly when he indicated how a poet battles with himself, the text, and language to find the suitable name for a character or a place, since proper names have three peculiarities: the power of "essentialness" (*Pouvoir d'essentialisation,* since they refer to one single referent), the power of "quotation" (*pouvoir de citation,* since uttering a name means evoking its essence) and the power of "exploration" (*pouvoir d'exploration,* since they raise memories).[2] Therefore, a proper name crystallizes individuality which is unique and unclassifiable (unlike scientific truths). More generally it enfolds the story and all the narrative potentialities of a person. It is probably beside the point here to call attention to the Latin etymology of 'persona', that is the dramatic character represented by the mask; however, it is very much to the present purpose to point out that a proper name affects the representation of the self, to the extent that it encloses the human being in an identity which is then 'worn' throughout life.

The naming of a child is so crucial a concern that it is recognised by the United Nations in the Declaration of the Rights of the Child which states: "Principle 3 – The child shall be entitled from his birth to a name and nationality".

The International Covenant on Civil and Political Rights[3] states in similar terms: "Article 24:2 – Every child shall be registered immediately after birth and shall be given a name".

---

[1] M. Ragussi, *Acts of Naming* (Oxford: Oxford University Press, 1987), 4.

[2] R. Barthes, "Proust et les Noms", in *Le degré zéro de l'écriture* suivi de *Nouveaux essais critiques* (Paris: Seuil, 1972).

[3] See The United Nations High Commissioner for Human Rights, http://www2.ohchr.org/english/law/ccpr.htm (December 2009).

According to Italian legislation, from birth a 'person' holds certain rights. In particular, individual rights *and* those pertaining to the family code: the right to life, to physical integrity, to be properly considered, to a name. Art. 6 of the Italian civil code states:

> Right to a name
> All persons have a right to a legally given name
> This name includes a first name and a family name
> No changes or additions are allowed, unless properly ratified by law.[4]

Similarly, British law requires parents to register a child's birth and name within six weeks of his/her birth. A person's legal name consists of the name given by the parents and established by reputation and usage, plus their surname or family name.

Around the world, all cultures have their own naming ceremonies, which are also social events. For example, in Hindu tradition a particular name is chosen according to the date and time of birth of the child: the father whispers the name four times in the right ear of the baby, the baby then receives blessings from all those present, and a feast is organized for the priest and the guests. In Jewish tradition both girls and boys are given names during their *zeved habat* and their circumcision ceremony respectively. The Christian practice of giving a personal name with baptism comes from a prior Jewish custom[5]. It is interesting to note that when baptized in the name of the Father, the Son, and the Holy Ghost, the Christian baby is identified with the power exerted by the Trinity over his/her life (Matthew 28:19). This infers that naming implies the idea of "belonging": from St. Paul's affirmation "you belong to Christ", we can, in fact, say that having a family name means that the new individual is brought not only into the community, but also into his/her cultural system, so that he/she acquires all the rights and duties established by family and society. As a result a proper name not only suggests uniqueness (the mask of the person), but also places the child within the *polis*,

---

[4] *Codice Civile*, Art. 6. "Diritto al nome. Ogni persona ha diritto al nome che le è per legge attribuito. Nel nome si comprendono il prenome e il cognome. Non sono ammessi cambiamenti, aggiunte o rettifiche al nome, se non nei casi e con le formalità dalla legge indicati." In the text my translation.

[5] In *Genesis* we are told that, before the creation of Eve, Adam occupied his days in naming the animals as they passed by: 01:002:019 "And out of the ground the LORD God formed every beast of the field, and every fowl of the air; and brought them unto Adam to see what he would call them: and whatsoever Adam called every living creature, that was the name thereof." 01:002:020 "And Adam gave names to all cattle, and to the fowl of the air, and to every beast of the field; but for Adam there was not found an help meet for him."

that is in a complex of beliefs, symbols, and rules, while suggesting ethnic, religious, historical, sexual, linguistic signs that remain valid for existing classificatory groups[6] (the mask of identity): proper names negotiate social relations. In Aravind Adiga's novel *The White Tiger*, the main character does not have a name till his first day at school since his parents and relatives have been too busy to give him one. Thus, it is his school master who names him Balram, the "sidekick of the god Krishna"[7]. This episode strikes the reader's attention as a tangible sign of the character's loneliness and also uniqueness. Later in life, disregarding his sense of belonging to the community, the boy consciously decides to become "a pervert of nature", a monster, in order to assert his independence and freedom from the family and cultural systems which have enslaved and prevented him from following his true nature:

> (…) only a man who is prepared to see his family destroyed – … – can break out of the croop. That would take no normal being, but a freak, a pervert of nature.[8]

He decides to kill a rich man and, in doing so, condemns his family to a cycle of revenge while he himself escapes far from the village using a different name. By denying the boy a name, the family has morally detached him from both an affective context and a network of legal and social interaction. This is, certainly, not the only example in literature, but it is probably the most recent since *The White Tiger* was awarded the Booker Prize in 2008.

So, if on the one hand a name is valid for purposes of legal identification, on the other it encloses a cultural supra-structure which defines social relationships. A name becomes a cognitive necessity for any person who wants to feel him/herself as part of a society, since it converts "anybody" into "somebody".[9] Names not only delineate the boundaries of social status, but they also bridge our comprehension of the Other: they have the capacity both to fix identities and to detach from those same identities.[10]

What happens when my relationship with an "other" is beyond the realm of familiarity, that is when I encounter someone whom I do not recognize as part of my natural and cultural milieu? What do I call him/her? What type of

---

[6] See B.Vecchi, ed., *Zigmunt Bauman, Intervista sull'identità* (Milano: Laterza, 2009), 80.

[7] Aravind Adiga, *The White Tiger* (London: Atlantic Books, 2008), 13–4. Balram comments: " … what kind of place is it where people forget to name their children?", 14.

[8] Adiga, *The White Tiger*, 176

[9] See C. Geertz, *The Interpretation of Cultures. Selected Essays* (New York: Basic Books, 1973), 363.

[10] See B. Bodenhorn and G. Vom Bruck, eds., *An Anthropology of Names and Naming* (Cambridge: Cambridge University Press, 2006), 2.

stories can I tell about subjects which I do not recognize as being "like my-self" or which do not belong to my linguistic universe? What I am trying to do is connect the generality of "otherness" to the specificity of a proper name, the species to the person and the self. Postcolonial culture has taught us to interpret the idea of "otherness" in many different ways – foreigner, nomad, exiled, mimic man, hybrid, mongrel, mirage – but all these ways imply both the "normal" vs "different" opposition and the concept of "be-longing" which, if taken to extremes, could either lead to silence or violence.

Today when we talk about hybridity, we refer to an "in-between" identity situated in a "third space", that is someone who lives suspended between East and West[11]. This is the typical condition of first and second generation postcolonial migrants and it generates a sense of uncertainty, ambivalence, and dissatisfaction, since it suggests the need to create a new space for sur-vival in the dominant political structures of the so-called host community. Since art and literature have repeatedly shown us the way when dealing with this area, they should currently guide our considerations on bio-technologi-cal hybridism. Mary Shelley, H.G. Wells, Frank Baum, Philip Dick, Shelley Jackson, Hanif Kureishi, Kazuo Ishiguro, Dean Koontz, Stanley Kubrick, Steven Spielberg, Tim Burton, Jan Švankmajer spring to mind in this context. We ought really to consider as hybrids all those creatures made of flesh and technological devices or science-bred which emerged from the recombi-nation of numeric, genetic and atomic codes. The first problem is to assess/decide whether posthuman-otherness of this kind is to be considered as a "person" and consequently subjected to juridical evaluation[12], that is deemed a "legal entity"; or whether it is to be seen as a simulacrum, or simply a living paradox. If it is true that the body is the privileged, archetypical site of the human, it appears legitimate to probe into the extent to which the posthu-man body falls within the natural boundaries of the "human being". I myself suggest that s/he/it might be seen as a de-territorialized affiliate to humanity who conceptually inhabits the interstices of two opposite ideas: continuity vs. discontinuity. All hybrids are crossbreed of nature, culture, expertise and knowledge, and thus resistant to specific identification. I am not saying that the hybrid-nomad has no identity, subjectivity or legal liability, but rather that in order to be publicly recognized within a single "standard" (that is accord-ing to a principle of continuity with humanity), s/he/it first of all needs to be

[11] See H. K. Bhabha, *Nation and Narration* (London: Routledge, 1990) and H. K. Bha-bha, *The Location of Culture* (London: Routledge Classics, 1994).
[12] See S. Rodotà, "Il corpo e il post-umano", *Pólemos, Rivista semestrale di diritto, politica e cultura* 2 (2008), 13.

linguistically identified. What is the difference between a complex biological organism and a human being as a generic and genetic entity? On what bases do we recognize human subjectivity? These are only two of the questions literature should help us to answer.

My question is: why does Frankenstein's monster have no name and why does Shelley Jackson's Patchwork Girl likewise remain nameless? Readers frequently mistake Dr. Victor Frankenstein's family name for the monster's given name[13], so that this patronymic has not only become common to all nameless monsters in literature but is also familiar as the metonymic form of address for monstrosity and biological deviance. If, generally speaking, a name is a sign which refers to but is not equivalent to a description or a meaning[14], in this specific case it becomes a monstrous tool of extraordinary semantic power since it apparently designates an alien[15] like a linguistic sign without a referent, a signifier without the signified because it avoids all concepts, intentions and contents which a common speaker could add to it. In the past, a monster[16] was considered as both a creature afflicted with a defect from birth and as a portent (prodigy), something extraordinary which, according to St. Augustine, was the manifestation of God's will; later, either a

---

[13] From *Oxford English Dictionary*: Frankenstein's monster: n. – a thing that becomes terrifying or destructive to its maker. – ORIGIN from Victor *Frankenstein*, a character in a novel (1818) by Mary Shelley, who creates a manlike monster which eventually destroys him.

[14] See R. Barthes, "Proust et les Noms", 122. In *A System of Logic* (1843) J. S. Mill also defines names as "meaningless markers.", since proper names do not allow us to conjecture anything with respect to people's real features.

[15] The word "alien" not only designates something that is not part of our normal experience including a creature from outer space, but in legal terms it also refers to someone who is not a legal citizen of the country in which s/he lives.

[16] See *Online Etymology Dictionary*: monster c.1300, "malformed animal, creature afflicted with a birth defect," from O.Fr. *monstre*, from L. *monstrum* "monster, monstrosity, omen, portent, sign," from root of *monere* "warn". Abnormal or prodigious animals were regarded as signs or omens of impending evil. Extended c.1385 to imaginary animals composed of parts of creatures (centaur, griffin, etc.). Meaning "animal of vast size" is from 1530; sense of "person of inhuman cruelty or wickedness" is from 1556. In O.E., the monster Grendel was an *aglæca*, a word related to *aglæc* "calamity, terror, distress, oppression." Monstrous: 1460, "unnatural, deviating from the natural order, hideous," from L. *monstruosus* "strange, unnatural," from *monstrum*. Meaning "enormous" is from 1500; that of "outrageously wrong" is from 1573. *Monstrosity* "abnormality of growth" is from 1555, from L.L. *monstrositas* "strangeness," from L. *monstrosus,* a collateral form of *monstruosus* (cfr. Fr. *monstruosité*). Sense of "quality of being monstrous" is first recorded 1656. Noun meaning "a monster" is attested from 1643. http://www.etymonline.com/index.php?search=monster&searchmode=none (June 2009). See also *Oxford English Dictionary*, 1036–37.

person of inhuman cruelty or wickedness or else a deviation from the natural order which embodies both the angelic and the demonic. Currently, in line with neo-gothic taste, it is mostly seen as mental deviance[17]. Within a species, the identity of a person is unique, but a monster has few common characteristics with any living species either physically or mentally. Monsters such as Mary Shelley's creature do not have models (extra-linguistic referents) and they exhibit their being as discontinuous with the environment through their physical appearance and looks:

> I couldn't be mistaken. A flash of lightening illuminated the *object*, and discovered its *shape* plainly to me; its gigantic stature, and the deformity of its aspect, more *hideous than belongs to humanity* [...][18] [my italics]

> I perceived, as the *shape* came nearer (sight tremendous and abhorred!) that it was the *wretch* whom I had created.[19] [my italics]

At first Dr Frankenstein identifies this living thing with its form and consistence in so much as to call him "the spectre", "the object" and "the shape". As the narration continues, the creature is also called "monster", "demon", "fiend", "evil", "wretch": these terms carry (in different ways) a strong connotative meaning referring to inner peculiarities and moral judgments linked to human and non-human beings. If it is true that names are discursive practices involving the process of naming, by using these common names Victor does not only deny his "living thing" a proper name, but he patently dispossesses his creature of a human identity while denying him human paternity[20]. This stigmatizes Frankenstein's unnatural, monstrous and guilty act: neither a parent nor a novelist would ever behave in this way. He makes it clear that the "spectre" lives at the threshold of language: it is commonly acknowledged that as soon as an object is named, it is immediately integrated into the cultural-linguistic system. But Frankenstein's monster is so "different" and so "discontinuous" that even his creator is terrified and unable to position his identity within a species or a social net. From the moment it exists, Shel-

---

[17] It might be useful to underline that a cyborg or, more generally, a posthuman body is rarely configured as a "disordered body". On the contrary, it is usually represented artistically as an attractive "thing", or at least fascinating because endowed with extraordinary powers.

[18] M. Shelley, *Frankenstein, or the Modern Prometheus* (Harmondsworth: Penguin Books, 1992), 76.

[19] Shelley, *Frankenstein*, 99.

[20] In this context see the "activation" of DAVI as a "virtual son" in Steven Spielberg's *AI*. The "mother" pronounces a list of words that function as a naming ceremony on the character's consciousness, his family and society. In so doing, the artificial intelligence recognizes the woman as its/his mother.

ley's character is a *filius nullius*, nobody's son, an illegitimate son, because he cannot be part of any family: he is without tradition or normative precedent[21]. This condition denies him not only relationship with other people but also subjectivity as a culturally-mediated entity: pragmatically, he sees a phenomenology denied along with the experience of being procreated.

From a linguistic point of view, a unique individual such as this is self-referent in the same way as the poetic function of language. If the poetic word lives in form more than in content and its peculiarity is to de-familiarize what is normal, common, obvious and known, then what role can the nameless ogre with a powerful physique play? The Monster makes the world look "different" and this should lead to awareness, knowledge, culture, light. Rather than a product of scientific progress, the monster should be seen as a tool for cultural progress. Dr Frankenstein maintains:

> Life and death appeared to me ideal bounds, which I should first break through, and pour a torrent of light into our dark world: a new species would bless me as its creator and source [...][22]

Contrary to this, Mary Shelley's creature apparently causes the world to fall into darkness.

Certainly, he does not belong to any order or standard and therefore cannot be classified. The word "monster" itself embodies all those proper names which cannot be pronounced because they contain the fears, tensions, anxieties and mysteries connected to a referent which is still in the process of linguistic identification: the lack of a name is implicit in the monstrous. Plato claimed that name is a phantasmagoria, a chaotic set of things, concepts, elements, thoughts and images mixed up in the mind of the speaker/observer: to quote or explore it means delving into what Freud has taught us to recognize as the relationship between *Heimlich* and *Unheimlich*[23]. The monstrous is disturbing because it does not mask its strangeness; on the contrary, (negative) physical diversity (nature, shape) is so clearly visible to the observer that it instantly ostracizes the creature from all human communities even if he is endowed with a soul glowing "with love and humanity".[24] As a

---

21 *The Universal Declaration of Human Rights* states: Article 6. "Everyone has the right to recognition everywhere as a person before the law."

22 Shelley, *Frankenstein*, 54.

23 This point has been widely studied and there are many interpretations of the novel as a metaphor for psychological, social and political terror. Almost all adaptations of the novel to the cinema deal with this topic.

24 Shelley, *Frankenstein*, 101. We see here an undeniable echo of the Noble Savage myth, misunderstood or corrupted by society.

consequence, he is denied all social interaction (even the Doctor flees from his own moral and legal responsibilities).

If relationships flourish through proximity, since all communities live on the pattern of a neighbourhood, the monster is (dis)placed "far" from all social systems[25], and is unable to initiate responsible reciprocal interaction with anyone. Frankenstein's monster reveals what codified social relations, determined by custom and law, usually keep under control. Controlling means recognizing the boundaries of the Other and, within this space, of defining the categories of good-and-evil, rights-and-duties. In the novel, the people around the creature fear what does not belong to their *interpellative* milieu, that which cannot be named since it is unknown or which springs from their deep unconscious and is thus beyond control. The acceptance of the "other" depends on our capacity to stay in control of it.

The fact that the monster is unnameable generates a fracture which forcibly destroys serenity, harmony and all certainties so that each individual eye-witness of his presence can create his/her own particular, horrifying narration of the creature. Quite clearly, the problem here is not the monster it/himself but the incapacity or failure of the observer to recognize the "mask" of this different living entity. Paradoxically, he is too similar and too contiguous to man to be simply frightening: he is overwhelmingly threatening because he conceals infinite potential mystery and unlimited over-determination. Before he actually becomes dangerous, it is the very presence of Frankenstein's ogre that contradicts all rules of harmony and coherence in society: if we are all naturally and genetically the same in view of the fact that we all have the same biological origin, the monster manifests it/himself as intangible evil simply by the fact of its/his not "belonging to" the species. If it is possible for us to accept cultural differences, it is extremely difficult to accept that which undermines our vital process: even Jesus Christ was born of a woman even if he was the Son of God.

The monster embodies the derridian "presence of the absence": he is the presence of a non-existing origin, that is, he is the supplement *sous rature* in society. But he defies the aporia since he cannot be linguistically codified as he does not belong to any cultural system. The question is: how does the monster *interpel* (negotiate) the consciousness of those who view him? As a phenomenon, not as a living being, he is first of all an aesthetic construction (a meaningless one): his body appears as a simulacrum of human identity, al-

---

[25] See B. Vecchi, ed., *Zigmunt Bauman, Intervista sull'identità*, 16. In contemporary society, the principle *cuius regio, eius natio* is less and less significant, since identity loses its social and territorial roots.

though he reveals only what is implied (or hidden) in the linguistic interstices of the concept of "humanity". On the one hand, he reacts to stimuli, makes decisions, controls his behaviour (if only at the beginning), perceives and registers personal experiences; on the other, he is "different from" any natural model or any other literary character wearing a specific identitarian mask. When I state that "he is not a man", I am implying that he is "anything but" a man: nonetheless, the hermeneutical boundaries he crosses are only those connected to the idea of "human being", although it is not clear if he is a real human being or simply an anthropoid. Symbolically, his uniqueness lets the "generality" of human nature emerge. The aesthetics of the Monster opens up a transition towards a new kind of awareness of the human standard.

But the question is: how can we accept a standard for mankind while any single individual struggles to see his/her peculiarities accepted? The replicants in *Blade Runner* have been produced to live within society, they are endowed with a name, a story, a beautiful and powerful body and a keen intelligence, but their social identity (personhood) is not enough to guarantee safe cohabitation. While Shelley's monster is ready to undertake human responsibilities[26] towards his neighbours since *others exist* for him (though, conversely, not even his creator recognizes any responsibility towards him), the replicants try to take advantage of the humans in the name of their own "identity". This fundamental opposition lies at the core of all literary works on the subject, and it leads to a double, totally different, conception of posthuman entities even if, as simulacra, they all provide an – extremely accurate – mirror image to human beings.

As Zygmund Bauman suggests, a generic identity is a battle cry used in defensive wars, it is a fight against dissolution and fragmentation, against the fear of being devoured or the intention of devouring. In *Frankenstein*, we have a monster who demands to be imagined by the community, to be part of an "us"[27] and against this we have a community trying to defend itself using identity as a weapon (the very same principle which gives leverage to nationalisms, the holocaust and modern dictatorships). At first the monster tries to learn how to behave so to create his/its own social mask through a process of homologation towards a "solid identity", that is through imitation

---

[26] See K. Tester, *Il pensiero di Zygmunt Bauman* (Trento: Erickson, 2005), 182–4. Consider also the monster's words: "I swear to you, by the earth which I inhabit, and by you that made me, that with a companion you bestow I will quit the neighbourhood of man, and dwell, as it may chance, in the most savage of places. My evil passions will have fled, for I shall meet with sympathy! My life will flow quite away, and in my dying moments I shall not curse my maker.", M. Shelley, 148.

[27] Tester, *Il pensiero di Zygmunt Bauman*, 74.

(he turns into a mimic hybrid). He searches for a lasting identification with the "us" around him, while it is, tragically, that very "us/we" which prevents him from learning a code of communication. With this the monster sees his possibility of being part of the community violently denied along with his chance of telling everyone his story. He is, therefore, obliged to live half-way between silence and the sound of his brute yell[28]. He almost pleads for Victor's attention:

> Let your compassion be moved, and do not disdain me. Listen to my tale: when you have heard that, abandon or commiserate me, as you shall judge that I deserve. *But hear me. The guilty are allowed, by human laws, bloody as they are, to speak in their defence before they are condemned.* Listen to me, Frankenstein.[29] [my italics]

With these words the monster calls on the law to assert his presence and his right to life, just the same as that of any human being. He asks to tell his story and to have the right to be judged in the same way as any "common" guilty-"man" would. He tries to establish one single standard of identification between himself and the community based on moral principles, that is he tries to root himself within the community. Nevertheless, as soon as he realises that he cannot succeed, he asserts his body and individuality as his "legal boundaries": *he* decides what is good and evil, *he* legislates and *he* applies the law, even though we see this only as violent revenge. It is extremely interesting to note that in the above passage he is asking Frankenstein to follow a procedure, which is exactly what the doctor failed to do when he broke the law of nature. Victor's manipulation of the body condemns his living creature to solitude[30]. It would appear that the absence of a name turns him into a mysterious exoticism, while ugliness turns him into an exotic weirdness – anything but a person.

A century later, Mary Shelley redeems the monster when she eventually agrees to create a companion for him: I am referring here to Shelley Jackson's

---

[28] We could underline how the Golem, usually described as an artificial creature in human form, but devoid of human soul, speech, and reproductive capacities, was animated through manipulations of Hebrew letters called *zerufei ha-otiyot,* often including the Tetragrammaton, the four-letter name of God. The communion between the creature and its creator is manifested through its passive, mute, and obedient behaviour.

[29] Shelley, *Frankenstein*, 101–2. Consider also *The Universal Declaration of the Human Rights*, Article 10: "Everyone is entitled in full equality to a fair and public hearing by an independent and impartial tribunal, in the determination of his rights and obligations and of any criminal charge against him."

[30] Referring to a monster, we cannot talk about 'exile', because it/s/he does not usually have a father/motherland to leave.

Patchwork Girl. *PG, or a Modern Monster* is currently one of the most popular literary hypertexts, probably the only digital literary work to have influenced our reading of a traditional linear novel. From the title page of the hypertext we learn that the authors are "Mary Shelley, Shelley Jackson & Herself", where Herself means the patchwork girl herself who, at a certain point, decides to leave her mother and create her own life abroad. So, she leaves home – England – for voluntary exile[31] in the United States. More than a manipulation of the body, PG reminds us of an act of creation, an act of writing. The author says:

> I had made her, writing deep into the night by candlelight, until the tiny black letters blurred into stitches and I began to feel that I was sewing a great quilt, as the old women in town do night after night, looking dolefully out their windows from time to time toward the light in my own window and imagining my sins while their thighs tremble under the heavy body of the quilt heaped across their laps, and their strokes grow quicker than machinery and tight enough to score deep creases in the cloth. I have looked with reciprocal coolness their way, not wondering what stories joined the fragments in their workbaskets. [WRITTEN]

Moreover:

> My birth takes place more than once. In the plea of a bygone monster; from a muddy hole by corpse-light; under the needle, and under the pen. Or it took place not at all. But if I hope to tell a good story, I must leapfrog out of the muddle of my several births to the day I parted for the last time with the author of my being, and set out to write my own destiny.
> [BIRTH in M/S]

This hybrid and monstrous creature soon develops a wish, i.e. to tell her own story, because PG knows that she can only meet reality through linguistic constructions of her own "self". In this she is aligned to a postmodern sensitivity which conceives the subject as a linguistic and cultural construct, but there is more to it than that. PG tells us how we can give her life:

> I am buried here. You can resurrect me, but only piecemeal. If you want to see the whole, you will have to sew me together yourself. [GRAVEYARD]

From a narrative point of view, she exists only when we stitch together all her identities, that is all those single human masks which in the past belonged to all the women and men who are now parts of her body. Each mask (text,

---

31 Note that Victor's monster does exactly the opposite. Furthermore, one of the commonest names given to Frankenstein's monster is "wretch". In Old High German *(w)recch(e)o* meant exile, adventurer, knight errant (German *recke warrior, hero*), in Old English miserable being, despicable person. See C.T. Onions, *The Oxford Dictionary of English Etymology* (Oxford: Oxford University Press, 1996).

lexia) has a voice, and tells its own personal story, taking precedence over the others from time to time, until in the end they all succeed in achieving their complete original self-sufficiency. This is one of the main differences between Frankenstein's and Jackson's monsters: the former wants to be a whole while it remains a juxtaposition of parts mimicking nature (a natural form). The latter is an identity in progress, as is inter/intra-textuality. The former is a forced nomad seeking for a stable territory, the latter is a living, impermanent nomad. However, like the monster, PG has no one name, she has many: Margaret, Susannah, Flora, Geneva, Walter, Judith, Angela, Roderick, Bella, Thomasina, Mistress Anne, Helen, Tristessa, Eleanor. PG is clearly conscious of her anonymous multiplicity:

> (... Every part of me is human and proportional to the whole. Yet I am a monster – because I am multiple, and because I am mixed, mestizo, mongrel.); [WHY HIDEOUS?]

> I am made up of a multiplicity of anonymous particles, and have no absolute boundaries. I am a swarm. "Scraps? Did you call me Scraps? Is that my name?" [SELF SWARM]

If in Mary Shelley's novel, the void of the signifier gave light to an infinite universe of would-be stories told by those who had been unlucky enough to see the creature, here all proper names are first and foremost human referents which condemn the character (PG) to a communicative entropy: she is so pregnant with information that she cannot signify. PG is monstrous because she contains excessive information in perpetual transition. Her body is an amplified medium of her Self, which lives its androgynous[32] nature beyond any concept of sex and gender, nationality and citizenship, race and class. PG is a finite set of narrations sewn together partly by Mary Shelley and partly by the reader who is asked to weave together both the body and the hypertext.[33]

Her physical sutures, highlighted in Jackson's drawings, become the metonym of her constant over-crossing, the proof of her (deleuzian) nomadic

---

[32] Androgyny was one of the main features of all archaic deities. In the ancient past there were many divine couples with the same name: Aphrodite-Aphrodito, Libera-Liber, Fauno-Fauna, Freya-Freyer. See L. Bearné, *Le vergini arcaiche, ovvero di come le antiche Donne custodissero la Libertà, l'Ebbrezza e la Gioia* (Milano: Edizioni della Terra di Mezzo, 2006), 80.

[33] We already know that identity emerges from relationships, including relationships people have with books. This hypertext bonds us not only to *Frankenstein* but also to all those works implicitly or explicitly quoted by the author, such as: *Patchwork Girl of Oz* by F. Baum, *Grammatology* by J. Derrida, *Mille Plateaux* by G. Deleuze e F. Guattari, etc. I personally think that Mary Shelley's novel is today ideally connected to all the monsters who came after hers.

transition. PG is *certain* of being a human being and proud to live in the stitches of her body, unlike her forerunner who suffered because his stitches made him look ugly (and implicitly evil). It was due to his stitches that Frankenstein's creature was seen by the community more in the light of an object or simple shape than a person. On the contrary, PG calls her sutures her skeleton [DISPERSED], that is to say a net which prevents her from collapsing, like the lexia of a hypertext which need to be filled by links:

> I am like you in most ways. My introductory paragraph comes at the beginning and I have a good head on my shoulders. I have muscle, fat and a skeleton that keeps me from collapsing into suet. But my real skeleton is made of scars: a web that traverses me in three-dimensions. What holds me together is what marks my dispersal. I am most myself in the gaps between my parts, though if they sailed away in all directions in a grisly regatta there would be nothing left here in my place. [DISPERSED]

So to PG sutures [CUT] do not represent fractures but connections. This creature is a social being who lives in interaction: unlike her mate who suffered solitude, she is an entire "neighbourhood" in herself. To be more precise, PG is a structure of possible structures, and she herself is a scarred synthesis. The author appears to have overcome the problems of nihilism as a consequence of an absent origin, thanks to a re-formatted and re-assembled fragmentation supported by hypertextual sutures. Shelley Jackson's creature is a typical posthuman subject, as stated by Katharine Hayles:

> (1) The posthuman view privileges informational pattern over material instantiation, so that embodiment in a biological substrate is seen as an accident of history rather than an inevitability of life. (2) It considers consciousness, regarded as the seat of human identity in the Western tradition long before Descartes thought he was a mind thinking, as an epiphenomenon, an evolutionary upstart that tries to claim it is the whole show when in actuality it is only a minor sideshow. (3) It thinks of the body as the original prosthesis we all learn to manipulate, so that extending or replacing the body with other prostheses becomes a continuation of a process that began before we were born. (4) Most importantly, by these and other means the posthuman view configures human being so that it can be seamlessly articulated with intelligent machines. In this view there are no essential differences between bodily existence and computer simulation, cybernetic mechanism and biological organism, robot teleology and human goals.[34]

Information implicit in her proper names is embodied in the girl's flesh and encoded in her mask: those men and women are DNA and persons, that is to say they are codes of genetic information and codes of behaviour. Franken-

---

[34] N. K. Hayles, "The Posthuman Body: Inscription and Incorporation in *Galatea 2.2* and *Snow Crash*", *Configurations* 5.2 (1997), 241–266.

stein's monster struggles to learn what is already embodied in PG. While a posthuman subject can be considered as a biological, chemical, psychological, information system, as a person PG corresponds to a constantly renewed set of semiotic exchange. The protagonist is a monster not because she is multiple, but because she lives in a multiple impermanence, that is she lives in transition and in the continuous evolution of her single multiple personas. She is a whirlwind of voices, of shouted words even though we perceive her as a unity, as one single mask, a complex of aporias. She is, of course, as coherent as a patchwork; whereas Mary Shelley's creature was an example of incoherence, Jackson's is more like the attempt to achieve a harmonious quilt, in which the single parts go together well and are in proportion. Frankenstein's monster perceived it/himself as a human being in a living hybrid body who was obliged by circumstances to behave contrary to its propensities[35]; by contrast PG does not distinguish her own Self (at least initially) from those of the men and women who make up her body. Moreover, her physical structure is driven by the peculiarities and emotional characteristics of their original owners. For example:

> My foot belonged to Bronwyn, who had extraordinary balance. Modest Bronwyn never said a word on her own behalf, but kept what she had; when pushed, gentle Bronwyn never budged. She outlived her immediate family, held the house and grounds for herself against a torrent of creditors, suitors, and poor relations.
> My foot drags a little in walking, but excels in standing still.
> [FOOT]

---

[35] A comparison with Hanif Kureishi's *The Body* might be fruitful here, since in this short-novel the transplanted brain and consciousness of the sixty-year-old Ralph lives in the body of a young man. In neurological terms, we realize that Ralph's and the monster's bodies and consciousness do not harmonize with each other, since the abstract ideals from their brains almost never correspond to the particular things they happen to meet. See H. Kureishi, *The Body* (London: Faber and Faber, 2002).

With reference to the brain, it is interesting to note how James Whales' film *Frankenstein* (1931), produced by Universal Studios, starts with the faulty transplant of an "abnormal brain" which immediately turns the creature into a criminal. All psychological and social motivations implicit in the novel are here eliminated in favour of a biological justification. The consequences of this *hamartia* are curiously reconsidered in *Ghost of Frankenstein* (1942) directed by Erle C. Kenton where Ludwig, the scientist, says "What will he think when he resumes life in that body? Will he thank us for giving him a new lease on life? Or will he object to finding his ego living in that human junk heap?" Moreover the same focus on the brain is evident in *The Curse of Frankenstein* (1957) by Terence Fisher, and, ironically, in *Frankenstein Junior* (1974) by Mel Brooks.

My hands are a cabal. A twitching finger, and I suspect my hands of thievery (Dominique, ambidextrous pickpocket, had already lost her right hand to punitive justice but later extracted a silk purse from the judge with her left.) The callous on the middle finger of my right bespeaks scholarship (a renowned essayist had an unknown collaborator: Livia, his wife, who wrote his books as well as dusted them.) One of my fingers is comfortable enough with a needle, another seems easier on the handle of a knife.

[HANDS]

The postmodern glorification of ambivalence and the doubts it casts on solid identities lead one to maintain that the only space for relationship lies in accepting the Other as a *fluid face* not as a *mask*: the Other cannot be "different from" me because the concept of Myself is not definitive. The monster is a face without a mask, and his immediate neighbourhood is simply not prepared to accept this face (that is to say his anomalous truth) unsupported as it is by any ideological or cultural paradigm. Perhaps, it is his lack of a proper name which engenders his lack of homologation within a social and cultural role.

There is a further difference between the novel and the hypertext: whereas the monster's namelessness underscores the impossibility of his reaching a sense of belonging (which generates his desperation), PG embodies the tradition and normative precedent of those men and women who once belonged to a community and who now want to reaffirm their inscribed membership (they were natural persons with juridical responsibilities). In a lexia called I AM the protagonist claims: "I am never settled. I belong nowhere. This is not bizarre for my sex, however, nor is it uncomfortable for us, to whom belonging has generally meant, belonging TO." Rather than a monster or a mimic man, PG is a true hybrid, an in-between identity since we cannot precisely say what her sex is, nor can we say what or who she is: she is a fluid subject, who decides to accept all her possible Others. Before being accepted as a patchwork, she recognizes herself as such a patchwork. Unlike a puzzle, which tries to conceal its multiplicity below a flat figurative surface, a patchwork reveals itself as a harmonious compound of different pieces. Mary Shelley created a subjectivity which feels uncomfortable with its difference, Shelley Jackson creates

> a new kind of self which doesn't fetishize so much, grounding itself in the dearly-loved signs and stuff of personhood, but has poise and a sense of humor, changes directions easily, sheds parts and assimilates new ones. Desire rather than identity is its compositional principle. Instead of this morbid obsession with the fixed, fixable, everyone composing their tombstone over and over. Is it that we want to live up to the dignity of our dead bodies?[36]

---

[36] S. Jackson, "Stitch Bitch: the Patchwork Girl", §"Everything at Once". Transcript of Jackson's presentation at the *Transformations of the Book Conference* held at MIT on October 24–25, 1998.

Yet living in the nextness of her multiple identities, she realizes that her consciousness needs a body as well as a memory. We must remember how, according to Roland Barthes, names are the linguistic forms for reminiscence: memory works as a means to localize the subject in a cultural context. At the same time the condition for memory is the acknowledgment of a story to tell, a name to evoke and, perhaps, a sense of belonging. In *Blade Runner*, Tyrell explains that by gifting replicants with a past, they were creating a pillow for their emotions so they could be more efficiently controlled. It consists in implanting human memories usually associated with family and function as a replacement for actual historical interpellative experience. It means that replicants are gifted with one single virtual story, that is with one distinct identitarian mask. Paradoxically, compared with Frankenstein's monster, these creatures appear less complex and more pathetic since the spectator perceives an inescapable sense of artificiality. The anonymous monster becomes a mass product with a proper name. This fact throws the creation vs. production opposition into high relief. Despite their perfect appearance, most of the replicants are more similar to Azimov's robots or an artificial intelligence. This is very different from the condition of Jackson's character.

One day PG decides to buy an identity and a past [BUYING PAST]. Hence she asks an umpteenth real woman to sell her her story and from that moment on PG is Elsie[37], which is to say a fictional repertoire, a net of social relations, a role, a name, that is, a narration; this latter is, however, historical (that is rooted in cultural and temporal circumstances), unique and personal. This strikes the reader's imagination mainly by virtue of the hybrid's determination to break all standard practices. If a name is a right recognized by law, this act undermines both the relationship between authenticity and fiction, and the concept of identity connected to the "possession" of one's own identity. PG buys her name – Elsie – legally, and from that moment she does not only have a name but she is the "owner" of it. In common law, she could have given herself a name but, in the light of the common law *habeas corpus* principle, she chooses to buy the name because she demands public recognition[38], namely civic duties and rights: she is looking for a relationship me-

---

37 It might be just a vague possibility to associate the name "Elsie" with the idea of "else", "someone else", "somewhere else", "whatever else", etc. in order to create a suggestion as to what "other", "another", "different", "additional", etc. refer to. With this in mind, the name not only refers to a character, but it also describes and enhances/carries forward the narrative potential of the character.

38 Once again in literature a proper name defines a subjectivity more than a generic subject. In Tim Burton's *The Nightmare Before Christmas*, Jack takes possession of Santa Claus's name for a short time: in doing so, he wears a different mask for a

diated by law between herself as a citizen (body, consciousness and memory) and the State. Frankenstein's monster and Patchwork Girl both want to be considered as juridical persons and, to a certain extent, they (in their 'paradoxical' nature) succeed since the former is generally recognized as a criminal and a wretch, while the latter is able to establish a legal transaction with a woman. This implies that they can lawfully interact with other people, and that they are treated (however involuntarily) by the community as "legal entities" incorporated within the "legal system" even if they are not always perceived as sharing the same "natural identity". They are consequently "cultural outsiders" while remaining "juridical insiders" just as long as their "difference" needs to be controlled. Since they communicate mainly through their bodies, it is clear that the bodies themselves are bearers of rights for both monsters, as is implicit in the *Habeas Corpus ad Subiciendum*. Ironically, it is their unnatural physical structure which renders them juridical subjects as without their soma they would never exist.

What happens to the original Elsie? Generally speaking, from a posthuman point of view, she has disembodied her information. I believe that this is possible only in legal terms: Elsie erases neither the information encoded in her DNA since she does not erase her physical body nor in her non-biological self, that is her mnemonic conscience, she simply divests herself of the information archived in a registry office. She becomes another living paradox: beyond the confines of common sense, she is a natural mature human body without historical identity. This means that she is not allowed to use her mask any longer, and so the question is: is she still a person? We can presume that, just like Frankenstein's creature, she is allowed to live a responsible, reciprocal relationship with her neighbours based more on common sense and emotions[39] than on civil rights and duties. But, unlike the monster, she is accepted by other people because of her natural appearance. This makes it clear that personhood is more than the bio-cultural entity theorized by Donna Haraway[40]. At the same time it is more than a role or an identity: a

---

while. "That identities can be stolen, traded, suspended, and even erased through the name reveals the profound political power located in the capacity to name; it illustrates the property-like potential in names to transact social value; and it brings into view the powerful connection between name and self-identity.", B. Bodenhorn and G. Vom Bruck, eds., *An Anthropology of Names and Naming* (Cambridge: Cambridge University Press, 2006), 2.

[39] We will discover that Elsie does not leave PG, but does stay with her till her final dissolution.

[40] See P. Ferri, *La rivoluzione digitale. Comunità, individuo e testo nell'era di Internet* (Milano: Mimesis, 1999),142–156.

person is a bio-cultural-legal *complex*. PG (like the replicants) becomes a passive archive of stories so as to guarantee, in line with the postmodern perspective on life, the character's capacity to develop[41].

The possession of property rights and a legal contract does not correspond to our idea of identity since a social mask is a puzzle whose end pattern remains hidden. Biological, cultural and social (family) legacy give us the pieces and the main rules needed to build one or more semantic orders. Whereas the absence of a name impeded both action and the possibility of the character's evolving, PG overcomes the problem with a deed of sale. For PG the possession of a name means acquiring an identity which is single and free from her patchwork multiplicity. In this way she becomes more and more similar to a human being or a modern subjectivity, but at the same time her narrative potentiality decreases. It is as if through the authentication of her own name, PG has asked the State to have her identity recognized over the autonomy of her multiple voices. As a consequence, her body reacts to uniformity and starts to dismember itself, so that the single pieces take the upper hand over the whole.[42] All her single fragments of soma appeal to their specific writ of *Habeas Corpus* to ratify their own individual liberty at the expense of PG's "arbitrary exercise" of power. Thus, the protagonist chooses to reject the notion of fluid identity, typical of our digital society. A fluid identity perceives cohesion as a burden, a constraint to freedom and to all the opportunities of change. PG is the attempt to make a woman from a patchwork, but she dissolves in the tub, liquefying like any other contemporary identity.

I MADE MYSELF OVER/ ELSIE TRIUMPHANT
My parts bobbed in changing patterns in a warm reddish slurry of bathwater and blood. Elsie was immersed in me, surrounded by fragments, but somehow she held me. I was gathered together loosely in her attention in a way that was interesting to me, for I was all in pieces, yet not apart. I felt permitted. I began to invent something new: a way to hang together without pretending I was whole. Something between higgledy-piggledy and the eternal sphere.
I became supple. My furniture parts became mellow as wax and the joints and junctures, long turned to proper purposes, bent past their right angles into impossible obliquities, or found curves not known to their before-uses. I was think-

---

[41] Any act of naming is an act of possession, even if the family name is the only relationship that a person has with his/her ancestors. See Ragussi, *Acts of Naming*, 7.

[42] In the movie *The Nightmare Before Christmas*, it is less important for Jack to know the names of the children than the fact of his being Santa Claus. The children do not wonder what his name is, because they are too surprised to meet a completely different Santa Claus from the stereotyped appearance.

ing of Mary and speaking in tongues of places and names, Bronwyn and Roderick, Judith, Susannah, Flora and Bella and Anne, Eleanor, Tristessa, Geneva, Thomasina, Agatha, Constance, Jennifer, Jane. All disassembled, I made myself over, forgetting not to remember.

I was many things before I became something like human again, and all the while Elsie was magnificent, like a woman in a fairy tale, holding her true love tight, though she turn badger or wildcat or asp.
[I MADE MYSELF OVER]

In the tub, PG dissolves but she is practically "held" both by Elsie, who lies down in the tub with her, and symbolically by her legal name. Surprised by her body's dismemberment, she starts speaking in "tongues of places and names" that is with the single voices of her different pieces. Not even her proper name seems to warrant her consistency and unity, as Elsie gets angry and says:

NAME
"You'll have to find another name, because I'm taking mine back," said Elsie when she left.
[NAME]

We can only make assumptions as to Elsie's moral responsibilities and arbitrary expropriation of the main character's "property"[43], but, after the legal relinquishment of her name, this cynical and final abandonment works on our imagination as a denial of the right to life for PG: she is not a person, she has no name with which to assert her identity, she has no individual legal rights, therefore she is worthless for humankind. For the reader, this lack of value is not reason enough to justify Elsie's behaviour, since we have developed feelings towards the main character. Instead of exploding into myriad pieces, PG implodes thus gaining an absolute literary dimension which only exists beyond time and space. Neither the puzzle metaphor (modern identity), nor the patchwork one (postmodern identity) and not even man's genetic submission to nature can delete her infinite complexity.

Summing up, the comparison of the two literary works reveals two different ways of conceiving human subjectivity. The earlier monster wants to be accepted for "what he is" in his quest for belonging to the neighbourhood. He conceives identity as "sameness", as an entity codified by morality, society, and religion. The later monster lives in inter/intra-textual relationships; it/she is an essence perpetually re-codified in interacting patches. Her identity lives in the flouting of boundaries and endless transmutation of

---

[43] *The Universal Declaration of the Human Rights*, Article 17 states: "(1) Everyone has the right to own property alone as well as in association with others. (2) No one shall be arbitrarily deprived of his property".

meaning, as she has been created as an act of writing and keeps on living as an act of reading. Her identity is pure pervasive energy, the mask of which can never materialize over her face because, in the meanwhile, that face has changed. Its essence cannot be manifested in attributes but only in a stream of energy or meaning. Both monsters, as "spaces beyond", highlight the fact that if, on the one hand, only culture can help us to accept diversity, on the other, a manipulated body is neither different nor the same (aporia) as the Other, because it eludes classification of any kind, whether of identity, or name.

Paul Cheung
University of Sydney

A Serious Reading of Biotechnology in Japanese Graphic Novels: Weak Thoughts Regarding Ethics, Literature and Medicine

> A novel examines not reality but existence. And existence is not what has occurred, existence is the realm of human possibilities, everything that man can become, everything he's capable of. Novelists draw up *the map of existence* by discovering this or that human possibility.[1]

In the collection of texts published as *The Art of the Novel,* novelist Milan Kundera reflects on his ideas of the novel, particularly those concerning its identity and its construction. Stressing the novel's origins in the spiritual and cultural entity called "Europe", Kundera refers frequently to a range of novelists such as Miguel de Cervantes and Herman Broch on the one hand, and philosophers such as Edmund Husserl and Martin Heidegger on the other. With passion he argues in favour of "the wisdom of the novel"[2] in the present age, where he believes the populace remain passive towards broadly disseminated but a narrow range of conceptions of human existence. Kundera prefaces his volume by stating that his ideas and writings are merely those of a novelist, not a philosopher. This essay revisits the vexatious question of the relationship between literature and philosophy, drawing its illustrations primarily from the unlikely source of Japanese graphic novels. Known as *manga* in Japan, and featuring an enormous array of biotechnology, these graphic novels raise important questions of the basis of literary and philosophical value. The essay will conclude with a consideration of the search for such value in the fields of ethics and medicine.

The battle for ascendancy between philosophy and other endeavours which Kundera alludes to has had a long history, notes classicist and philosopher Alexander Nehamas.[3] In Book X of *The Republic*, for example, Plato has Socrates say:

---

[1] M. Kundera, *The Art of the Novel* [1968], trans. Linda Asher (London: Faber and Faber, 2005), 43.

[2] Kundera, *The Art of the Novel*, 158.

[3] A. Nehamas, "Serious Watching", *South Atlantic Quarterly* 89.1 (1980), 157–80.

> [...] we must remain firm in our conviction that hymns to the gods and praises of famous men are the only poetry which ought to be admitted into our State. For if you go beyond this and allow the honeyed muse to enter, either in epic or lyric verse, not law and the reason of mankind, which by common consent have ever been deemed best, but pleasure and pain will be the rulers in our State.[4]

Plato's Socrates showed a clear preference for "law and reason" over "epic and lyric verse". So strong was the preference that poets were categorically excluded from the perfect state in *The Republic*. Nehamas points out that a similar orientation of wholesale dismissal exists among some critics of contemporary popular culture in the United States. As poetry had long since moved to the higher echelons of the hierarchy of literary and philosophical value, modern critics target instead one of the most popular forms of today's entertainment, namely, television. Nehamas mounts a convincing case for the considerable philosophical and literary sophistication of one form of television, that of serialised drama. Addressing the concerns of the critics over issues such as the alleged lack of depth allowed for character development and the singularity of perspective, Nehamas paints a picture of the dramatic television series which is strikingly similar to that of Kundera's European novel. As the novelist draws up a map of human existence by discovering possibilities of what it is to be human, so do the producers of television drama discussed by Nehamas. However, such a map may not be legible or even visible to the viewer who categorically rules out any value. Value is discernible to those who make the effort, argues Nehamas, to engage the wide-ranging nuances of being human in television drama. The possibility of "seriously watching" television drama, suggests Nehamas, renders the ascendancy of canonical tradition rather problematic, and possibly even less important.

It is not only poetry or television that has been found wanting in philosophical value. Literature is often liable to serious charges of deficiency in philosophy. For example, in the dialogue between continental philosophers Jürgen Habermas and Jacques Derrida, the former accused the latter of reducing philosophy to literature.[5] Following on from Nehamas' attempt at "serious watching", I would like to examine two related claims of ascendancy by a "serious reading" of Japanese graphic novels or *manga*. The first claim is that literary value resides primarily in literature, not in philos-

---

[4] Plato, "The Republic" in *The Collected Dialogues of Plato* [1961], eds. E. Hamilton and H. Cairns (Princeton: Princeton University Press, 1985).

[5] J. Derrida, "Is There a Philosophical Language?" in *The Derrida-Habermas Reader*, ed. Lasse Thomassen (Chicago: University of Chicago Press, 2006), 35–45.

ophy, and further, not in the products of mass media. Interlocked with this is the second claim that philosophical value resides primarily in philosophy, not literature. When combined, these two claims would accord *manga* merely the status of popular entertainment, not unlike poetry in the words of Plato's Socrates. As such, *manga* serves as an extreme case for the understanding of the manner by which Kundera's "human possibilities" might be mapped.

Sometimes considered equivalent to American comics and Italian *fumetti*, *manga* is in many respects emblematic of post-1945 Japanese popular consumer culture.[6] It remains one of the most widespread forms of entertainment for children and adults alike. Outside of Japan, *manga* has been mainly marketed to a narrower segment of the population, namely that of school-aged children and young adults. Presented and enjoyed in greater quantities than a single strip or page, *manga* narratives are typically packaged into palm-sized volumes of framed images and words that may extend up to one or two hundreds pages. Popular *manga* invariably become animated either into televised series or feature films, known as *anime*. Adaptation from or into other media is practised extensively and creatively, forming the basis of the sizable symbolic economy that is Japanese popular culture. In this economy, the adaptation of elements from graphic novel into animated features and merchandising is only of many routine practices. Furthermore, many narratives found in graphic novels have originated from written novels from within Japan or beyond. Redrawing boundary relations is one of the most dominant characteristics of *manga*.

Post-war Japanese *manga* has seen more than the redrawing of boundaries between *genres* and *media*. Very few existing boundaries have remained unchallenged. A pioneer of this subversive entertainment is Osamu Tezuka (1928–1989), who remains an iconic figure in Japanese society. Although not the originator of *manga*, the name Tezuka is almost synonymous with *manga*.[7] The era of steadily rising demand among readers and the resulting increase in the scale of serialisation early in the post-war period provided Tezuka ample opportunities to conduct bold experiments, which included not only the

---

6   For a well-illustrated art-historical introduction to *manga*, see B. Koyama-Richard, *One Thousand Years of Manga*, trans. David Radzinowicz (Paris: Flammarion, 2007).
7   A thoughtful collection of essays on Tezuka can be found in The National Museum of Modern Art – Tokyo. *Tezuka Osamu*, edited by The National Museum of Modern Art – Tokyo, Yoshikazu Iwasaki, Kunio Motoe, Yukio Kondo, Atsushi Tanaka, Takeshi Mizutani and Koji Takahashi, Asahi Shimbun, Tokyo, 1990.

incorporation of film-narrative techniques into his *manga* but the creation of a robot protagonist of reversible and neuter gender in *Metropolis* (1949), inspired by a photo of Maria in Fritz Lang's 1927 film of the same title. Instead of Maria, Tezuka invented Mitchy, who is transformed between the genders and neuter status by means of a switch located in the throat. Mitchy is followed shortly by another robot protagonist, robot *Astro Boy* (*Tetsuwan atomu*, 1951–1968), who was intended to be a neuter character modelled on the female gender. Somewhat like Mitchy in the other narrative, Astro Boy is a human creation who wonders more than occasionally about its human identity. Through these early serialisations, Tezuka had developed a flair for the cautious assessment of science and technology. The focus on biotechnology in his narratives as opposed to other areas of rapid scientific advances in the exploration of questions of existence is particularly noteworthy. Considering that science-fiction had popular appeal among Tezuka's readers, the other areas of science could well have been used to create stories resulting in record *manga* rental figures, begging the question of the focus on biotechnology.

As if to redefine his burgeoning reputation as the author of futuristic science fiction *manga*, one of Tezuka's subsequent offerings is that of *Black Jack*, serialized between 1973 and 1983. In the first chapter of *Black Jack*, *Is there a doctor?*, the protagonist surgeon by the same name performs a facial transplantation, a technology that arguably did not exist in 1973 and did not come into being until the first recorded procedure in France in 2005. The surgical procedures in *Black Jack* are a mixture of the realistic, possible and improbable. From one perspective, surgery in *Black Jack* and robotics in *Astro Boy* are little more than the "thrills and spills" which form the core of entertainment. If entertainment is the only function of deploying biotechnology in these narratives, then nuances of meaning would be unnecessary, even distracting. However, *Black Jack*, perhaps more so than *Astro Boy*, is replete with such nuances. Consider for example Black Jack's practising of surgery without a license, the exorbitant fees he extracts, at least from wealthy patients, his outdated attire and his scarred, two-toned face. The surgery performed on the patient in the first story was without consent and indeed involved deception. And yet, the fee charged, the deception and the subsequent escape had one purpose, that of saving the vulnerable in the face of certain death callously caused by the powerful. In a style typical of the Black Jack character, and perhaps of Tezuka, the act of saving life through medicine was not entirely deliberate and certainly not bound by anything resembling the Hippocratic Oath of doing no harm. In the space of a mere 23 pages of manga-style frames, Tezuka has managed to raise some profound questions of the nature of the demand for medicine, shattering the mirror reflecting the image of the benevolent doctor while also bringing the axiomatic right to treatment under the spotlight. These disconcerting possibilities of the human are highlighted long before "bioethics" became a distinct entity in written discourse. Kundera says in an interview that:

> The novel discovered the unconscious before Freud, the class struggle before Marx, it practised phenomenology [...] before the phenomenologists. What superb 'phenomenological descriptions' in Proust, who never even knew a phenomenologist![8]

When read "seriously" then, it becomes clear that biotechnology in Tezuka's *manga*, and in the works of many other Japanese graphic novelists, often serve the purpose of exploration as well as popular entertainment. The mapping of previously uncharted segments of existence would qualify *manga* for the designation of "the novel" despite vast differences in form and content compared to their written, "European" counterparts. To the extent that "the novel" possesses literary value, such value can also be found outside the traditional boundaries of literature, even in the unexpected location of mass media.

Tezuka – the prolific cartographer of human existence – did not himself transcend the effects of shifting or contested boundaries. In an interview given in 1979,[9] he claimed that he had originally envisioned *Black Jack* to be an increasingly sinister character over the course of the series. He described as unfortunate readers' interpretation of the first few chapters as the beginnings of a tale of a medical rebel with a cause. Although later chapters were supposedly created to accommodate this interpretation, they are not less nuanced than earlier chapters in the ways described above. The value of saving lives remains a point of contention throughout the series. The readers of *Black Jack* could accept a range of possibilities, but not that of an ethically equivocal protagonist. Apart from the morality of characters, other contested boundaries included visual style and affective tone. The 1960s saw a surge in the supply of and demand for seriousness in *manga*. Tezuka's drawings were judged by some to be lacking in realism, while the frequent use of humour was interpreted as the sign of childish entertainment. In this line of thought, serious subject matters ought to be coupled with a serious way of storytelling. Characterised by a more realistic visual style and bereft of comic relief, the genre *geikiga* emerged partly out of dissatisfaction with main-

---

[8] Kundera, *The Art of the Novel*, 32.
[9] A. Tanaka, "Black Jack", in *Tezuka Osamu*, 214–25.

stream, *suttorii manga* (or story *manga*) typified by Tezuka's Astro Boy.[10] Tezuka's response is another act of creative subversion of the distinction between *geikiga* and story manga. He created *Eulogy to Kirihito* (1970–1971) in which the protagonist, a junior medical doctor, falls victim to the intrigue of his supervisor and ends up developing a disfiguring and fatal disease affecting patients in a remote area. The disfigured face has a dog-like appearance, as do the limbs. Dr. Kirihito eventually accepts his disfigurement, succeeds in preventing those affected from dying before bringing the perpetrators to justice. In terms of visual style and affective tone, *Eulogy to Kirihito* is clearly an example of *geikiga*. The genre boundary had been redrawn, however, with the fantasy of animalizing disease and the unrealistic plot this specific device supports.

Appearing as it did after *Eulogy to Kirihito*, the return to the style and tone of story *manga* in *Black Jack* could be seen as a defiant attempt by Tezuka to decouple the seriousness of appearance from the seriousness of substance. This was partly stymied by readers who expected Black Jack to be another vigilante healer after the mould of Dr. Kirihito. For the rest of his career Tezuka continued to give voice to questions of human existence in the polyphony of *geikiga* and story manga. The villainous protagonist finally appears in MW (1976–1978), named after the biological and chemical weapon forming the backbone of the narrative. The questions of existence expected of Tezuka can of course be found in MW, but here they are posed in relation to the doers of evil in a relentless fashion throughout the series.

In Tezuka's oeuvre, the influence of the readership upon style and characterisation is discernible. In the commercial environment in which *manga* was read, was this influence ever reciprocated? Did any aspect of the production of *manga* change the way *manga* was read, seriously or otherwise? A number of writers, including philosopher and literary critic Azuma Hiroki, have claimed that a new generation, with access not only to *manga* and *anime,* but also the computer and the internet, display less interest in the narrative elements of *manga* and *geikiga*, the staple of their forebears.[11] Drawing on Derrida's exposition of deconstruction, Azuma paints a picture of the progressive "animalisation" of obsessive Japanese fans of popular culture against the background of the demise of meta-narratives from the mid-1990s

---

[10] P. Brophy, "Tezuka's Gekiga: Behind the Mask of Manga" in *Tezuka: The Marvel of Manga*, ed. Philip Brophy (Melbourne: National Gallery of Victoria, 2006), 123–33.

[11] H. Azuma, "The Animalization of Otaku Culture" in *Mechademia*, ed. Frenchy Lunning (Minneapolis: University of Minnesota Press, 2007),175–87.

and the rise of the internet. These fans, derogatively called *otaku* in Japan, have become fixated, not on narrative *per se*, but on the corporeal aspects of character design. Azuma illustrates this with reference to Dejiko, a so-called "character"[12] which first appeared in popular culture magazine in 1998. The name "Deijiko" is a play on the syllables "deiji", taken from "digital" and fused with "ko", from a common compound feminine name in Japan. Drawn to child-like but exaggerated proportions; with large sparkling eyes, cat-like ears, and bells, Deijiko embodies the aesthetic of different generations of *manga* dating back to Tezuka. Its resulting appeal to popular culture fans led to its adoption as the corporate brand of Gamers, a Toyko department store specializing *manga*, *anime*, related soundtracks, computer games and merchandising. In its figurine form, Deijiko is not a stable identity but a series of possibilities, with accessories and body parts[13] that consumers mix and match at will. Instead of her own personality, Deijiko is at any given time a particular permutation of corporeal features. This consumer-driven "assemblage" is also exploited in so-called "novel games" in which character design interacts with story development. In a reversal of the usual course of cross-media adaptation, Deijiko became a character in the sense of being part of a narrative shortly afterwards in the televised *anime* series *Digi Charat* (1999–2000), and a series *manga* by the same title (2000). Azuma claims that *otaku* are led to find sustenance in the minutiae originating in data quite independent from narrative elements. This supposed contrast between *otaku* and Japanese consumers of popular culture, including those of earlier generations, raises the possibility of the subversion of the literary affordance in a piece of work by the intentions divergent from authorial design. The mere availability of literary value in *manga* does not guarantee its serious reading.

The location of value exclusively within a poem, novel, manga or a treatise but not the reader ultimately leads to the same problem of underestimating the dynamic, interactive nature of value.[14] The brief search for literary value in *manga* above demonstrates that distinctions, frequently of the dichotomous kind, breakdown with a predictability that calls for a thoroughgoing reassessment of common approaches to the question of value. The second claim to ascendancy, implicit in Habermas' criticism of Derrida, and in the positioning of philosophy above literature, has been an issue within philo-

---

12 Azuma, "The Animalization of Otaku Culture", 182.
13 The available accessories or body parts recall similar designs in *manga* and *anime* narratives in preceding decades. As such, Deijiko was primarily an intertext or source of intertextual references when considered analytically.
14 P. Bartoloni, "The Value of Suspending Values", *Neohelicon* 34. 1 (2007), 115–22.

sophical circles. Contributors to the broader debate over philosophy's identity, form and scope are numerous, as are the positions put forward. A group of positions associated with philosopher Gianni Vattimo is especially pertinent to the examination of this claim. Under the rubric "weak thought" (*pensiero debole*), Vattimo developed an approach to metaphysics likened to the reduction of the ultimate to the indefinite, and the absolute to the relative.[15] Metaphysical thought has produced strong, self-evident fundamentals often manifesting themselves as dichotomies: reality versus fiction, repetition versus novelty, the sacred versus the profane, and the centre versus the periphery. Rather than promising a cure for the ills of metaphysics, an enticing but self-contradictory option, Vattimo advocates the unambiguous acknowledgement of the continuing influence exerted by metaphysics. For Vattimo, it is the accepting of the tendency to gravitate towards the transcendent that precariously opens the way to eventually overcoming this tendency. Concerns over the return to the reliance on metaphysics in attempts to overcome it prompted a monograph by Vattimo entitled *Beyond Interpretation: The Meaning of Hermeneutics for Philosophy*.[16] The weakening of ontology into hermeneutics written about by Heidegger, of whom Vattimo is an exponent, is not in itself valuable or sufficient. By not acknowledging the historical and social conditions in which weakening has been taking place, hermeneutics became, ironically, the means to transcendent truth for some. It may be argued that it is such a science of interpretation which pursues the distinction of philosophy from other endeavours such as literature and, in turn, literature from popular media. Overcoming metaphysics involves the difficult process of weakening these distinctions, not with a foreseeable final destination, but with a view of discovering more possibilities of human existence, as Kundera states in the *Art of the Novel*.

In *Beyond Interpretation*, Vattimo discusses in some detail ethics as another trend in philosophy which also runs the risk of falling into the trap of metaphysics. In here and a subsequent article,[17] Vattimo outlines his preference for an ethics that is characterised by dialogue, including dialogue with tradition, by which he means the Judeo-Christian tradition. Based on negotiation and consensus, this is an ethics not so much of the transcendental "Other", but of the "other", or more precisely the plural "others". Such a relational

[15] G.Vattimo, *The End of Modernity: Nihilism and Hermeneutics in Post-Modern Culture.* [1985], trans. Jon R. Snyder (Cambridge: Polity Press, 1988).

[16] G.Vattimo, *Beyond Interpretation: The Meaning of Hermeneutics for Philosophy* [1994], trans. David Webb (Cambridge: Polity Press, 1997).

[17] G. Vattimo, "Ethics without Transcendence?", *Common Knowledge* 9.3 (2003), 399–405.

ethics weakens in that it involves the "endless attenuation of the self".[18] The emphasis on dialogue that weak thought brings to ethics is valuable in considering the relationship between philosophy, literature, ethics and medicine. The relationship between these endeavours has been a concern in medical education and to a lesser extent, continuing professional development. This is seen for example in calls for the study of literature to be incorporated into the training of medical personnel not only during but also following the qualification process. Rita Charon and colleagues, addressing this process in the United States, identify a broad range of uses of literature for the purposes of training medical students and medical doctors since the 1970s.[19] They include the teaching of the illness experience to medical personnel; examination of the role and impact of medicine in society; deepening of the understanding of patients and doctors through their own narratives; the practice of ethical reasoning, and the critique of medicine itself. For Charon et al., literature encompasses the texts of "the traditional literary canon" and "works of contemporary and culturally diverse writers, novels, short stories, poetry, and drama".[20] To these they also add personal accounts of illness and of professional practice, as well as various methods of literary analysis. One of Charon's co-authors, Anne Hudson Jones, had earlier argued for the use of literature in the ethical education of medical students above analytic philosophy.[21] Jones identifies the value of literature in ethical education with the former's function of "providing a context that demonstrates how much more than rationality is involved in human belief and moral behaviour".[22] This approach to the relationship between ethics, literature and medicine is characterised by the emphasis on literature's utility. It advocates the serious reading of a broad range of texts, including serious ones, for their philosophical value.

An alternative approach has been put forward by Neil Pickering,[23] drawing upon the ideas of author Iris Murdoch. Accordingly, the reading of literature is akin to the encountering of the other at the expense of the persistent focus on the self. Rather than the intentional search for philosophical or literary value, this alternative focuses on the reader's imaginative engagement

---

[18] Vattimo, "Ethics without Transcendence?", 405

[19] R. Charon et al., "Literature and Medicine: Contributions to Clinical Practice", *Annals of Internal Medicine* 122. 8 (1995), 599–606.

[20] Charon et al., "Literature and Medicine", 599.

[21] A. Hudson Jones, "Literary Value: The Lesson of Medical Ethics", *Neohelicon* 14. 2 (1987), 383–92.

[22] Hudson Jones, "Literary Value", 391.

[23] N. Pickering, "Imaginary Restrictions", *Journal of Medical Ethics* 24 (1998), 171–75.

with a piece of work and the unpredictable ethical developments that may or may not result. The emphasis here is therefore not on ensuring growth in one's own ethical reasoning capacity through the investment of seriousness in the reading of literature.

A third approach is found in a series of three novels by British author Hazel McHaffie published in 2005: *Vacant Possession*,[24] *Paternity*[25] and *Double Trouble*.[26] Conceived of as medical ethics novels and carrying the intention of "bringing ethics to life", these works represent intentionality in the composition of ethical issues in the novel. In *Paternity*, for example, McHaffie addresses the relatively recent issues raised by the use of assisted reproductive technology and genetic testing through the story of her characters Declan and Judy. Devastated by a failed attempt at assisted reproduction, this couple is led by a subsequent chain of events to the shocking discovery of closely-linked genetic lineages which may have contributed to the death of their first child. These developments compound the strains already present in their relationship. Eventually, further tests reveal that their first child died due to a cause unrelated to their genetic heritage. Intricately constructed, McHaffie's novels are clearly examples of novelised bioethics as opposed to novels in which bioethical issues may be discerned irrespective of authorial intentions, such as Kazuo Ishiguro's *Never Let Me Go* (2005).[27]

A fourth approach can be identified with the high-circulation medical thrillers produced by Robin Cook, the US author of *Blindsight* (1992)[28] and *Chromosome 6* (1997) and many other volumes.[29] Written primarily to evoke suspense and provide entertainment, Cook's novels offer detailed, complex portrayals of biotechnology in use. For instance, in *Chromosome 6*, recent advances in molecular biology and immunology are projected forwards, fictionalised and applied to organ transplantation and forensic pathology. While there is no overt concern with ethical issues, some of the biotechnol-

[24] H. McHaffie, *Vacant Possession: A Story of Proxy Decision Making*, Living Literature Series, (Oxford: Radcliffe, 2005).

[25] H. McHaffie, *Paternity: A Story of Assisted Conception*, Living Literature Series (Oxford: Radcliffe, 2005).

[26] H. McHaffie, *Double Trouble: A Story of Assisted Conception*, Living Literature Series (Oxford: Radcliffe, 2005).

[27] K. Ishiguro, *Never Let Me Go* (New York: Alfred A. Knopf, 2005). Ishiguro has commented that despite founding the plot on reproductive human cloning and organ transplantation, the novel is primarily not concerned with the bioethics of advanced medical technology. He acknowledges that readers might nonetheless arrive at such an interpretation.

[28] R. Cook, *Blindsight* (London: Macmillan, 1992).

[29] R. Cook, *Chromosome 6* (London: G.P. Putnam's Sons, 1997).

ogy in Cook's novels predates what was available at the time of publication and also any ethical controversy that arose at their introduction.

A number of observations regarding these approaches in a spirit of dialogue consistent with weak thought might be made at this point of the essay. In medical education and professional development, a sharp focus on reading serious works of fiction could be accompanied by the cultivation of relevant reading practices. How this could be achieved in specific education systems and curricula remains a challenge considering the intensity and longevity of medical training. And as pointed out by Charon et al., the intended impact of an instrumental approach to literature in medicine would benefit from being assessed.[30] It would not be unreasonable, following Kundera, to hope that this wider approach to medical training would encourage serious engagement with philosophy not found in the form of narrative. The same observation can be made in relation to the imaginative engagement with literature discussed by Pickering.[31] A similarly creative engagement with explicitly philosophical thought is possible and desirable. In terms of the lesser emphasis on intentionality and the self, this approach is conceptually very similar to Vattimo's ethics. The advent of the intentionally bioethical genre, exemplified by McHaffie's novels, is relatively recent. Interest in this genre has separately been shown in Japan.[32] Given their ethical specificity, novels of this genre may be a particularly appropriate complement to those of other genres among motivated readers. Not written for any ostensibly serious purpose, Cook's best-selling medical thrillers present a challenge not found in the other three approaches. Perhaps it is relevant to note here that the readers of these novels include those who will at some stage interact with medical personnel as patients, and even join their ranks some day. Again by applying weak thought, Nehamas' reminder of Plato's Socrates is worth bearing in mind. To the extent that medicine is not only a matter of its practitioners but also that of patients and their relations, reading on a broader scale, well beyond the discourses of the healer and the healed, would also be desirable. As educators operating in medical facilities have little control over what future trainees will have read by the commencement of training or the impact of the reading material, there is a larger issue of how and why literature is read in the earlier years of life. By extension, this is also the issue of

---

[30] Charon et al., "Literature and Medcine".
[31] Pickering, "Imaginary Restrictions".
[32] N. Kaneyasu, "The Ethics of 'Brightness of Life' (4): The Possibility of The 'Abortion Novel' and The 'Bioethics Novel'", *Bulletin of Seisenjogakuin Junior College* 24 (2005): 39–50.

the media and genres that accompany young people in their development. In this respect, writers, philosophers and artists have much to offer. The increased availability of their offerings across national and geographical boundaries raises a possibility not extensively discussed by Vattimo. In maintaining dialogue with tradition in the overcoming of metaphysics, it will also be necessary to include not only the Judeo-Christian tradition, but other traditions as well. In doing so, it may well be discovered that Kundera's European novel, with its spirit of complexity, its yearnings for the possibilities of the human, have their provenance also in many locations far removed from the European continent. As shown with reference to Japanese graphic novels, such a location includes Japan, but surely others and those not defined rigidly in geographical terms.

Laura Apostoli
University of Verona

Fulfilling Personhood at the Margins of Life: Anna Quindlen's *One True Thing*

"Personhood is a talisman that confers status, respect, and moral worth".[1] It is widely assumed to be an essential concept for grounding moral and legal intuitions about personal rights and values. Far from being obvious, it is rather a Protean and fluid notion, lacking determinate boundaries. Distinct components may in fact be included in the definition of what constitutes a person, depending on the perspective we assume. They vary from the plain biological idea of the human being to the philosophical conception of a rational and moral agent. Psychological connotations are also put at stake, including the ability to exhibit self-consciousness, autonomy and self-determination. In this latter sense what is reflected is the self-image of the individual in contemporary culture, in which rationality, individuality and freedom are central values.[2]

The current scenario displayed by increasingly complex technological progress in biomedicine has further problematized the notional development of the concept of personhood and our understanding of it.[3] The new realm of artificial interventions on human life has undoubtedly created many possibilities for enhancing individuals' wellbeing. Nonetheless, it has drastically altered representations of ourselves at the same time, questioning the fundamental traits of our human condition.[4] Radical advances in technology and health care practice have in fact led to the existence of different "degrees

---

[1] J. D. Ohlin, "Is the Concept of the Person Necessary for Human Rights?", *Columbia Law Review* 105 (2004): 209–249, 211.

[2] R. H. J. ter Meulen, "Towards a Social Concept of the Person" in *Personhood and Health Care*, eds. D. C. Thomasma, D. N. Weisstub, C. Hervé (Dordrecht: Kluwer Academic Publishers, 2001): 129–135, 130.

[3] Using the term "biomedicine" I follow the account given by Beauchamp and Childress, who suggest that we employ it as a "shorthand expression for many dimensions of modern biological science, medicine, and health care". T. L. Beauchamp, J. F. Childress, *Principles of Biomedical Ethics* [1979] 3rded. (New York: Oxford University Press, 1989), 10.

[4] L. Séve, *Pour une critique de la raison bioétique* (Paris: Odile Jacob, 1994).

of life", namely prenatal, terminal and marginal.[5] In the light of such crucial changes in society, a modern re-evaluation of pivotal terms such as "person" and "human" has become inevitable. Besides, our cultural recognition of the boundaries defining life and death has become problematic as well. According to Lars Reuter, who clearly outlines the point at issue,

> [t]he variety of ways in which human life can be reproduced, genetically altered, and medically sustained, prolonged and terminated, have instigated an extensive debate on the very nature of human existence.[6]

As a primary outcome, the borders distinguishing ethical and legal arguments have blurred quite radically. Indeed, both disciplines have come to closely and explicitly converge in biomedicine, eventually leading to the birth of biolaw.[7] In spite of this, the majority of international legislations is still

---

[5] I am referring to: embryos and fetuses (prenatal); comatose, brain-damaged and terminally-ill patients (terminal); clones, cyborgs, chimeras and artificial intelligences (marginal) – as they no longer pertain to the sole realm of science fiction.

[6] L. Reuter, "Human is What is Born of a Human: Personhood, Rationality, and the European Convention", *Journal of Medicine and Philosophy* 25.2 (2000): 181–194, 182.

[7] The coin of this term is relatively recent. It defines a field of knowledge obviously closely connected to bioethics, as they both address social tension and conflict caused by emerging bio-scientific research and application. As an interdisciplinary resource, biolaw correlates several disciplines with the law, primarily biology, medicine and medical ethics, philosophy, humanistic and social studies. Its main purpose is the achievement of a firm legislative framework able to regulate controversial and complex ethical issues rising from contemporary biomedical advances, and to enforce patients' rights. Bio-legal concern covers the different levels of application on which recent developments in medicine and biology are considered. It aims at protecting the individuals from their improper use, namely against instrumentalization in both therapeutic and research contexts. Further readings on biolaw: D. Beyleveld, R. Brownsword, *Human Dignity in Bioethics and Biolaw* (Oxford: Oxford University Press, 2001). F. D'Agostino, *Bioetica nella prospettiva della filosofia del diritto* (Torino: G. Giappichelli Editore, 1996). J. L. Dolgin, L. L Sheperd, *Bioethics and the Law* (New York: Aspen Publishers, 2005). P. Kemp et al. eds., *Basic Ethical Principles in European Bioethics and Biolaw, Vol. I, Autonomy, Dignity, Integrity and Vulnerability* (Barcelona: Centre for Ethics and Law, Copenhagen and Institut Borja de Bioética, 2000). J. Menikoff, *Law and Bioethics. An Introduction* (Washington DC: Georgetown University Press, 2001). S. C. Poland, "Bioethics, Biolaw and the Western Legal Heritage", *Kennedy Institute of Ethics Journal* 15.2 (2005): 211–218. J. D. Rendtorff, "Basic Ethical Principles in European Bioethics and Biolaw: Autonomy, Dignity, Integrity and Vulnerability – Towards a Foundation of Bioethics and Biolaw", *Medicine, Health Care and Philosophy* 5 (2002): 235–244. W. Van Der Burg, "Bioethics and Law: a Developmental Perspective", *Bioethics* 11.2 (1997): 91–114.

missing a regulatory framework capable of managing these new levels of life. Hence, it can be assumed that ethical and moral quandaries will increasingly test the flexibility of the law in the next decades. What is called into question is firstly the legal recognition of "who counts as a person", in order to determine the attribution of the full bundle of rights and values consistent with such notion. Are embryos and fetuses *already* persons? Are comatose and terminally-ill patients *still* persons? And, in the near future, in the matter of clones, cyborgs, chimeras and artificial intelligences: *what* are they? The attainment of a bio-legal guardianship able to deal with these new forms of "liminal beings" is thus felt as an urgent need, so as to prevent abuses and unfair discrimination. The fact that legal theory is loath to cope with such conflictual ethical matters should cause no wonder. What is at stake is in fact no less than the regulation of our proper humanity projected in the future, which will inevitably affect the structuring of society as a whole. As maintained by Francis Fukuyama in his recent and well-known work, *Our Posthuman Future*,

> [t]he most significant threat posed by contemporary biotechnology is the possibility that it will alter human nature and thereby move us into a "posthuman" stage of history.[8]

Therefore, exploring the issue of personhood intrinsically relates to one of the most essential questions lying at the core of contemporary thought, as argued by the philosopher Lucien Sève, that is: what do we want humanity to be like?[9] This vital dilemma is left open more than ever in current technological times and in aching need of in-depth reflections in the light of different perspectives. Literature undoubtedly offers precious contributions to widen the discussion, providing insights to unexplored scenarios and contexts, so as to plumb the depths of such contentious ethical and legal quandaries. As a matter of fact, this discipline has always dealt with the investigation of basic themes defining human experience, looking for a holistic elaboration of the peculiar features that constitute individuals' subjectivity. By all means it has always served as a fundamental stage upon which the bodily and moral character of our human nature have been strenuously debated.

In the humanistic perspective, being a person does not simply mean to be alive in a biological sense or to be able to fulfil certain cognitive actions, but to lead a meaningful human life, namely to be in a communicative relation-

---

[8] F. Fukuyama, *Our Posthuman Future* (New York: Picador, 2002), 7.
[9] Sève, *Pour une critique de la raison bioétique*, 26.

ship and exchange with others.[10] Hence, one of the most positive benefits that the intervention of literature in bioethical and bio-legal matters can reach is that of reconsidering the individual not only as such, but in close connection with his fellow humans, as a relational creature capable of inter-relating with other creatures in concrete acts. In other words, literature brings us to account human beings as full entities, as the harmonious sum of body and mind in a *worthy* life.

As a matter of fact, a wide spectrum of literary works dealing with thorny bioethical dilemmas promotes the necessity of an enlargement of the concept of personhood, including new rights, values and attributes, embracing all relevant functions of humanity. By the same token, a great part of the narratives that intrinsically refer to bio-juridical issues supports a rethinking of what constitutes a person in moral and legal terms, oriented towards the concepts of *quality* and *dignity.* Following this perspective, the fictional display of boundary but possible situations concerning genetic engineering, cloning, abortion or euthanasia encourages the necessary updating of some of the ancient enquiries rooted in the history of moral philosophy. The previously mentioned questions – such as "Who is a person?" or "How should a person be treated?" – although still fundamental in any ethical consideration, are in some circumstances no longer sufficient to face the moral and legal implications stemming from the most recent advances in biomedicine.

The fact that biotechnology has achieved the ability to sustain life well past the point where natural forces would have brought certain death in earlier times, gives rise to the questions: are we prolonging it unnecessarily? And for which and whose benefit? As a result, new troublesome dilemmas are put at stake in contemporary debates, thanks to the incisive intervention of humanistic disciplines exploring such contentious matters, literature in first stance. *When is life worth being lived and who should decide for it? Are we the masters of our own body, mind and destiny?* These are some amongst the current quandaries that should enhance bio-juridical discussions, shifting the attention more accurately towards the individual's peculiar life.

Tackling such questions becomes remarkably salient when we deal with literary works that focus on the controversial issue of euthanasia, and on voluntary euthanasia or physician-assisted suicide in particular.[11] That is, when

---

[10] D. Solies, "The Crisis of Personhood. Towards a Bioethical Theory of Compassion", paper presented during the Conference of Metanexus Institute in Madrid: *Subject, Self and Soul: Transdisciplinary Approaches to Personhood,* July 2008. Available at: http://www.metanexus.net/conference2008/articles/Default.aspx?id=10527.

[11] The word "euthanasia" derives from the Greek *eu,* meaning well or good, and *thanatôs,* meaning death. It covers a number of end of life scenarios. A preliminary

we face the grief of competent terminally-ill patients[12] strongly wanting to conclude their agony through medical "mercy killing". The fictional portrayal of such painful experiences of extreme sufferance can undoubtedly instil a certain feeling of discomfort and queasiness. Nonetheless, it can also lead up to a significant process of discovery and self-improvement. Reflections over persons forced to confront the advancing of death day after day, living with the persistent awareness of their finitude and vulnerability, can offer precious hints to a reconsideration of the concept of personhood from moral and legal viewpoints.

A passionate depiction of this sort of experience, where pain, self-discovery and determination indissolubly merge together is offered by the compelling novel *One True Thing*, written by the American journalist, essayist and novelist Anna Quindlen, first published in 1994.[13] Like the majority of Quindlen's narratives, this work focuses on the complex description of family relationships destabilized by a distressing crisis that affects the moral dimension of mutual duty and responsibilities. It frankly analyses the ambiguities and mysteries that make up individuality related to marriage, family

---

definition of euthanasia is that of being a deliberate act aimed at ending the life, or hastening the death of individuals who are ill or injured beyond hope of recovery, for that person's benefit. The taxonomy of euthanasia includes active or passive, and voluntary or non-voluntary. Active euthanasia occurs when the patient's death is brought about at his own request, providing him with lethal drugs or other means. Passive euthanasia concerns withdrawing life-sustaining treatments, letting nature take its course. Non-voluntary cases include individuals who are incapable of indicating a preference, for examples those who have descended into a persistent vegetative state. "Mercy-killing" or assisted-suicide is described as the intentional killing with benevolent motives of a consenting patient who requires help to perform the act (presumably with a doctor's direct intervention). The category which is most vigorously contested is active voluntary euthanasia, which includes cases where a life is one of agonising and unrelenting pain, and the patient is fully competent. Active euthanasia is criminalized in almost every Western democracy, considered as a form of homicide. It is legal in the Netherlands, Belgium, the American State of Oregon, Japan and Colombia (Switzerland should be added, Swiss law punishes only those who assist suicide for "selfish motives"). This is perhaps the strongest category of claimants for the "right to die".

12   Following the Walton Report's account, patients are "competent" if they are able to understand the available information about their conditions, to consider with medical advice the risks, benefits and burdens of different treatments or courses of action, and thus to make informed decisions. See *Report of the Select Committee on Medical Ethics* (London: HMSO, 1994).

13   A. Quindlen, *One True Thing* (London: Arrow Random House, 1996). [All the following quotations are taken from this edition. Further references and page numbers will be indicated in text, abbreviated as OTT.]

and fate. *One True Thing* gracefully deals with pivotal themes defining human experience, such as love and death, identity and self-realisation, growth and change. The narrator of the novel, Ellen Gulden, is an ambitious young woman, a rising, talented journalist for a glamorous magazine in New York. When her mother Kate is diagnosed with terminal cancer, the disease is already far advanced. Her father, George Gulden, an esteemed English professor, asks her to put her promising job on hold and return to the house in which she was raised to take care of her dying mother. Initially reluctant and resentful she eventually accepts, not moved by real compassion towards her mother but primarily to please her beloved father.

The story is told retrospectively and we already know from the prologue that Ellen will later be accused of the mercy-killing of her mother, suspected of having voluntarily murdered Kate with a lethal overdose of morphine. She will be arrested and in the end discharged for lack of evidence against her. Actually, in the thought-provoking conclusive part of *One True Thing* we discover that no mercy-killing has taken place. Ellen is innocent, as she always protested. Rather, we become aware of the fact that Kate has committed suicide, overdosing herself with the morphine she has secretly hidden over time. Since voluntary euthanasia is forbidden by law, she has bitterly opted for an ultimate plan of self-annulment and self-destruction, overwhelmed by pain and by the fear of the loss of dignity in her dying. Besides, deciding to conceal her distressing intentions to her caregivers, she has implicitly forced herself to deceive her own family.

The two main characters weaved in the fabric of the narrative, Ellen and Kate, have to confront an unsettling drama, even if from different positions. Illness and impending death have entered their lives, upsetting the coherence and integrity of their existence. Disease in fact amounts to "dis-integration": it represents a rupture not just of the body, but of the unity of the person in his wholeness. According to the classic definition maintained by Eric Cassel,

> suffering is a specific state of severe distress induced by the loss of integrity, intactness, cohesiveness or wholeness of the person or by a threat that the person believes will result in the dissolution of his integrity.[14]

Such affliction leads both Ellen and Kate to a significant change involving the deep structure of their personality and their will. Quindlen's novel therefore offers the opportunity to explore the ramifications of the issue of personhood in end-of-life contexts focusing primarily on a moral perspective. Moreover, the portrayal of Kate's disturbing final act gives the clue to an in-

---

[14] E. Cassell, "Pain and Suffering" in *Encyclopedia of Bioethics* [vol. IV], ed. W. T. Reich (New York: Macmillan, 1995), 1899.

vestigation of personhood from a legal point of view. Her plight raises in fact poignant dilemmas concerning the ability of any person to act as a truly autonomous agent within contemporary society.

The definition of what is meant by fundamental principles such as dignity, autonomy and self-determination becomes ambiguous when they are weaved into grievous medical boundary situations. It is no longer clear who holds the effective power to decide *how* and *when* life is worth living: is it a prerogative of the physician, a duty of the jurist or a right of the patient himself? Do persons living experiences of great despair have the obligation to live in pain? If their real and informed wish as legal persons is to stop suffering through life-terminating choices, should their will be honoured? And, finally, is it possible and desirable for an alleged "right to die" to come into existence in a future bio-juridical legislation? These vexed questions are extensively compelled by the reading of *One True Thing*. They precisely reflect the serious disagreement existing amongst physicians, bioethicists, lawyers and philosophers over questions related to end-of-life decision-making within the Western bioethical and bio-juridical framework.

Illness as Disclosure: Exploring Moral Personhood

Medicine is constituted and justified as an art of eradicating human suffering. Nonetheless, in many cases it has to accept its own defeat, having nothing left to offer except a prolongation of the dying process. In the past we had good reason to fear a painful and uncomfortable aging and dying. Nowadays, with modern advances in the management of chronic illness and palliative medicine, we are more likely to be fearful of *living* in pain and discomfort, with the perceived degradation of physical decline and mental incompetence.

This is the case of Kate Gulden. She has undergone chemotherapy but her disease is too far progressed to positively respond. Physicians continue to provide her mere palliative treatments to gently attend her last journey towards the end, where dependence and decay inevitably follow. Unfortunately cancer is not easy to soothe, rather it deforms and impairs the body, caging the mind as well. Observing her mother, Ellen clearly perceives the burdens she is stoically hiding. She reflects:

> [...] her face was calm but empty; her smile was bleak, without light or warmth. And for the first time I thought of what it must be like to know that you were going to die, that the trees would bud, flower, leaf, dry, die, and you would not be there to see any of it. It was like standing too close to the fire. [OTT, 35]

Critical illness takes its travellers to the margins of human experience, undermining the relationship between the body, the self and the outer world. The
emotions of fear, frustration and loss are indissolubly involved with being ill.
Panic, uncertainty, denial and disorientation are the most common feelings
that compose the everyday life of a person continually confronted with
death. Kathy Charmaz talks about a "constant struggle" that people who are
suffering from an enduring disease have to fight for their own identity, ad-
justing themselves to their pain and loneliness.[15] A similar view is expressed
by David C. Thomasma in his account of how serious illness forces a confrontation with the impact of disability, pain and death on the hopes and aspirations we all collect during our lifetime. It challenges us with the possibility of a substantially altered image of ourselves or even of non-existence.[16]
"Pain is like the mist invading a house with its doors and windows open to
the elements", argues Guillermo D. Pintos. He further states that in its dee-
pest sense, it substitutes the radical nature of the self that it penetrates at will.
Like the Cartesian *cogito*, during illness the level of the "I think" becomes the
same as that of the "I suffer".[17] Arthur W. Frank passionately resumes:

> [p]ain is about incoherence and the disruption of relation with other people and
> things; it is about losing one sense of place. [...] Loneliness also enters, then
> doubts about who you are and what your life is worth, hope mixed with depres-
> sion, anger mixed with a desire for contact with others, dependency mixed with a
> need to continue to do things for yourself.[18]

Illness and pain therefore have many dimensions. They proclaim the presence of the body through its experienced disintegration. Corporeality comes
to be imposed as a prison that affects the person's identity, threats intimacy,
and alters self-understanding. At the same time, sufferance obliges to re-as-
sess personal values, assumptions and interpersonal interactions previously
taken for granted. In this sense "illness" greatly differs from the idea of "disease", where the latter plainly relates to an objective, physical malfunction.
On the contrary, illness involves a process of perception, affective response
and cognition. It can be viewed as the shaping of disease into behaviour and

---

[15] K. Charmaz, "Struggle for a Self: Identity Levels of the Chronically Ill" in *Research
in the Sociology of Health Care* [Vol. VI], eds. J. Roth, P. Conrad (Greenwich CT: JAI
Press, 1980).

[16] D. C. Thomasma, "Moral and Metaphysical Reflection on Multiple Personality
Disorder" in *Personhood and Health Care*, 221–241.

[17] G. D. Pintos, "The Medical Interpretation of Pain and the Concept of a Person" in
Personhood and Health Care, 364.

[18] A. W. Frank, *At the Will of the Body, Reflections on Illness* (New York: Mariner Books,
2002) 52.

experience.[19] The consuming impact of dying necessarily compels to an exploration of the deep meaning and purposes of one's existence. Such challenging terrain can provide the opportunity to start a process of reflection and self-discovery that can lead to both positive or negative outcomes. Illness can thus become the vehicle to start a journey from an unexamined life into the awareness of one's true potential. It opens up an entire new realm of experiences that impacts on individual self-awareness and self-knowledge, bringing about a more authentic, honest and frank understanding of ourselves.

By becoming the caregiver of her mother, Ellen becomes the caregiver of herself as well. The intricacies of illness involve in fact the caregiver's emotional and intellectual personality almost in the same way as they affect the ailing person's one. She undertakes an overwhelming journey made of revelations, losses and self-understanding. The sorrow and frailty of the prospect of death bring Ellen through a gritty rite-of-passage to her real maturity. It entails the reassessment of the values she has adhered to her entire life. Ellen is compelled to re-evaluate everything she thought she knew about her family, her past and most importantly, about her own self. She concisely admits in the opening passage:

> [i]t was Thursday, and *I was still my old self*, smug, self-involved, successful, and what in my circles passed for happy. [OTT, 14, my emphasis]

While growing up Ellen never considered herself close to her mother. The reader can perceive a tangible unspoken distance between them, coming to think that they even hardly know each other. Ellen carries a deep resentment for not having anything in common with Kate. The figure of the mother can in fact be understood as the *alter-ego* of Ellen. She represents the sensitive and uncontrived side that Ellen is unable to express. Kate is a gentle person who finds joy in establishing happy relationships, modest, tender and calm. However, the reader can perceive right from the beginning that her personality is far more complex and stratified than it seems on the surface. As the events unfold in fact we come to appreciate the hidden strengths she hides, the secret inner core she keeps, made of acuity, shrewdness and determination. Instead Ellen has always regarded her mother as a simple person, quite an uninteresting homemaker with small-town hobbies, devoted to domestic duties and community works. Kate is thus unable to stand comparison with Ellen's

---

[19] A. Kleinman, *Patients and Healers in the Context of Culture: An Exploration of the Borderland between Anthropology, Medicine and Psychiatry* (Berkley and Los Angeles: University of California Press, 1981).

worshipped and idolised father, a brilliant intellectual, impatient with medi-
ocrity, who nurtured her own writing aspirations. Ellen plainly states:

> [a]ll my life I had known one thing for sure about myself, and that was that my life
> would never be [my mother's] life. I had moved as far and as fast as I could; now I
> was back at my beginning. [OTT, 24].

However, the drama that has affected the family obliges Ellen to question
and wipe out stable assumptions concerning her surrounding reality. Kate
plays the role of a sort of catalyst to reach such new level of knowledge. It becomes apparent how fundamental the role of the mother is to the Gulden
family when she starts dying. The fabric that holds them all together slowly
begins to tear. Her husband is unable to cope with her condition and unfairly
unloads his responsibilities on his daughter, placing himself more and more
distant from the situation. On the other hand, Ellen bonds with her mother,
their intimacy and closeness constantly grow, leading her to a heightened
understanding of herself in the end. The first meaning that we may associate
to the title of the book – the one true thing – is thus the outcome of Ellen's
strenuous quest for self-awareness, in search of her authentic personhood.
In the epilogue we discover that years after the death of her mother she has
become a "professional" carer, a doctor, leaving the journalistic career to her
old cynical and judgemental self. As in a sort of purgatory, she has to pass
through the burdens of illness, death and jail to achieve her own re-birth,
that ironically and mournfully springs from her mother's death.

Following the development of the disease we may perceive in Kate a clear
distinction between the reactions of her body and those of her mind. Their
responses to annihilation work in inverse proportion. The more the devastating effects of the disease develop and physical pain becomes unbearable,
the more her inner strength, perseverance and self-awareness increase. Although her body no longer answers her will, her mind strongly refuses to accept such disrespectful numbness. As reflected by the moving lines by Dylan
Thomas "[d]o not go gentle into that good night [...] rage, rage against the
dying of the light", Kate reacts to darkness, decay and frustration.[20] Her resistance cannot be violent or aggressive, as it would not be consistent with
her personality, rather it is an intellectual one. Kate in fact proudly proposes
to her daughter to start a book-club together. They pass the time during
chemotherapy sessions reading classical literature and comparing their own
feelings and sensations. They do not follow academic lines or theoretical ap-

---

[20] D. Thomas, "Do not go gentle into that good night", in *Selected Poems 1934–1952*
(New York: New Directions, 2003), 122.

proaches to analyse the books, as George would have done, they instead read classics in the light of their shared felt-experience of pain and self-discovery. Precious masterpieces such as *Pride and Prejudice*, *David Copperfield*, *Middlemarch* and *Anna Karenina* become intellectual tools to challenge physical discomfort, letting themselves get lost in the narrative world. At the same time, dipping into the imaginative scenarios helps Ellen and Kate appreciate a more authentic and honest relationship between them, which paradoxically passes through metaphorical identification with the fictional characters.

If sufferance and pain reflect the "unmaking of their world",[21] the display of its chaotic and incoherent features, then literature represents its reconstruction, a ploy for meaning-creation between mother and daughter. Through the examples of great literary women, Kate educates her daughter to come to terms with her feelings about her own nature, emotions and relationships, counterbalancing the over-emphasised cerebral dimension of her life. Conversely, Ellen is compelled to re-establish her thoughts about her mother's intellectual acuteness and capability, having to recognise that she has always greatly undervalued her potential. In this perspective another meaning may be associated to the title of the book, as "the one true thing" may be interpreted as the discovery of the "one true Kate", through the gradual surfacing of her pure and most concealed personality: sharp, bright, proud and emotionally intense. Following Ellen's new perception of reality, Kate extraordinarily comes to resemble the character traits of Anna Karenina, the elegant and literate woman that surprises the reader for her passionate spirit and determination.

Nonetheless, as the course of the narrative progresses, the perception that the disease is prevailing over Kate and that pain is becoming burning, beating and intolerable is almost tangible. Quoting again Frank's mind, he tries to explain the sort of distress that illness causes referring metaphorically to the idea of *silence*. Pain is inexpressible, it renders the ill persons mute.[22] More than that, physical pain does not simply resist language but actively destroys it, bringing about an immediate reversion to a state anterior to language.[23] Such a grievous sort of isolation is fully mirrored in Kate's figure, since her silences actually increase as the story unfolds. Her dialogues with Ellen lose the brightness and acuity they had at the beginning, her language becomes more and more fragmentary and "medicalized". She is not able to participate in the

---

[21] E. Scarry, *The Body in Pain. The Making and Unmaking of the World* (New York: Oxford University Press, 1987).

[22] Frank, *At the Will of the Body*, 34.

[23] Scarry, *The Body in Pain*, 4.

book-club any longer. As the contents of her world disintegrate, so do the contents of her language. In the end the only words she has the force to utter are those for asking for help, or more morphine. The reader can learn of her inner struggle and understand the grief she is bearing only by relying on an emphatic interpretation of her reactions and behaviour. Kate's world, voice and essence are in fact slowly fading through the thick agony of her body.

As in a sort of modern gothic novel in *One True Thing* the main sources of fear, discomfort and anxiety come from the inside. Kate's own body smothers the most dreadful of woodworms, her own skin smells of chemicals. Her corporeal unity is shattered, as she has almost lost the capacity to eat, sleep, walk, read and talk. Her sense of belonging is undermined for the unavoidable impossibility to make future plans and accept responsibilities. Even her dreams are compromised, filled with nightmares and hallucinations due to the assuming of morphine. The home she has revered all life long has turned into the metaphorical transfiguration of her inner decline, the theatrical stage of her decay. The living room has been emptied out to make room for the hospital bed and the wheelchair. The air is stale, it no longer perfumes of wild flowers and cheesecake, rather it stinks of medicines and body fluids. The things once familiar, safe and reassuring, now mirror the new reality imbued with distress and queasiness. Such gloomy condition involves Ellen as well, as she is obliged to come to terms with the loss of the stable assumptions that had founded her own existence until then. She is in fact compelled to revise her idealised view of her father, having to recognise his insensitivity and selfishness and how these faults have affected her own personality. Certainties and stability are thus indissolubly lost in this strenuous path through existential anguish, towards new levels of self-consciousness.

Due to the withering effects of the disease, Kate finds herself entirely denuded before her vulnerability. She has lost the one first fundamental right consistent with the notion of personhood, namely that to the integrity of her physical and psychical characteristics. According to Peter Kemp, respect for integrity in the bio-juridical perspective accounts for the inviolability of the human being in its corporeal and spiritual dimension. It refers to the wholeness and coherence of the internal and external order of a person that should not be touched and destroyed.[24] In its dynamic sense, it is synonymous with the right to a holistic idea of health, referring to bodily, mental and existential intactness.[25] It expresses the untouchable core, the inner nucleus of life that

---

[24] Kemp, *Basic Ethical Principles in European Bioethics and Biolaw.*
[25] The term "integrity" comes from the Latin words *integer* and *integritas*, which refers to an intact, harmonious state in a person. The very core of its meaning probably

must not be subjected to external interventions. Kate's integrity is fully damaged and compromised. The relations between the physical, rational, moral and emotional elements of her self are by now fragmented.

Even if Kate has hardly learnt to tolerate her bodily deterioration and her decaying, slow passage towards death, she cannot bear to passively witness her mental decay, the loss of control and the ability to reason that the last stage of illness implies. She perceives such condition as too hard to contemplate, almost intolerable. Hence, she gradually achieves the resolution to stop her wearing journey by herself, through an extreme and calculated act, killing herself with an overdose of painkillers. One last meaning may thus be linked to the title of the book. The "one true thing" comes to indicate Kate's "one true choice", where *true* amounts to final and uncompromising. If illness, pain and suffering are so very true for the level of self-awareness they bring about and force us to deal with, then this is Kate's one true decision: reasoned and craved, although fundamentally irrational in its nature.

Kate's Ultimate Act: Exploring Legal Personhood

It is relatively uncontroversial to state that while death is an event, dying is a process that happens contemporaneously with living. The processes of living and dying are necessarily intertwined, and involve characteristics shared by all of us. Therefore perfectly healthy people and people with proximity to death all share the common status of being "*locus* of rights" in equal measure.[26]

Placed in the bio-juridical framework, the character of Kate Gulden represents a conscious and competent terminally-ill patient, bearer of the full bundle of rights associated with legal persons. In the modern legal system the role of the legal person is dominant and supreme as the basing building block of society. This subject is granted fundamental rights, such as those to life, health and liberty, followed by political, economical and civil ones.[27] Alongside this conception, personhood is generally assumed to have gained a central place in morality as well. According to Jean Delumeau this notion

---

coincides with one of the Platonic cardinal virtues, namely the harmony of the soul (dikaiosyne).

[26] J. Coggon, "Could the Right to Die with Dignity Represent a New Right to Die in English Law?", *Medical Law Review* 14 (2006): 219–237.

[27] These rights have been recognised in documents such as the *Magna Charta*, the *United States Constitution*, the *Canadian Charter of Rights and Freedom*, the *UN Declaration on Human Rights*, even if these are subject to interpretation by courts and cultural values.

has in fact come to signify the human being in terms of what is fundamental to it, and of what is singular and irreducible about each member of humanity.[28] The combining of these perspectives brings to mind the forensic dimension of the term "person" as depicted by John Locke, as identifying "intelligent agents, capable of a law, and happiness and misery".[29]

Following these accounts Kate has to be fully understood as a "person" in moral and legal terms. Once one is identified as a person, no right is held more sacred and carefully guarded by any legislation than that to the possession and control of his own individuality, free from all restraint or interference from others.[30] This doctrine has been extensively supported by one of the most influential jurists of the twentieth century, Benjamin N. Cardozo, who famously stated: "Every human being of adult years and sound mind has a right to determine what shall be done with his own body".[31] In this sense, the very notion of liberty comes to be inextricably intertwined with the idea of physical freedom. In the medical context this basic precept is reflected by the legal recognition of a person's right to self-determination, which includes the ability to exercise personal autonomy, to act as a sovereign individual and to exercise independent choices.[32]

Such crucial notions – *self-determination*, *autonomy* and *choice* – are, in turn, the pivotal factors blending into the concept of personhood in the contemporary bio-legal perspective. Indeed, these are the very rights that Kate Gulden feels she is not entitled to. The reader cannot directly become aware of Kate's thoughts, as the story is narrated by Ellen. However, during an unexpected moment of anger, Kate lets herself go to liberation, and plainly reveals her mind:

> [...] Rotten. That's what I look like now, like a peach when it's all rotten. Like bad fruit. Why can't I just die and be done with it? It's a *crime* for a human being to have to live like this, rotten like this. [OTT, 163, my emphasis]

What is worth pointing out here is the choice of the word "crime". Kate sees herself as the victim of an abuse, caused by the constriction of her freedom.

---

[28] J. Delumeau, "The Development of the Concept of Personhood" in *Personhood and Health Care*, 13.

[29] J. Locke, *An Essay Concerning Human Understanding* [1690] (London: Penguin, 1997), 211.

[30] Statement of the U.S. Supreme Court in *Union Pacific Railway Co. v. Botsford*, 141 U.S. 250, (1891), at 251.

[31] In a 1914 New York Court of Appeals decision, *Schloendorff v. Society of New York Hospital*, 211 N.Y. 125, 129–130, 105 N.E. 92, 93.

[32] H. Biggs, *Euthanasia. Death with Dignity and the Law* (Oxford and Portland, Oregon: Hart Publishing, 2001).

What she feels she has lost is the authority to determine the conditions under which she deems her life to be intolerable. She misses the opportunity to secure herself a peaceful death, a good death, as the etymological meaning of the word "euthanasia" suggests. In fact, even if Kate's belief is that her existence is significantly lacking in value, she is not allowed to be overtly helped in dying in a medical context, in a humane and effective way, assured that there will be no mistakes and no return from the brink of death. She is not legitimized to act following such autonomous exercise of deliberate and informed choice. The availability of this option is constrained by both social mores and the criminal law.

Kate perceives herself as defrauded of her ability to choose and to maintain a sense of control over her life. She therefore considers suicide her last and only chance to escape her unbearable suffering, and, paradoxically, a means of self-preservation. The only opportunity she has to fulfil her will and continue to be the master of her destiny is by secretly erasing her own existence. Her inner torment therefore concerns the loss of another fundamental right, that of self-determination, namely to the capability to make choices with her own free will, to determine her own fate, and to engage in goal-directed and self-regulated behaviour.

Self-determination involves persons' rights of pursuing their own lives' projects and preferences. It includes the capacity to determine and value the parameters of their own existence, and even to disvalue them. In the medical setting it deals with respecting persons' decisions concerning their medical care. Under some circumstances it includes patients' rights to remain ill or even to endanger their lives.[33] Legislations allow the competent patient who is being kept alive by some form of life-sustaining technique to order that treatment to be discontinued. It is a right about protecting oneself from bodily intrusion. However, such right does not extend to physicians' role, as they cannot be asked to assist a patient to die. The performance of this act is in fact clearly contrary to professional ethics and can lead to criminal prosecution.[34] As a result, the patient that does not need artificial means to survive obtains no benefit from the negative right to be free from unwanted care. In this way the law produces the unfair result that a person's possibility to end his life with medical assistance is determined by the type of disease that affects him.

---

[33] C. Delkeskamp-Hayes, "Respecting, Protecting, Persons, Humans, and Conceptual Muddles in the Bioethics Convention", *Journal of Medicine and Philosophy* 25.2 (2000): 147–180, 155.

[34] Biggs, *Euthanasia*.

Michael L. Wehmeyer defines self-determination as:

> [t]he ability of acting as the primary causal agent in one's life and making choices and decisions regarding one's quality of life free from undue external influence or interference.[35]

Kate is no longer the primary causal agent of her life, illness is. She therefore needs to demonstrate, mostly to herself, that she is *still* a person in classical Kantian terms, namely an autonomous and self-determining being. She feels that the defining properties of her personhood must be honoured.

According to the well-known assumptions advocated by Immanuel Kant, a person is a rational and free being who can determine his own actions. In his view, persons are the only earthly beings that are free of "causal determination". As moral agents, we are free to act upon the deliverances of reason, rather than merely from natural causes. He famously stated that human beings have an intrinsic worth that makes them valuable above all price, which derives from their ability to make their own decisions and set their own goals.[36] Autonomy forges the moral subject's personality, ensures his dignity and gives him the capacity to legislate for himself. This constitutes the freedom of the person in Kant's perspective. It is not surprising then, that autonomy is heralded as being pivotal in biolaw as the watchword that symbolises patients' moral and legal claims to make their own decisions, without constraint and in accordance with their own value system. Relying again on Kantian terminology, a future bio-legal legislation will be intended to be rooted on the centrality of the individual's self-realisation, self-respect and self-esteem, in order to guarantee, at its best, the universal protection of the persons with intrinsic value as end-in-themselves. As Jacob D. Rendtorff precisely resumes, bio-juridical principles:

> [...] express the necessary protection of humanity and of the human person as guidelines for a future politics in bioethics and biolaw, [...] they indicate the political morality of the medical and legal systems [...]. They can be understood as the foundation of a new generation of human rights, or bio-rights.[37]

Nonetheless, in the case of Kate Gulden these overarching concepts that assume the idea of wholeness, harmony and balance of the person at their core have completely lost their intrinsic meaning. They are totally incompatible with her actual wounded and humiliated self. As Charles Taylor writes,

---

[35] M. L. Wehmeyer, D. J. Sands, eds., *Self-determination across the Life Span: Independence and Choice for People with Disabilities* (Baltimore, MD: Paul H. Brookes, 1996), 22.

[36] J. Rachels, *The Elements of Moral Philosophy* (New York: McGraw-Hill College, 2003).

[37] Rendtorff, "Basic Ethical Principles in European Bioethics and Biolaw", 235.

"a person is a being who has a sense of self, has a notion of the future and the past, can hold values, make choices, adopt life-plans".[38] The self then is the awareness a person has of his "ownness", it reflects the cohesive character-istics that constitute and identify a person.[39] As far as Kate's reality is con-cerned, such cohesiveness is entirely disrupted. Having lost her sense of per-sonal integrity due to impairment, disability and dependence, together with the possibility to act autonomously, due to social and legal restrictions, she perceives almost definitely that her own dignity is going to crumble.

The concept of dignity appears to be an absolute value: it has been uni-versalised as a quality of the person as such. It expresses in fact the intrinsic worth and fundamental equality of all human beings and the inter-subjective value of the individual in his encounter with the other. As such, this principle seems to anchor our modern-day notions of personhood.[40] Respect for the dignity of human beings entails that nobody may be treated with less respect than anybody else with regard to basic human rights, such as the right to life, fairness and equal treatment.[41] Furthermore, it emphasizes the outstanding position of the human individual in the universe as being capable of free-dom, autonomy, moral reasoning and involvement in a good life. Kate Gulden is definitely not involved in a good life, as she is experiencing unac-ceptable suffering and cannot fulfil her will to avoid it. In this sense she can-not be considered in a position of equal standing in the human community. Rather, her intrinsic human worth is greatly impaired and devalued since her life has degraded into a mere painful biological existence.

Following the philosopher Harry G. Frankfurt's perspective, what defines personhood is the capability of achieving a critical consciousness of one's own desires and of translating them into a motivation which can lead to ac-tion. In simpler words, persons are the masters of their desires, able to enjoy a free will, and to act accordingly.[42] Although Kate can no longer be the master of her own life, she is determined to be the master of her own death, exer-

---

[38] C. Taylor, "The Concept of a Person" in *Human Agency and Language, Philosophical Papers* (New York: Cambridge University Press, 1999), 97.

[39] L. Benaroyo, "Suffering, Time, Narrative, and the Self" in *Personhood and Health Care*, 376.

[40] Kemp, *Basic Ethical Principles in European Bioethics and Biolaw.*

[41] The United Nations' *Declaration of Human Rights* has tried to capture the notion of dignity as the idea of an equal human value. The first article declares: "All human beings are born free, equal in dignity and human rights. They are endowed with reason and conscience and should act towards one another in a spirit of brother-hood". (General Assembly of the United Nations, 10th Dec. 1948).

[42] H. G. Frankfurt, "Freedom of the Will and the Concept of a Person", *The Journal of Philosophy* 68 (1971): 5–20.

cising choice over the time and circumstances of her dying. At the margins of life, in the light of the new self-awareness that illness has forced on her, Kate learns that dying should reflect the quality of the life lived until that time. She needs a death where she is still able to maintain control and exert a similar influence over dying as experienced through her lifetime. She perceives such death as inherently dignified. To put it in Ronald Dworkin's words, "it is important that life ends *appropriately*, that death keeps faith with the way we want to have lived". He further suggests that the intrinsic value of human life lies also in the fact that it is in part our own creation. The values by which we have lived our life and the things we think fundamental about it should thus be honoured. Sometimes this could mean ending it when we are facing a death that would be demeaning or degrading.[43] In other words, it is the very intrinsic value of life itself that mandates ending it when we feel that to continue living would devalue it. Following this account, euthanasia and death with dignity are inextricably linked. For this reason Kate voluntarily and consciously ends her life. This could be read as an ultimate expression of her will to re-shape her own destiny, choosing to bring about her own demise.

Questioning the Limits of Personal Freedom

The profound ethical and legal controversies consistent with end-of-life decisions result so hard to unravel because of their substantial involvement with fundamental notions concerning human experience. Questions of death inevitably bring us back to essential questions of life, since how we die may be as important as how we live. Once again Dworkin brilliantly illustrates the issue, stating that the nature of decision-making at the end of life

> [...] involves decisions not just about the rights and interests of particular people, but about the intrinsic, cosmic importance of human life itself. [...] Opinions divide [...] because the values in question are at the centre of everyone's lives.[44]

Paraphrasing Peter Singer, our new reality necessarily imposes a "rethinking of life and death" in order to provide socially acceptable answers for the serious dilemmas that arise from a rapidly changing medical environment.[45] And the law should be reshaped in response, since any evaluation over these contentious issues strongly depends on the questioning of basic individual rights

---

[43] R. Dworkin, *Life's Dominion. An Argument about Abortion, Euthanasia, and Individual Freedom* (New York: Vintage Books, 1994), 179.

[44] Dworkin, *Life's Dominion*, 217.

[45] P. Singer, *Rethinking Life and Death* (Oxford: Oxford University Press, 1995).

and values. Is life something at our disposal and over which we may have control? Might the *right to life* imply or include a *right to die*?

The argument is not necessarily one about institutionalizing euthanasia. What is more poignant is rather the recognition of the essential exercise of personal autonomy – literally "self-governance" – achieved through the respect for one's considered choice. What is at stake is the possibility to enable individuals to retain control throughout the entire span of their lives, including death. It means acknowledging individual sovereignty over all purely self-regarding acts that might extend to respect a person's choice in favour of death over life. A mentally competent adult, even if strangled into the unfavourable circumstance of pain and disability, is without doubt still a full legal person. His right to shape his own destiny in accordance with his beliefs, values and convictions should therefore be honoured. Besides, it can be suggested that by refusing a right to die, the law creates a duty to live in pain, as the affected parties are not allowed to exercise any choice.[46] Liberty itself can be seen as the right to define our own concept of existence, of meaning, and of the mystery of human life.

Nonetheless, the right of self-determination cannot hold unlimited sway, if humans are to live in a community. John Stuart Mill's famous "harm principle" declares that individual autonomy should be respected and that the state should criminalize only conduct which is harmful to others.[47] However, the limits of self-possession are set by the demands of social life. One agent's right to choose is to be exercised only in ways that are compatible with respect for the rights of fellow agents.[48] The right to life has traditionally been taken as a negative right, that is a right that others do not deprive us of life. The notion of a hypothetical right to die is much more obscure. It actually moves to a positive right, not just to remain alive, but to be enabled to dispose of our existence. It entails the right to be killed by having a lethal dosage of medication administered or by having something necessary for life withheld.[49] Euthanasia is therefore no longer a matter only of self-determination, a self-regarding act, rather it implies a mutual, social decision between rational persons. Furthermore, over-emphasising patient autonomy in end-of-life decisions may compromise the dignity of others, namely of those who are instrumental in performing euthanasia.

---

[46] W. Grey, "Right to Die or Duty to Live? The Problem of Euthanasia", *Journal of Applied Philosophy* 16 (1999): 19–32.

[47] J. S. Mill, *On Liberty* [1859] (Cambridge: Cambridge University Press, 1989).

[48] Beyleveld, Brownsword, *Human Dignity in Bioethics and Biolaw.*

[49] S. B. Chetwynd, "Right to Life, Right to Die and Assisted Suicide", *Journal of Applied Philosophy* 21.2 (2004): 173–182.

The investigation over the limits of individual autonomy through the display of its possible negative implications is well depicted in *One True Thing*. It is Kate herself in fact, that during one of her last moments of despair begs her daughter to help her die. She whispers:

> [h]elp me, Ellen. I don't want to live like this anymore. You must know what to do. Please. No more. You're so smart. You'll know what to do. [OTT, 172]

This request profoundly plagues Ellen. After such anguished plea, Ellen comes to the ultimate stage of her understanding of the experience of pain. She almost physically perceives its overwhelming power to affect and strangle the whole reality of one's life, till the last boundary desire to crave for death. She reflects:

> I woke to the ugly fluorescent brightness of a world deep in fallen snow, covered with pitiless whiteness. It was a world changed forever, a world in which I found it difficult to meet my mother's eyes. [OTT, 173]

Although Ellen truly would be able to satisfy her mother's appeal, she resents missing the strength and courage necessary to perform this ultimate act of responsibility, compassion and mercy. Ellen's incapability to accomplish her mother's will greatly upsets her poise. She is overwhelmed by guilty feelings, as if her denial would mean abandoning Kate to sink into tragedy alone. She feels embarrassed and anxious, as if her lack of determination would imply a betrayal towards her mother. Besides, the fact that she is unjustly arrested further impairs her personal integrity. She is subjected to the indignity and ordeal of a criminal prosecution. While imprisoned she must endure the odium of the whole town boosted and sustained by the media, since her case of suspected mercy killing has become notorious. During the trial she is constantly torn apart by the press and constrained by the humiliation of having the details of her private life exhibited to the world by the reporters. Nonetheless, while in jail Ellen paradoxically admits to "feel free". Her newly sense of freedom stems from her relief of knowing her mother at peace, liberated from the burdens and ignominy of her disease. On her side, she can free herself from the feelings of inadequacy and blame that have so deeply distressed her for not having helped her mother fulfil her ultimate wish.

Advocates of voluntary euthanasia perceive the indignity of enduring "living death" as more harmful than death itself. So, ending the harm by bringing the sufferer's life to a dignified end could be considered caring.[50] In this perspective, it is respect for persons' lives rather than for life in itself that is

---

[50] Biggs, *Euthanasia*.

fundamental. And the dignity of persons entails their right to choose and to act upon that recognition accordingly. On the other hand, opposition to and condemnation of suicide are consistent themes of our philosophical, legal and cultural heritage. We are confronted with an almost universal tradition that has long rejected a so-called right to die, and implicitly continues to reject it today. There are different reasons that explain this situation from ethical and legal viewpoints. They primarily deal with the nature of the medical profession. A great part of the medical community agrees that to hasten a patient's death is fundamentally incompatible with the physician's role as a healer. It in fact violates the central precept of medical ethics of non-maleficence, namely the duty not to harm patients. Furthermore, legislatures have the duty to protect human life. Particular interest goes to vulnerable groups such as the poor, or elderly and disabled persons that might be pressured into a "duty to die", for example to avoid becoming a burden for their families. In this sense, individual rights to autonomy are seen as not sufficient to overrule fundamental collective interests. Appeals to human dignity serve to confine rather than to support individual freedom. Opponents of euthanasia argue in fact that no legislative framework could provide sufficient protection to save potentially vulnerable individuals from abuse or coercion and, as a consequence, society from a decline into moral decay.

The arguments for and against the legalisation of euthanasia therefore continue to create contentious dilemmas, which apparently defy social and legal resolution. What is certain is that in our age of sophisticated life-sustaining technology some elements of the Hippocratic oath, such as the injunction to preserve life at all costs, have undergone an in-depth process of re-examination. This has happened also in response to the contributions given by literary works that explore the elusive, complex human condition tangled into contentious contemporary medical circumstances. Novels such as *One True Thing* and others of the kind undoubtedly offer challenging insights to the understanding of the ethical and legal dimension of medical practice. Quindlen's passionate exploration of alienation and disability, of free-will and self-determination, suggests that in contemporary times it is no longer credible to promote the sustaining of life regardless of its *quality* and *dignity* as the primary and paramount concerns of medical ethics. Especially at the margins of life, where compassion is a natural response for a person who seeks relief from the anguish of intractable disease, reason and emotion necessarily need to converge. Since this is partly an autobiographical narrative, the novel witnesses the veracity of personal experience. Quindlen therefore embeds her very thoughts in Ellen's harrowing and poignant words:

[w]hen people wonder how I survived being accused of killing my mother, none of them realizes that watching her die was many, many times worse [OTT, 276] – and again – […] the truth is that I did not kill my mother. I only wish I had. [OTT, 10]

The fulfilment of personhood in end-of-life contexts, intended as the ultimate honouring of individual autonomy and self-determination is one of the most compelling challenges imposed by our modern times. The critical question of whether a right to exert control over our death *should* broaden the concept of person in the legal perspective is to be considered in detail. The call for human dignity is confronting medical science in the courts, and the law is being called on to re-interpret the threshold between life and death.[51] Paradoxically, in the name of the same argument of "human dignity" some people advocate more permissive legal frameworks, whilst others oppose any relaxation in the law.[52] Legislatures are thus bound to face such ethical crisis as soon as possible, although courts manifest a certain anxiety in manipulating legal personhood. And such an impasse is likely to become subject to additional sources of pressure in the future, as technological progress promises to further muddy the waters of personhood.

One of the most forceful contributions of literature to ethical and legal matters in their many facets is given by the display of such boundary and compelling experiences that inevitably confronts us to take a stand. Although critical evaluations of basic moral and legal norms can be irritating, or even frightening, sometimes discomfort is a good thing. Queasiness challenges the (un-common) reader to reassess personal stable assumptions and values. The undermining of certainties obliges to re-evaluate accepted behaviours and modes of thinking of contemporary society. Such representations deepen a sense of responsibility, empathy and identification towards the displayed situation and, consequently, towards the fragility of the human condition as a whole. The combining of these elements is essential to motivate and sustain the ethical quest for equity and fairness. A quest that is more necessary than ever when it deals with crucial judgements about the paramount disposal of our own lives, and deaths.

---

[51] Biggs, *Euthanasia*.
[52] Beyleveld, Brownsword, *Human Dignity in Bioethics and Biolaw*.

Valentina Adami
University of Verona

"So what is a human being?"
An Exploration of Personhood Through
Jeanette Winterson's *The Stone Gods*

Jeanette Winterson's 2007 novel *The Stone Gods*[1] is a multilayered socio-environmental cautionary tale about repeating worlds and human self-destruction. It alternates history and science fiction, bending time and moving between new and old worlds. It explores the past while anticipating potential futures, and represents humanity as doomed to endlessly repeat the same mistakes over and over again.

Billie Crusoe is our guide through the novel, which is divided into four parts, set in three different end-of-the-world scenarios. The stories are kept together by a common theme, that is, the inhumanity of humans to themselves and to their planet, as well as by the love story between Billie, a human being, and Spike, a Robo *sapiens*.

Part One, entitled "Planet Blue", begins on Orbus, a planet very much like our own Earth, policed by robot traffic cops and inhabited by genetically-engineered, bio-enhanced humans who look all alike "except for rich people and celebrities, who look better. That's what you'd expect in a democracy" [SG, 23], states the narrator. Orbus is near collapse, and humans are running out of resources and suffering from the effects of ecological disaster, climate change and wars. A new planet perfect for human life has just been discovered, and, after the success of a first exploratory mission, a second mission leaves Orbus to start colonizing Planet Blue. The mission is also an opportunity for the government to get rid of some "troublesome" people. This is why Billie – a disillusioned scientist suspected of terrorism because she has rebelled against the autocratic regime of Tech City (for example, by choosing to live in a traditional farm at a time when farms no longer exist and people only eat meat cloned in a lab because they think natural food is diseased) – is part of this mission. Along with her are Captain Handsome – a

---

[1] All the quotations and references to the text will be taken from the following edition: J. Winterson, *The Stone Gods* [2007] (London: Penguin Books, 2008), hereafter indicated by the abbreviation SG.

space pirate defined as a "space privateer" or "freelance predator with semi-official sanction" [SG, 56] – and Pink McMurphy, a woman who has been fighting to be genetically "fixed" at age twelve to please her paedophile husband. Spike, a beautiful Robo *sapiens*, is also part of the Planet Blue mission: she has been taken on the space ship by Captain Handsome, who has rescued her from being dismantled and now wants to teach her what love is. During the mission, Spike will actually evolve and fall in love, but not with Captain Handsome.

The second part of the novel is set in the eighteenth century on Easter Island, a pristine and balanced environment, until humans arrived and made it barren by cutting down all the trees to transport stones and build the stone gods of the novel's title. Like Orbus, Easter Island is a dying world, killed by human hubris: as the narrator observes, "mankind [...] wherever found, Civilized or Savage, cannot keep to any purpose for much length of time, except the purpose of destroying himself" [SG, 132].

Parts Three and Four, entitled "Post-3 War" and "Wreck City", are set in the near-future, in a bombed-out city run by MORE, a private corporation which has created Spike, the first Robo *sapiens*. In this post nuclear war dystopia, Spike is a disembodied robot-head who is being programmed (or educated?) by Billie so that she will one day be able to take objective decisions in place of humans and guide the world rationally.

The novel offers several possibilities for reflection on bioethical issues, such as the ethics of human enhancement and genetic engineering, environmental ethics and ecology, the relationship between humans and robots, biopolitics and biopower. However, in this paper I will focus exclusively on the problems concerning the definition of Spike, the Robo sapiens, asking whether she should be entitled to personhood and moral status.

The questions raised by Winterson's novel recall quite closely the concerns about the definition of humanness in the context of modern technology expressed by Ridley Scott's 1982 movie *Blade Runner* (and, of course, by Philip K. Dick's story *Do Androids Dream of Electric Sheep?*, from which Scott's movie derives): "if the creature is virtually identical in kind to the creator, should not the creature have virtually all the same rights and privileges as the creator?".[2] In *Blade Runner*, the creature in question is Rachael, who initially does not know that she is an android because she has been given fictional

---

[2] M. Gwaltney, "Androids as a Device for Reflection on Personhood" in *Retrofitting Blade Runner*, ed. J. B. Kerman, (Madison: The University of Wisconsin Press, 1997), 32.

memories of her past and has always been treated as a human being: it is therefore difficult for the reader/spectator not to think of her as a person. After discovering that she is actually an android, Rachael goes through an identity crisis and starts questioning her humanness and personhood. This prompts us to ask whether we should extend the concept of person beyond the realm of the human.

Similarly, Winterson invites us to reflect on the concept of person-hood and analyze the status of Spike: can she be considered as a person? Of course, she is not biologically human, being made of silicon, but what is her actual moral status? In order to try and find a possible answer to these questions, I will examine some of the descriptions and definitions of Spike in Winterson's novel with reference to various theories of moral status.

Spike is first introduced to us by the narrator, Billie Crusoe, with these words:

> [...] a Robo *sapiens*, incredibly sexy, with that look of regret they all have before they are dismantled. It's policy; all information-sensitive robots are dismantled after mission, so that their data cannot be accessed by hostile forces. She's been across the universe, and now she's going to the recycling unit. The great thing about robots, even these Robo *sapiens*, is that nobody feels sorry for them. They are only machines.
> [...] It's a kind of suicide, a kind of bleeding to death, but they show no emotion because emotions are not part of their programming.
> Amazing to look so convincing and be nothing but silicon and a circuit-board. [SG, 6–7]

As the quoted paragraph reveals, at the very beginning of the novel, Billie seems to consider Spike as nothing but a robot: "nobody feels sorry for them. They are only machines" [SG, 6], "they show no emotion because emotions are not part of their programming" [SG, 7], and they are "nothing but silicon and a circuit-board" [SG, 7]. However, there are already some hints at Spike's (possible) humanness: she is "incredibly sexy, with that look of regret" [SG, 6], her dismantling is "a kind of suicide, a kind of bleeding to death" [SG, 7], and she looks "so convincing" [SG, 7]. Thus, the question I am trying to answer in this paper is: which are the characteristics that Spike should have in order to be considered as a person, as a being with moral status?

Some philosophers claim that there is only one property that confers moral status (for example, human dignity), while others argue that more properties are necessary in order to have moral status. According to Tom Beauchamp and James Childress,

the properties identified in prominent theories of moral status will not, by themselves, resolve the main issues about moral status, but [...] *collectively* these theories
can be used to provide us with a general, although untidy, framework for handling
problems of moral status.[3]

I will thus analyze Spike's features through a brief excursus on the main
groups of theories of moral status, following the latest edition of *Principles
of Biomedical Ethics* (2009) by Beauchamp and Childress and Mary Anne
Warren's essays on moral status.[4]

The first theory identified by Beauchamp and Childress argues that an individual has a moral status if it has distinctively human properties: according
to this kind of theory, being a living member of the Homo sapiens species is a
necessary and sufficient condition of moral respect. However, as Beauchamp
and Childress underline, supporting such a theory is problematic because the
properties that qualify humans for moral status are only contingently connected to being human: they could be possessed by members of nonhuman
species or by other entities such as God, chimeras, robots and genetically
manipulated species.[5] Therefore, the concept of person cannot presuppose
biological membership in the Homo sapiens species.

Literature has long shown a belief in the claim that personhood cannot be
confined to human beings. In particular, Mary Anne Warren highlights how
the existence of non-human persons has often been postulated in children's
books and science fiction:

> Children's books often depict animals as persons, who speak, wear clothes and live
> exactly like human beings. While this is fantasy, it demonstrates that the existence
> of non-human persons is not inconceivable. [...] Other evidence that the term
> "person" does not necessarily apply only to human beings can be found in science
> fiction. A popular theme is that of an initial encounter between human beings and
> extraterrestrials. At first, the members of one or both species fail to recognize
> members of the other species as persons. Often this recognition comes about
> through the determined efforts of a few humans or extraterrestrials who have
> learned to sympathize and communicate with members of the other species.
> Sometimes the aliens are not extraterrestrials but machines that have been con-
> structed by human or other beings, who initially fail to recognize the personhood
> of their creations.[6]

---

[3] T. Beauchamp and J. Childress, *Principles of Biomedical Ethics* [6th ed] (Oxford: Oxford University Press, 2009), 66–67.

[4] M. A. Warren, *Moral Status: Obligations to Persons and Other Living Things* (Oxford:
Oxford University Press, 1997); and M. A. Warren, "Moral Status" in *A Companion
to Applied Ethics*, eds. R. G. Frey, and C. H. Wellman (Oxford: Blackwell Publishing, 2003), 439–450.

[5] Beauchamp and Childress, *Principles of Biomedical Ethics*, 69.

[6] Warren, *Moral Status*, 92–93.

This is precisely what happens in *The Stone Gods*, where Winterson describes a robot with distinctively human characteristics: "[t]he first artificial creature that looks and acts human, and that can evolve like a human" [SG, 17]. Thus, if we agree with Warren that "just as a being need not belong to one's own sex, race or tribe to be a person, neither need it be biologically human or of terrestrial origin", and that "[p]ersonhood is a psychological concept, not a biological one"[7], then we should analyze the moral status of Spike according to her mental and behavioural capacities.

A second kind of theory of moral status is based on cognitive psychological properties, that is, on processes of awareness such as perception, understanding, memory and thinking. According to this theory, individuals have moral status because they are self-conscious, free to act, able to communicate through language, and rational. The following quotes from Winterson's novel show how Spike displays all of these characteristics.

"The Robo sapiens had the idea herself. [...] Call it her Last Request. This is a poignant personal moment for her. They are draining her data right now." [SG, 32]

"If your data can be transferred, as it is happening now, then why must we dismantle you when you cost so much to build?"
"It is because I can never forget"
"What? I don't understand. We take the data ..."
"And I can recall it"
"But you can't – it's vast, it's stored computer data. When it's downloaded, the host, the carrier, whatever you are, sorry, can be wiped clean. Why aren't you a machine for re-use?"
"Because I am not a machine [...] Robo sapiens were programmed to evolve ..."
"Within limits"
"We have broken those limits." [SG, 35]

"The Robo sapiens has escaped!" [...] "She'll go to the Border. She must be defecting"
"Robots can't defect. They aren't made to think for themselves"
"This one was." [SG, 46]

"In any case," she said [...] "is human life biology or consciousness? If I were to lop off your arms, your legs, your ears, your nose, put out your eyes, roll up your tongue, would you still be you? You locate yourself in consciousness, and I, too, am a conscious being." [SG, 76]

"I have disabled my Mainframe connection" said Spike. "I have chosen to live as an outlaw." [SG, 209]

---

[7] Warren, *Moral Status*, 93.

These passages clearly illustrate that Spike perceives, understands and re-members things; she can obviously think and communicate through lan-guage; she is indeed rational, self-conscious and, from a certain moment on-wards, even free to act. Therefore, if we followed a theory based on cognitive psychological properties, we should conclude that Spike is a person and has a moral status. However, this theory does not provide a fully satisfactory defi-nition of personhood, because it fails to protect vulnerable human beings (such as young children and brain-damaged humans), who do not possess some of the required properties.

The third category of theories of moral status derives mainly from Kant and is based on moral agency: according to these theories, an individual has moral status if she can act as a moral agent, that is, if she is able to make moral judgments about the rightness and wrongness of actions. As Warren under-lines, the great strength of such theories is that they give "full and equal moral status to all individuals who are capable of moral agency".[8] In the novel, Spike seems to know very well what is right and what is wrong, although her understanding is probably not based on moral judgment, but rather derives from her ability to make connections: "These are robots who join the dots. Ask them for advice, and they will give it to you: impartial advice based on everything that can be known about the situation" [SG, 17]. However, Beauchamp and Childress argue that "[b]eing a moral agent is not the only way to acquire moral status".[9] Similarly, Warren doubts the adequacy of moral agency theories because they exclude non-human animals, and even young children and brain-damaged humans from moral consideration.

The fourth group is constituted by theories based on sentience, that is, the capacity to feel pain, pleasure and other conscious mental states. In Winter-son's novel, a particularly interesting conversation that takes place on the space ship about sentience as the defining characteristic of human beings:

> "What are you girls talking about?"
> "The fact that Spike isn't a girl", I said, "We're trying to work out the differ-ences between Robo sapiens and Homo sapiens"
> "You think too much", said Pink, "I'll get you a drink. It's obvious – cut me and I bleed"
> "So blood is the essential quality of humanness?" said Spike.
> "And the rest! The fact that you had to be built – I don't know, like a car has to be built. You were made in a factory"
> "Every human being in the Central Power has been enhanced, genetically modified and DNA-screened. Some have been cloned. Most were born outside

---

8 Warren, "Moral Status", 440.
9 Beauchamp and Childress, *Principles of Biomedical Ethics*, 75.

the womb. A human being now is not what a human being was even a hundred years ago. So what is a human being?"

"Whatever it is, it isn't a robot," I said.

[…] Spike wasn't giving up. "But I want to know how you are making the distinction. Even without any bioengineering, the human body is in a constantly changing state. What you are today will not be what you are in days, months, years. Your entire skeleton replaces itself every ten years, your red blood cells replace themselves every one hundred and twenty days, your skin every two weeks"

"I accept that" I said, "and I accept that you are a rational, calculating, intelligent entity. But you have no emotion"

[…] "So your definition of a human being is in the capacity to experience emotion?" asked Spike. "How much emotion? The more sensitive a person is, the more human they are?"

"Well, yes" I said. "Insensitive, unfeeling people are at the low end of human – not animal, more android." [SG, 77–78]

This conversation highlights the main problem inherent in a theory of moral status based solely on sentience, that is, the fact that such a theory risks disallowing moral status for certain human beings (for example, those who have lost the capacity for sentience because of serious brain damage). However, if we believe, as Beauchamp and Childress do, that sentience is a sufficient condition of moral status (although not a necessary one),[10] then we should conclude that Spike does have moral personhood. In the following passage, Spike describes the moment in which she felt an emotion for the first time. Very interestingly, this happened after reading one of John Donne's poems, "The Sun Rising":

"On the official space mission, when we hung in our ship over Planet Blue, Handsome came aboard for the celebrations. […] Underneath the digital images of Planet Blue, he wrote, *She is all States, all Princes I, Nothing else is.*"

"I can read several languages and I can process information as fast as a Mainframe computer, but I did not understand that single line of text."

"I went to Handsome and asked him to show me the book. He sat beside me, our heads bent over the page, his hair falling against mine, and he explained first of all the line, and then the poem, then he put the book into my hands and looked at me seriously, in the way he does when he wants something, and he said 'My new-found land'"

"He left, and I went back to my data analysis, and I thought I was experiencing system failure. In fact I was sensing something completely new to me. For the first time I was able to feel." [SG, 80–81]

The fifth type of theory examined by Beauchamp and Childress, as well as by Warren, argues that moral status derives from being part of a net of relations (either social or biological/ecological). As Warren suggests, relationship-

---

[10] Beauchamp and Childress, *Principles of Biomedical Ethics*, 78.

based theories can indeed offer interesting insights about moral obligations, but they risk denying moral status to sentient beings who do not have social or ecosystemic relationships.[11] In any case, even according to these theories Spike could be considered as a person worthy of moral status, because she definitely has social relations, and even a couple of love affairs (a heterosexual one with Captain Handsome, whom she doesn't really love, and a lesbian one with Billie):

> "He's so strong, so romantic. [...] I just don't understand why he's in love with a robot – no offence intended to you, Spike, I'm not prejudiced or anything, it's not your fault that you're a robot – I mean, you never had any say in it, did you? One minute you were a pile of wires, and the next thing you know you're having an affair" [said Pink].
> "I don't love Handsome," said Spike.
> "Well, of course not – y'know, like I said, you're a robot"
> "That isn't why I don't love him," said Spike [...]. [SG, 69–70]

> Spike leaned forward and kissed me. "Bend the light."
> "You're a robot," I said, realizing that I sounded like Pink McMurphy.
> "And you're a human being – but I don't hold that against you"
> "Your systems are neural, not limbic. You can't feel emotion"
> Spike said, "Human beings often display emotion they don't feel. And they often feel emotion they do not display" [SG, 75–76]

Warren identifies a further type of moral status theory, which she calls "the organic life theory": according to it, organic life is the only legitimate criterion of moral status and all living organisms have equal moral status.[12] However, as Warren highlights, this theory is inapplicable on a practical level: every day we destroy millions of micro-organisms to protect ourselves, and thus we cannot claim to accord the same moral status to microbes as to human beings. But can we consider Spike as a form of life? Apparently not, because she is not made of organic material and does not display human bodily functions:

> My lover is made of a meta-material, a polymer tough as metal, but pliable and flexible and capable of heating and cooling, just like human skin. She has an articulated titanium skeleton and a fibre-optic neural highway. She has an articulated

---

11 Warren, "Moral Status", 444–445. To support her claim that the absence of prior social relationships cannot justify unprovoked aggression, Warren proposes the example of E.T. in Spielberg's movie: he is a good and gentle extra-terrestrial accidentally stranded on earth. The children who find him (with whom spectators are invited to empathize) understand that he is a good and gentle person and protect him, while adults want to kill him without determining if he is hostile or dangerous. Warren sees this story as a metaphor for meetings between human groups that are initially alien to one another.
12 Warren, "Moral Status", 443.

titanium skeleton and a fibre-optic neural highway. She has no limbic system because she is not designed to feel emotion.

She has no blood.
She can't give birth.
Her hair and nails don't grow.
She doesn't eat or drink.
She is solar-powered.
She has learned how to cry. [SG, 83]

Yet, later on, it is the narrator herself who acknowledges Spike as a living entity:

> I lay beside Spike and thought how strange it was to lie beside a living thing that did not breathe. There was no rise and fall, no small sighs, no intake of air, no movement of the lips or slight flex of the nostrils. But she was alive, reinterpreting the meaning of what life is, which is, I suppose, what we have done since life began. [SG, 99]

After having examined various theories of moral status, both Warren and Beauchamp and Childress close their analyses by stating that none of the uncriterial theories provides a satisfactory definition of moral status, because no single criterion can take into account all the relevant ethical considerations. Therefore, they suggest that we should combine the criteria proposed by different theories, taking into consideration both intrinsic and relational properties in order to provide a multicriterial account of moral status[13]. This is precisely what I have tried to do in my analysis of the character of Spike, who – I think – cannot be denied moral personhood simply on the basis that she is not made of organic material and that she has been created in a lab, because she actually displays most of the other characteristics required by different theories of moral status.

In conclusion, I would like to underline the fundamental role of literature in bioethical reflection: by broadening our imagination and enriching our understanding of moral concepts and ideas, narratives such as *The Stone Gods* have potential not only to illustrate personhood, but to help us *imagine* personhood. As Jeanette Winterson points out:

> Stone Gods isn't a pamphlet or a docudrama or even a call to arms, it is first and foremost a work of fiction, but I am sure that change of any kind starts in the self, not in the State, and I am sure that when we challenge ourselves imaginatively, we then use that challenge in our lives. I want the Stone Gods to be a prompt, but most of all, a place of possibility.[14]

---

[13] Beauchamp and Childress, *Principles of Biomedical Ethics*, 82; and Warren, "Moral Status", 445.

[14] J. Winterson, "The Stone Gods", http://www.jeanettewinterson.com/pages/content/index.asp?PageID=471, retrieved on April 28, 2009.

Therefore, thinking about the moral status of robots, androids and extrater-
restrials – which can of course only be done through fiction – is relevant not
only in itself but also and most importantly because it offers us a model for
thinking about the liminal stages of human life, and about moral status and
personhood in general, without all the emotional and political involvement
which characterizes discussions about abortion, euthanasia and other poignant bioethical issues.

Sidia Fiorato
University of Verona

The Problem of Liminal Beings
in Alasdair Gray's *Poor Things*

The novel *Poor Things* by Alasdair Gray is presented in its subtitle as a partial autobiography of the Scottish Public Health Officer Dr. Archibald McCandless.[1]

In the introduction, Gray assumes the fictional role of editor of the volume and pronounces his belief on the truth of the narrated events. His opinion is in contraposition to the historian who retrieved the manuscript from the local history museum of Glasgow and declared it a work of fiction, and to Victoria McCandless, wife of Archibald McCandless and female protagonist of the book, who contests the whole story in an afterword by denouncing it as a fiction and openly declares its literary sources. In this merging of narrative and authorial planes, which embodies the postmodern characteristics of fragmentation, crisis of authorial position and reliability, parody, pastiche and intertextuality,[2] the main source for McCandless'/Gray's novel is declared to be Mary Shelley's *Frankenstein*.

As a matter of fact, *Poor Things* can be considered as a re-writing or a parallel version of the Frankenstein story, presenting analogies with, as well as meaningful differences from, the original text. In particular, it brings the story to a further level by analyzing the social and legal consequences of the main deed of the creation of a human being. As many critics – as well as Gray himself – have underlined, the neglect showed by Victor Frankenstein towards his creature, more than the act of creation itself, led to the tragic events in Shelley's novel:

---

[1] All the quotations and references to the text will be taken from the following edition: A. Gray, *Poor Things* [1992] (Chicago and Normal, Illinois: Dalkey Archive Press, 2001) and the page number will be indicated parenthetically preceded by the abbreviation of the title PT.

[2] With regard to this aspect of the novel, see N. Bentley, *Contemporary British Fiction* (Edinburgh: Edinburgh University Press, 2008), 44ff; S. Malpas, *The Postmodern* (London: Routledge, 2005), 23ff; S. Berstein, *Alasdair Gray* (Lewisburg: Bucknell University Press, 1999), 109.

[t]he problem with Frankenstein is that [...] having brought his creature to life, he doesn't even bother to give it an education [...] It's just the idea of a creator being so careless of his creation.[3]

In the light of recent advances in the medical field of organ transplants and bio-engineering supported by corresponding advances in science and technology, when "Frankenstein's once fateful attempt might appear, retrospectively, likely to succeed and altogether desirable",[4] the novel *Poor Things* offers itself for a reflection concerning both bioethical and biojuridical issues.

The figure of the Promethean scientist-creator of Gray's novel is Godwin Baxter, who, however, appears at the same time to represent Frankenstein's monster. Multiple hints suggest that this character was created by his father, a renowned medical man: nothing definite was known about his mother, he is described as possessing an "ogreish body", a strange appearance and frightening voice and he says: "I was big from the start" [PT, 12, 15, 19]. Moreover, his father taught him to keep the values of his body under control because of a chemical imbalance which he has to attend to regularly.[5] McCandless once observed that Baxter's medical records "showed daily fluctuations too irregular, sudden and steep for even the strongest and healthiest body to survive" [PT, 73]. To McCandless's question of whether he ever asked his father where he originally came from (and the suggestion of this question already implies that he is different from average human beings), he answers that his father explained it to him by means of diagrams and models, giving him lessons.[6] Baxter enjoyed the lessons, because they taught him to admire his internal organization and preserved his self respect when, as he says, "I learned how most people feel about my appearance" [PT, 19], that is, when he gains consciousness of not being like other people, of being differ-

---

[3] A. Gray, "Frankly Gray", *The List*, (August 28–September 10, 1992): 63. See also G. Levine, U. C. Knoepflmacher eds., *The Endurance of Frankenstein. Essays on Mary Shelley's Novel* (Berkeley: University of California Press, 1979), 9.

[4] V. Poggi, "Mary and her Monster Revisited by Two Contemporary Scottish Writers" in *Mary versus Mary*, eds. L.M. Crisafulli, G. Silvani (Napoli: Liguori Editore, 2001), 279.

[5] "Having little or no pancreas, he made his digestive juices by hand, stirring them into his food" [PT, 72].

[6] This recalls Frankenstein's diary with the annotations of his studies and experiments, with the difference of the pleasure, pride and satisfaction of his creation on the part of Baxter's father, and the horror on the part of Victor Frankenstein. See M. Shelley, *Frankenstein* (London: Penguin, 1985), 175: "your journal of the four months that preceded my creation. [...] the minutest description of my odious and loathsome person is given, in language which painted your own horrors and rendered mine indelible".

ent in some way. He appears to be fully conscious of the nature of this difference, and he seems to have accepted it and to be able to manage it well.[7]

However, Godwin Baxter is mainly a rereading of Victor Frankenstein; as a matter of fact, Frankenstein's narrative of his life to Robert Walton ends with him entreating the latter not to pursue ambitious aims and commenting that "I have myself been blasted in these hopes, yet another may succeed".[8] This "another" is Godwin Baxter.

The narrator Archibald McCandless meets him at the Glasgow Medicine faculty. He is described as an expert surgeon; even though he never graduated, he keeps following an autonomous line of study, which his colleagues do not know anything about, and spends much time in the dissecting room. He tells the narrator that he is refining some of his father's techniques. He is convinced that "Medicine is as much an art as a science" [PT, 16] and that "the universe keeps nothing essential from us" [PT, 99].

Baxter's father had discovered how to arrest a body's life without ending it, so that no messages passed along the nerves; although respiration, circulation and digestion were completely suspended, cellular vitality remained impaired. When Baxter discloses the nature of his studies to the narrator, the latter thinks with enthusiasm of the application of such an ability to the dead bodies in dissecting rooms, pleading for the cause of organ transplants, and seeing the benefit for humanity: "If you can use their undamaged organs and limbs to mend the bodies of others you will be a greater savior than Pasteur and Lister – surgeons everywhere will turn a morbid science into immediate, living art!" [PT, 23]. However, Baxter seems intent on pursuing his studies alone and the next time the two meet he has successfully applied his methods for the resurrection/creation of a woman ("who owes her life to these fingers of mine", PT, 27); he says he has "cured her of death", and defines his action "a skillfully manipulated resurrection" [PT, 27].

This unknown woman had committed suicide by throwing herself into the river Clyde and her body had been retrieved by the Glasgow Humane Society. On examining the body, Baxter had noticed that rigor had not yet set in, and that the body had hardly cooled. He had also noticed that she was pregnant and showed the signs of a wedding ring on her finger. He had "cleared her lungs of water, her womb of the foetus, and by a subtle use of

---

[7] This is another difference from Shelley's novel: when the monster sees himself reflected in the waters of a river he does not gain self-respect but realizes he is a wretch and therefore loses the balance he had managed to build up to that moment.

[8] Shelley, *Frankenstein*, 260.

electrical stimulus could have brought her back to self-conscious life" [PT, 33]. However, he had dared not do so, because he "knew nothing about the life she had abandoned, except that she hated it so much that she had chosen *not to be*, and forever! What would she feel on being dragged from her carefully chosen blank eternity[9] and forced to be in one of our thick-walled, understaffed, poorly equipped madhouses, reformatories or jails?" [PT, 33]. At that time, suicide was treated as lunacy or crime. So he kept the body alive at a purely cellular level.

We can assume that Baxter managed to maintain the woman in a persistent vegetative state, a condition in which "a functioning brainstem permits the human animal to breathe spontaneously, maintain a heartbeat, metabolize, and continue other major biological functions except for the capacity for consciousness".[10] Such individuals have lost their cognitive neurological function and awareness of the environment, but retain non-cognitive functions; they are technically alive but their brain can be defined as "dead". As nobody had reclaimed the body, instead of transferring it to the dissecting rooms Baxter had brought it to his father's laboratory. Driven by the reflection that every year hundreds of young women drowned themselves because of poverty and social prejudices, and at the thought of the many unnatural born who cannot live without artificial help or cannot live at all, he had taken a socially discarded body and a socially discarded brain and had united them into a new life. "I did so, hence Bella" [PT, 34]. Here the focus is not on the act of creation and the related concept of the consideration of human intervention on human life as licit or illicit; the "birth" of Bella is given in a matter-of-fact way. Therefore the possibility to manipulate life and intervene in the vital process is taken for granted. Moreover, when Baxter once compares himself to God, McCandless tells him: "You grafted a baby's brain into a mother's skull. Very clever. It does not make you an all-knowing god" [PT, 99].[11]

Baxter executed a brain transplant from the foetus to the mother; precisely, he implanted the brain of the unborn child (eight months and a half of gestation) into the body of his/her mother. To McCandless' question "Could you not have saved the child?" he replies "Of course I saved it – the thinking part of it. [...] Why should I seek elsewhere for a compatible brain when

---

[9] This reminds us of the desperate cry of Frankenstein's creature taken from Milton's *Paradise Lost*: "Did I request thee, my Maker, from my clay/ To mould me Man, did I solicit thee / From Darkness to promote me?" (10.II/43–45).

[10] D. DeGrazia, *Human Identity and Bioethics* (Cambridge: Cambridge University Press, 2005), 32.

[11] This is very different from *Frankenstein* where Victor dreams of becoming the blessed creator of a new species (Shelley, *Frankenstein*, 101).

her body already housed one?" [PT, 41, 42], "a perfectly healthy little brain" [PT, 30].

What is hypothesized[12] here is a whole-body transplant/brain transplant, an operation that involves moving the brain of one being into the body of another. This new woman therefore has a fully grown body and the mind of an infant child. It is as if she experienced a new birth and, even though her body is in her mid twenties, she has to learn everything all over again, starting from language and the way to relate herself to the world, the people around her and to society. She starts a new life; the body of the mother lives the mental life that should have been of her child, until, one day, her brain will be as adult as her body [see PT, 35], at least this is what Baxter thinks.[13]

As I said before, this novel takes Frankenstein's story further in that it considers the social and juridical effects of the creation of a new being. In *Frankenstein*, the creature is nameless and denied any kind of individuality or recognition because he is a product of and belongs wholly to his creator. Moreover, he symbolizes the unknown, the never-before-perceived, and therefore raises questions about how he should be perceived and fit into the cultural codes of signification.[14] In *Poor Things*, the first questions that McCandless addresses to Baxter are:

---

[12] This is considered as an intuitive case, that is a hypothetical case which takes the conceptually necessary conditions for the survival of a person into consideration, treating physical impossibilities as irrelevant in the context of a reflection on the concept of persons (or subjects) and their identity. (See DeGrazia, *Human Identity and Bioethics*, 23) The case presented in Gray's novel relies on the medical hypothesis that the brain retains not only the capacity for consciousness via (primarily) the cerebrum, but also the capacity via the brainstem to power the vital functions of any body to which it is successfully attached. (See DeGrazia, *Human Identity and Bioethics*, 53). Cerebrum transplants are considered as extraordinary cases that have not occurred yet. One of the most significant barriers to this procedure is the inability of nerve tissue to heal properly: scarred nerve tissues do not transmit signals well. Even if all the nerves were successfully connected, they may not transmit the same information as the same nerve connection to the old body. Moreover, the age of the donated body must be sufficient; an adult-sized brain could only fit into the skull of a head of adult size.

[13] The only moment Godwin shows an attitude of remorse towards his act he says: "I deserve death as much as any other murderer. [...] That little nearly nine-month-old foetus I took living from the drowned woman's body I shortened her life as deliberately as if I stabbed her to death at the age of forty or fifty, but I took the years off the start, not the ending of her life – a much more vicious thing to do. [...] Selfish greed and impatience drove me" [PT, 67].

[14] See A. K. Mellor, "Making a 'Monster': An Introduction to *Frankenstein*" in *The Cambridge Companion to Mary Shelley*, ed. E. Schor (Cambridge: Cambridge University Press, 2003), 20.

How do you explain her to them [meaning the people in the household]? [...]
How do you explain her to society? [...] are they told she is a surgical fabrication?
How will you account for her on the next government census? [PT, 34–35]

The problem of the woman's social justification comes to the forefront.
We witness here how medical science outpaces the laws regulating the legal
persona, rendering the application of existing social and legal categories to
new and liminal beings such as Bella problematic. Even if not all bioethical
considerations can be translated into biojuridical issues, this is perceived as a
need for those ones which affect the institutions of society.[15]

In this case, Baxter completely carries out his role as creator and also
creates a social identity for Bella as Bella Baxter; he says she is a distant
orphaned niece who suffers from total amnesia due to a railway accident
in which she lost her parents: "An extinct, respectable couple will be better
than none. It would cast a shadow upon her life to learn she is a surgical fabrication" [PT, 35].

Moreover, after learning of her intention to marry McCandless, he decides to make a will bequeathing her everything he owns, thus enabling her
to be inserted into property relations. Property is strictly linked to the development of the self and to a person's growth, to conceptions of freedom and
individualism. The relationship between property and personhood has always been considered fundamental by the law.[16] Therefore, by making her his
heir, Baxter confers upon Bella the status of legal person and the possibility
to engage in economic transactions. The legal subject is defined as a possessor of rights over things, that is, a person with *dominium*, and this constitutes
his/her social being. *Dominium* defines the sense of being rooted in a certain
culture, the ability to enter into relationships of various kinds [social, economic, legal] with others: "To own, to convey; to be located in a kinship structure, where one can inherit property, and leave property to be inherited. [...]
Identity and authority come out of this social and civic context".[17]

When Baxter is asked by the notary about the exact relationship he has
with Bella, in order to avoid any possibility of contestation of the will on the
part of other members of the family, he is forced to face the codification of
the law, a codification from which Bella escapes in the absence of norms for
such liminal beings. Therefore he turns to a corrupt lawyer in order to avoid

---

[15] See F. D'Agostino, *Parole di Bioetica* (Torino: Giappichelli, 2004), 10.

[16] See D. Carpi, *Property Law in Renaissance Literature* (Frankfurt am Main: Lang,
2005), 131.

[17] A. Gearey, "The Voice of Dominium – Property, Possession and Renaissance Figures in The *Cantos* of Ezra Pound" in *Property Law in Renaissance Literature*, 180.

questions and sets Bella's legal identity: Baxter sort of creates her legal identity independently of the law but at the same time inserts her into existing laws, thus demanding that her "persona" be recognized and protected and sustained by them.

In order to achieve this, Bella must be inserted into a social role: in its original meaning the word "persona" indicated the mask ancient actors used to wear in a play (the *dramatis personae*). The mask hid, or rather replaced, the actor's own face and countenance, but in a way that would make it possible for the voice to resound through it. From this metaphor the term passed to the legal context: the law assigns somebody the part he is expected to play on the public scene, with the provision, however, that his own voice is able to sound through.[18] Baxter provides Bella with a social/legal mask, in order to enable her to exert her rights at law.

Admission into society is also important for another reason: man is ontologically relational, that is, the possibility to enter into relationship with another is an unavoidable condition of man's identity; alterity precedes the "I", that is, the individual dimension. The function of the law is to grant coexistence and to grant the external conditions for intersubjective relationality.[19] Therefore, through Baxter, Bella is granted the possibility to posit herself in relationship with others, and therefore to fully realize her own relational identity and humanity.[20]

This is also what Frankenstein's creature longed for:[21] during his encounter with his creator on Mont Blanc he pleads for the possibility of accessing a social form of existence. Watching the cottagers and listening to their conversations, the creature learns about social structures and feels excluded from them:

> I heard of the division of property [...] I learned that the possessions most esteemed by your fellow-creatures were high and unsullied descent united with riches. [...] And what was I? Of my creation and creator I was absolutely ignorant; but I knew that I possessed no money, no friends, no kind of property.[22]

---

18 See H. Arendt, *On Revolution* (Harmondsworth: Penguin, 1976), 106–107, quoted in J. Gaakeer, "Ishiguro's Legal Chimera", *Pòlemos* 2 (2007): 126.

19 Cfr. L. Palazzani, *Introduzione alla biogiuridica* (Torino: Giappichelli, 2002), 90–91. Cfr. L. Palazzani, *Il concetto di persona fra bioetica e diritto* (Torino: Giappichelli, 1996), 58 and 108.

20 See Palazzani, *Introduzione alla Biogiuridica*, 94.

21 By reading literary texts he confronts himself with their protagonists and realizes that he is related to none (Shelley, *Frankenstein*, 173). He reflects upon his social condition and sees that he is united by no link to any other being in existence (Shelley, *Frankenstein*, 175), man will not associate with him (Shelley, *Frankenstein*, 189).

22 Shelley, *Frankenstein*, 165.

He perfectly understands the personal theory of property and economic relations as the basis for a social identity.

Furthermore, he claims his right to a companion and to form his own family, because this would imply an achievement of social identity in the dimension of the community. However, the family is also the nucleus of social and political life, therefore the creature claims the right to be considered a man and a citizen as a form of inheritance from his creator: "When the monster does appropriate the capacity for 'human speech', he almost immediately uses the capacity to narrate a history and voice a demand",[23] thus staging a juridical context. Victor cannot comply with his request because it would lead to the creation of a parallel and autonomous social and legal system to the existing one. The monster's eventual social identity is felt by Frankenstein as a direct threat to his own.[24]

The monster also realizes he is beyond the protection of the law; as a matter of fact, human law would judge him a murderer, but would not punish Frankenstein for killing him, that is, for committing the same crime: murder.

> The laws that can only identify him as a perpetrator, but not as a victim, apply to embodied subjects; beyond that they apply specifically to human bodies, the humanly embodied subjects. It is not the dictates of reason, but the limitations of the empirically human that debar the daemon's recognition before the law.[25]

Bella leads a free and independent life until she decides to marry McCandless. The wedding ceremony is interrupted by a man who declares he recognizes Bella as his lawful wedded wife Victoria Blessington, née Hattersley. He is supported in his claim by the testimony of her father, confirmed by a

---

[23] D. Reese, "A Troubled Legacy: Mary Shelley's *Frankenstein* and the Inheritance of Human Rights", *Representations* (2006): 96 and 49.

[24] See Reese, "A Troubled Legacy", 49 and 53. Moreover, the creation of the monster, carried out in solitude and separation from society signifies a separation and divorce from his social bonds on the part of Victor, "a conquest over his own social and sexual being, fulfilled in a creature to whom social and sexual ties are denied". C. Baldick, *In Frankenstein's Shadow. Myth, Monstrosity and Nineteenth-Century Writing* (Oxford: Clarendon Press, 1987), 51. Therefore, Shelley's novel can be seen also as a "Warning against the detachment of science from social ties" (Baldick, *In Frankenstein's Shadow*, 28).

[25] Reese, "A Troubled Legacy", 53–54. This refers to the monster's speech where he confesses his murders to Victor: "The guilty are allowed, by human laws, bloody as they are, to speak in their own defense before they are condemned. Listen to me Frankenstein. You accuse me of murder, and yet you would, with a satisfied conscience, destroy your own creature. Oh, praise the eternal justice of man!" (Shelley, *Frankenstein*, 365).

doctor and a solicitor. In this situation, "Bella was unknowingly seeing the father of her brain in the first husband of her body, the grandfather of her brain in the father of her body" [PT, 201] but she cannot remember seeing either of them before.

Baxter sustains the identity of Bella Baxter as a woman with no memory who had come to him three years earlier and had since lived as his ward, heir to his patrimony. He underlines that she had freely engaged to marry McCandless. However, he urges the men to settle the question through rational discussion instead of bringing the case before a judge and a jury in a court of law.

Therefore, a discussion takes place to determine the identity of the woman who calls herself Bella Baxter. On the one side, her previous family sustains that she is Victoria Blessington, and therefore subjected to the duties connected to such social identity and position. General Blessington and his solicitor make repeated appeals to the law for Bella's/Victoria's situation and future life. They are convinced that Baxter saved Victoria and abducted her in order to make her his mistress, subsequently giving her to Wedderburn and to McCandless.[26] She has the duty to go back to her husband and be subjected to his authority: "if she is now sane she will come home with me. If she refuses she is still mad, and it is my duty as her husband to place her in an institution where she will be properly treated" [PT, 219].

Baxter replies that Bella is not the same person as Victoria after having lain seven days in a mortuary. As the woman's previous relatives insist by saying that given the advancement of science in Glasgow and the scientific knowledge of his father Baxter had managed to resuscitate her,[27] the latter claims that, as a result, she is not General Blessington's wife any longer and this for manifold reasons:

---

[26] "If this story is put before a British jury they will believe it, because it is the truth" [PT, 220], claims General Blessington's lawyer. However, the case will not be taken to court because Victoria's family wants to avoid the social scandal which would ensue, and therefore avoids a confrontation of the law with such a liminal case.

[27] See PT, 212, when Dr Prickett says: "The London medical world is aware that since the start of this century the Glasgow surgeons have been putting electric currents through the nervous system of dead bodies. It is on record that in the 1820s one of your sort animated the corpse of a hanged criminal, who sat up and spoke. Public scandal was only prevented by one of the demonstrators severing the subject's jugular with a scalpel. Your father was present at that demonstration. I have no doubt that he passed on all he learned to you, who were his only assistant".

she has not been your wife since she drowned herself [...] The marriage contract says the marriage lasts until death do you part. [...] Dr. Prickett suggests I gave her a new life. If so I am as much the father and protector of the revived woman as Mr. Hattersley was of the earlier, and as entitled as he once was to present her in marriage to the husband of her choice. [PT, 219]

The survival of the person to the death of the organism is possible only in the hypothetical case of brain transplant; however, in this case Victoria's brain dies and is taken away from her skull. There is no person with no brain structure. The brain is considered the only organ which cannot be substituted (neither in a mechanical way nor through a transplant) in the human body; it coordinates the whole organism. Therefore human life coincides with cerebral life.[28] Following this line of reasoning, Victoria and Bella are two different persons.

Moreover, the modern concept defines persons as beings with the capacity for certain complex forms of consciousness, such as rationality or self-awareness over time. As far as the question of personal identity is concerned, the criteria for a person's continuing to exist over time can privilege psychological continuity (thus referring to Locke's classic formulation of a person as "a thinking intelligent being, that has reason and reflection, and can consider itself, as itself, the same thinking thing in different times and places")[29] or the continuity of a biological life.

Baxter sustains psychological continuity, that is, the continuity of a mental history over time, where present and past transient moments of awareness are connected by memory. A brain that no longer supports any form of psychological continuity would not preserve a person. According to the embodied mind account, that is, the fact that we are essentially minded beings or

---

[28] See Palazzani, *Introduzione alla biogiuridica*, 193; Palazzani, *Il concetto di persona fra bioetica e diritto*, 59 and 127; and H. T. Engelhardt, *Manuale di Bioetica* (Milano: Il Saggiatore, 1991), 246, the brain is the main organ granting corporeal integrity as it grants the integration of the most important systems of organs.

[29] See D. P. Kaczvinsky, "'Making up for Lost Time': Scotland, Stories, and the Self in Alasdair Gray's *Poor Things*", *Contemporary Literature* 42.4 (2001): 777. John Locke, in his *An Essay Concerning Human Understanding* defines the self as the immaterial, continued "consciousness" of one's past actions and possible future: "For since consciousness always accompanies thinking, and 'tis that, that makes every one to be, what he calls self; and thereby distinguishes himself from other thinking things, in this alone consists personal identity, i.e., the sameness of a rational being: And as far as this consciousness can be extended backwards to any past action or thought, so far reaches the identity of that person; it is the same self now it was then; and 'tis by the same self with this present one that now reflects on it, that that action was done". J. Locke, *An Essay Concerning Human Understanding* (London: W. Tegg, 1849), 222.

minds, "the criterion of personal identity is the continued existence and functioning, in non-branching forms, of enough of the same brain to be capable of generating consciousness or mental activity".[30] This is also in accordance with the intuitive method: when consciousness parts ways with the body, we tend to think that the person goes where consciousness goes. Sameness of mental life seems to be the most salient or decisive condition for personal survival. According to this line of reasoning, Bella and Victoria are two distinct persons. As Victoria has had a brain transplant, the new brain, originally belonging to a still unborn baby, cannot possibly retain a form of psychological continuity with Victoria or with its own former self because it still had to have been born. The thinking part of the baby, the part capable of developing consciousness, was born into the mother's body. We came into existence when the fetus acquired the capacity for consciousness or sentience: "The organism is conscious in virtue of having a mind that is conscious".[31]

Victoria's family, on the contrary, seems to cling to the assertion that identity is a function of biological life, therefore, believing her to be suffering from amnesia, they still consider her as Victoria, wife of General Blessington.[32]

Moreover, a person is characterized by a narrative identity, that is, a conscious, deliberate shaping of one's own personality or life direction according to one's values, from which also derive projects of self-creation for one's future. The *psyché*, the animation of individuality, gives an identity to the *bios*, the empirical life inevitably linked to corporeity and temporality. And it is only in the *bios*, the physical corporeal reality, that the *psyché* can express itself in the world as a value dimension.[33] Considering these elements, Victoria and Bella appear to be widely different from each other.

Victoria seems to be a victim of "mirroring", that is, the imposition of another's view of oneself to one's self-narrative which undermine the autonomy of her choices. She accepts the Victorian view of her personality, which

---

[30] J. McMahan, *The Ethics of Killing: Problems at the Margins of Life* (Oxford: Oxford University Press, 2002), 68. See also DeGrazia, *Human Identity and Bioethics*, 68.

[31] McMahan, *The Ethics of Killing*, 120.

[32] See Kaczvinsky, "Making Up for Lost Time", 779. Also according to Locke, if a person committed a crime but suffered from amnesia regarding it, he was nonetheless liable to punishment and had to face the consequences of his act (Kaczvinsky, "Making Up for Lost Time", 780).

[33] The law protects the life of the psyche, but as it cannot reach it, it focuses on the bios, the locus of juridical and ethical experience. The human bios needs the *polis* because in it the *nomos* is given, thus expressing the link between life and the law. (See D'Agostino, *Parole di Bioetica*, 29–32).

condemned her attitude towards sexual life and subjected her to the authority of male figures such as her father and her husband.

On the contrary, Godwin takes care of Bella and of the formation of her
identity (also as a narrative identity). Unlike Frankenstein, who abandoned
his creature to his fate, Baxter fostered and watched Bella's progress and her
gradual achievement of articulate thought, speech, literacy, and judgment.
He takes her around the world with him, so that she may obtain life experiences and be responsive to a great variety of stimuli. As a result, she soon develops a firm and talkative manner, and the second time the protagonist
meets her she seems to be in full control of the situation and to take care of
Baxter. She acquires a conscience of herself through literature (in an analogous process to Frankenstein's creature); she is aware that she lost her
memory, but she claims an autonomy of decision and of thought for herself,
apart from what she has been told by Baxter. She says:

> I have eyes and a mirror in my bedroom, I see I am a woman in my middle twenties
> and but nearer thirty than twenty, most women are married by then. [PT, 51]

She starts asking herself questions about who she is and why she is like that.
She autonomously decides to marry McCandless because she can treat him
as she likes, while Baxter treats her like a child. She is also conscious that she
lacks a past:

> I am only half a woman Candle, less than half having had no childhood [...] A
> whole quarter century of my life has vanished [...] I need more past [PT, 61].

Therefore she elopes with Wedderburn (the corrupt lawyer) and completes
her formation through human engagement, which for her represents "the
sine qua non for learning".[34] Her travelling is a bildung-process, a search
for her personal identity, in which she gains experience in decision making
(as Baxter says, PT, 69) and the completion of her narrative identity:

> nearly everyone who has an ongoing self-narrative has sufficient decision-making
> capacity to qualify as a moral agent, as someone who can be morally responsible.[35]

She gains experience of the world and of social conventions, which she does
not follow. She proves that she has a logical and clear mind to the people she
meets, who are fascinated by her sincerity: she only speaks to say what she
thinks and feels, not to disguise her opinions, and she is incapable of hypocrisy and lying. She's an adult person, but possesses the "tabula rasa of the

---

[34] Bernstein, *Alasdair Gray*, 120.
[35] DeGrazia, *Human Identity and Bioethics*, 89.

child's open and enquiring mind".[36] She makes judgments on politics and trends of social behavior in a lucid and deep analysis of the vices of society: one of the main results of Bella's process of education is her belief in the necessity for social justice.[37]

She accepts herself as she is and has an affectionate relationship with Baxter:

> You made me strong and sure of myself, God, by teaching me about the fine and mighty things in the world and showing I was one of them. You were too sane to teach a child about craziness and cruelty. I had to learn about those from people who were crazy and cruel themselves. [...] You are to blame for nothing, God, nothing at all where I am concerned. [PT, 194]

She attains the dimension of self-creation when she decides to become a doctor to help little girls, mothers and prostitutes; Baxter immediately offers her the social basis to render her wish possible by suggesting that McCandless should become a Health Officer in order to support her when she eventually obtains charge of her own clinic. Baxter always tries to protect her by including her in social and legal structures and even paving the legal path for her achievements.

Therefore, Baxter can assert: "I never met your wife, General Blessington. The drowned woman who came to consciousness here is someone else" [PT, 221–222]. He gives strength to his thesis by adducing the medical opinion of specialists in diseases of the mind and nerves, thus saying that Bella is:

> sane, strong and cheerful, with a vigorously independent attitude to life. [...] Apart from that [the amnesia] her balance, sensory discrimination, recollective and intuitive and logical powers are exceptionally keen. [...] Bella Baxter's most striking abnormality is her lack of it. Such a woman cannot be General Blessington's former wife [PT, 222–223],

whom they describe as suffering from sexual disturbances and therefore, according to their opinion, mad. Personal identity is not exclusively genetic, but also and mainly biographic; the individual acquires his/her specific originality through his/her life, thanks to the numerous relational dynamics which characterize it.[38]

The men insist on taking a legal view of the matter, according to which the body takes precedence, especially in cases of marriage: in this case, Blessington's lawyer can claim that Bella is still Victoria, the General's wife: "neither he nor the laws of the land will allow her to commit bigamy and live happily

---

[36] Bentley, *Contemporary British Fiction*, 47.
[37] Poggi, "Mary and her Monster Revisited", 281.
[38] See D'Agostino, *Parole di Bioetica*, 62.

ever after" [PT, 223]. However, according to the same legal perspective, as Baxter underlines, she can also decide to divorce General Blessington. As Victoria or as Bella she is a legal person entitled to her own decisions. Baxter always advocates Bella's freedom of decision, as he considers her a subject with full rights, both social and juridical:

> All I have said has been to persuade you it is honourable and possible to let this woman freely choose whether she returns to England with you or stays in Scotland with us – honourable and possible. [PT, 223]

The final decision is taken by Bella herself who asserts she will stay in Scotland: "here I stay, said Bella calmly [...] I will continue living here" [PT, 232]. Moreover, she further asserts her individuality and independence by sanctioning her separation from Victoria. She thanks the people for her lives, her former one and her current one, but she states she is happy to choose a man she does not have to thank for anything, that is, a man with whom she can be in the position of a peer, and who will allow her to lead the life she wishes. It is as if Bella discarded the persona of Victoria, therefore becoming free to live an autonomous life[39] as Bella; she manages to reach a cohesive identity by going beyond the two parts of a damaged life.[40]

The novel makes us reflect on the classical idea of the *persona juris*, as it was developed in Roman law, that is, "an entity possessing legal rights as well as duties, conferred upon him by the law and defined as 'he who can act in law'. [...] [If] personhood is a construct of law, a legal fiction, what are the criteria with which we construct individuals as legal subjects, as persons at law?"[41]

Baxter faces this problem and solves it by creating a personhood for Bella: his creation is complete, biological and social. His "creature" acquires more and more independence and autonomy until she is able to manage herself in society. Even if, at the beginning, he had tried to create a being who was utterly dependent upon him, he recognizes her autonomy when he observes:

> It is wonderful for a creator to see the offspring live, feel and act independently. I read Genesis three years ago and could not understand God's displeasure when Adam and Eve chose to know good and evil – chose to be Godlike. That should have been his proudest hour. [PT, 99–100]

---

[39] See Kaczvinsky, "Making Up for Lost Time", 786: Bella is what the Victorians most feared in a wife: a sexually liberated woman who is free to decide for herself and refuses to be tamed.

[40] See Bernstein, *Alasdair Gray*, 129.

[41] Gaakeer, "Ishiguro's Legal Chimera", 124. See also D'Agostino, *Parole di Bioetica*, 150: persona is one who is recognizable and can be qualified as an agent, the one who determines/his/her own actions.

In the end Bella chooses her identity autonomously, rebelling against the process of mirroring[42] (that all the men in the novel try to impose upon her) and separating herself both from her previous family and from Baxter, thus asserting Locke's claim that "every man has a property in his own person". The Lockian notion of property rights applies to one's body, and eventually results in the freedom of contract as the power to undertake obligations.[43]

Traditional answers about the unity of the self and about man's soul or memory are no longer credible, especially in a postmodern world where organ transplants, genetic engineering, and cloning are becoming not only possible but commonplace. A new creative act seems necessary on the law's part, through bioethical and biojuridical reflection and philosophy of the law. In this way the Roman meaning of the law emerges in the contemporary context: *ex facto oritur jus*. *Jus* was born out of the concrete observable needs of life.[44] Baxter had asked for a rational discussion to settle the question of Victoria/Bella's identity; this is precisely the attempt of philosophy in bioethics and biolaw, i.e. the retrieval of the faith in reason and its ability to search for the meaning (the *logos*) of reality by means of dialogue (*dialogos*), the dialectical confrontation with the motives of others.[45]

---

[42] See DeGrazia, 87–88: the process of mirroring takes place "when one person sees his own "reflection" in another person's apparent image, conception or characterization of him. [...] significant distortions in mirroring can undermine the autonomy of an individual's choices".

[43] Gaakeer, "Ishiguro's Legal Chimera", 127.

[44] See D'Agostino, *Parole di Bioetica*, 25–26.

[45] See Palazzani, *Il concetto di persona fra bioetica e diritto*, 13.

Mara Logaldo
IULM University Milan

"Murderous Creators": How Far Can Authors Go?

By a significant coincidence bioethics – and its juridical counterpart, biolaw – appeared in the same years in which postmodernism started to be extensively theorised. It is true that while the appearance of the word "bioethics" is generally made to coincide with the foundation of the Kennedy Institute of Ethics in Madison (Wisconsin) and Washington D.C. in the 1970s,[1] the birth of postmodernism is much more difficult to determine. However, although the origins of the term can be traced to the end of the nineteenth century and some of its distinctive aspects even farther back – as Ihab Hassan wrote "we continually discover 'antecedents' of postmodernism"[2] – it is between the late 1950s and the 1970s that the idea of a "postmodern turn" entered the lexicon of philosophy and critical theory to describe the paradigm of an age.[3]

In spite of the acknowledged difficulty of defining bioethical positions, we can recognize some common assumptions. Both bioethics and postmodernism, the latter particularly in its poststructuralist version, rejected positivist epistemologies, a faith in progress and in the search for certain knowledge that originated with the Enlightenment.[4] At the same time, they also rejected a theological view, preferring to it, at most, what has been defined as a "negative", "deconstructive," or "eliminative" theology.[5] This entailed, for both, a radical rethinking of the notion of human nature. Bioethics looked at it from a secularized perspective, which stressed its being the result of chance and natural selection rather than of a metaphysical project. By ques-

---

[1] Cfr. W. T. Reich, "How Bioethics Got Its Name," *The Hastings Center Report*, 23 (1993); see also M. Chiodi, *Modelli teorici di bioetica* (Milano: Franco Angeli, 2005).

[2] Cfr. I. Hassan, *The Dismemberment of Orpheus* (Madison, Wisconsin: The University of Wisconsin Press, [2nd edition] 1982).

[3] Cfr. S. Best and D. Kellner, *The Postmodern Turn* (New York: Guilford Press, 1997).

[4] On this point, there were also dissenting voices: just think about the defence of Enlightenment by Jürgen Habermas. Cfr. G. Ritzer, *Sociological Theory, From Modern to Postmodern Social Theory (and Beyond)* (New York: McGraw-Hill Higher Education, 2008).

[5] Cfr. D. Ray Griffin, W. A. Beardslee, J. Holland, eds., *Varieties of Postmodern Theology*, (Albany, NY: State University of New York Press, 1989). See also K. Vanoozer, ed., *The Cambridge Companion to Postmodernist Theology* (Cambridge: Cambridge University Press, 2003).

tioning the teleological view, bioethics removed mankind from a central position in the universe, emphasising the *quality* of life rather than its *sacredness*.[6]
Similarly, postmodernism claimed that man has a decentred self and that life
has neither an objective nor a transcendental meaning, at least in the traditional sense. The crisis of humanism announced by Husserl, and assimilated
by the postmodernists, may have led to some critical misinterpretations,
namely a confusion between the refusal of objectivism and mere subjectiv-
ism; but its real innovation lay in the introduction of a new idea of transcen-
dentalism, as a non-idealized reading of the relationship between thought and
perception.[7] In this framework, the only form of humanism still possible was
one based on self-awareness. While bioethics stressed the importance of self-
perception,[8] postmodernism "in the criticism of the seventies, [...] began to
appear to refer to contemporary self-conscious texts."[9] In addition to this, the
idea that man is a language-based, narrative, cultural and social construction –
an idea that science, especially with psychoanalysis, had embraced well before
postmodernism – was the most blatant common presupposition.

But apart from these aspects the intents of postmodernism and bioethics
seemed to be radically different: the more postmodern *theory* questioned
man's position in the world, the more medical *practice* felt the need to fix rules
to safeguard human life. It is difficult to say whether bioethics was mainly a
reaction to the unethical scientific experimentations that were still being carried out in the US in the early 1970s,[10] or a discipline which tried to prevent
the actualization of theories based on a relativized view of human beings.[11]
Of course, there were also dissenting voices among the postmodernists sig-
nalling the problem of confusing individual beliefs and general practice.
When Fredric Jameson, to give a significant example, distinguished a postmodernist space that was "deeply suffused and infected by its new cultural
categories," from "the luxury of the old-fashioned ideological critique", he

---

6   Cfr. P. Singer, *Practical Ethics* (Cambridge: Cambridge University Press, 1979).
7   Cfr. G. Vattimo, *La fine della modernità* (Milano: Garzanti, 1985), 43.
8   Cfr. R. Silcock Downie and J. Macnaughton, *Bioethics and the Humanities. Attitudes
    and Perceptions* (Oxon-New York: Routledge-Cavendish, 2007),18, 136ff.
9   L. Hutcheon, *Narcissistic Narrative. The Metafictional Paradox* (Methuen, London, [2nd
    edition] 1984), 2. Indeed, the adjectives used by Hutcheon to describe postmodern
    metafiction always contain this idea of self-consciousness: "self-reflective, self-in-
    forming, self-reflexive, auto-referential, auto-representational" (L. Hutcheon, *Nar-
    cissistic Narrative. The Metafictional Paradox*, 1–2). See also D. Carpi, *L'ansia della scrit-
    tura. Parola e silenzio nella narrative inglese del XX secolo* (Napoli: Liguori, 1995).
10 Cfr. Chiodi, *Modelli teorici di bioetica*, 27.
11 There are, however, exceptions, like Singer's acceptance of abortion and even in-
    fanticide precisely on the same premises. Cfr. Singer, *Practical Ethics*, 169ff.

explained his point by mentioning Hegel's moral stand, based on the distinction between "the thinking of individual morality or moralising" and the "whole different realm of collective social values and practices."[12]

If we shift our focus from postmodernism as a cultural and social phenomenon to its literature, we notice a similarly ambiguous relationship. Bioethics and postmodern literature arose from comparable presuppositions: the former from the loss of a boundless faith in the power of science;[13] the latter from the loss of a boundless faith in the power of literature.[14] In a sense, deconstructive scepticism invested the world of science as well as the world of literature. Both felt the need of a radical rethinking of their scientific or literary foundations, which often took the form of meta-discourse: metascience and metafiction respectively. Both reconsidered human life, no matter whether "real" or fictionally re-created, from an unidealized viewpoint. However, while science stressed man's ethical status – a position which could perfectly be reconciled with a secularized perspective – literature reconsidered characters from a critical standpoint which stressed their functional status and questioned their duty to embody moral integrity or a form of coherence that went beyond the (limited, in deconstructive terms) narrative logic of the texts they belonged to. As Gilbert Sorrentino wrote in *Imaginary Qualities of Actual Things*,

> These people aren't real. I'm making them up as I go along, any section that threatens to flesh them out, or make them "walk off the page," will be excised. They should, rather, walk into the page, and break up, disappear.[15]

While science brought to the foreground the right to physical and psychological integrity of human beings, literary characters became dislocated, dismembered, and unfathomable figures, "a block of impenetrable space in the form of a man", as Paul Auster wrote in *The Invention of Solitude* a few years later.[16]

Hence, at a superficial reading, the intents of bioethics and postmodern literature were symmetrically opposed. By claiming the fictional nature of characters, literature was allowed to escape any kind of "bioethical" control. While people could be conceived but not treated as "many pieces on a chess-

---

[12] F. Jameson, *Postmodernism, or the Cultural Logic of Late Capitalism* (London, New York: Verso, 1992), 46–47.

[13] Chiodi, *Modelli teorici di bioetica*, 11ff.

[14] Precisely on account of a blurring of the boundaries between the literary text and a textual construction of the world, literature lost its distinctive power as an aesthetic absolute.

[15] G. Sorrentino, *Imaginary Qualities of Actual Things* (University of Illinois: Dalkey Archive Press, 1992), quoted in Hutcheon, *Narcissistic Narrative*, 87.

[16] P. Auster, *The Invention of Solitude* (New York: SUN, 1982), 7.

board",[17] literary characters could be both conceived *and* treated as chess-board pieces that the omniscient narrator could freely move. While bioethics forbade any kind of unethical manipulation of real human beings, postmodern literature could claim the freedom of the artists to play with their human images and simulacra; in the case of the novelists, with their verbally-constructed creatures. All the pseudo-scientific or utterly "sadistic" attitude showed towards real human beings, which had been overtly condemned by bioethics, seemed now to be systematically reverted towards literary characters, whose intrinsic human representativeness was questioned and sometimes dismissed as a dangerous form of naivety.

At a deeper reading, though, what we take for playfulness and cynical disenchantment in postmodern literature is rather a sign of the need to reformulate the relationship between morality and creativity. It is true that, on the one hand, characters were "wordmasses" or "methods of composition",[18] on which authors could freely exert all their creative drive, playing god with the awareness of the impossibility of producing something new or really causing damage. Postmodernism considered aesthetic or critical distance[19] as a necessary caveat to avoid traditional mimesis. Even if an identification between fact and fiction was indeed acknowledged, it was in the name of an unexpected "short-circuit"[20] or ironic reversal caused by the fictionalization of life, rather than a trust in the capacity of literature to represent life. On the other hand, postmodernist novelists were conscious of the fact that, as René Wellek had pointed out, "the 'ontological gap' between a product of the mind [...] and the events in 'real' life" did not and could not "mean that the work of art is a mere empty play of forms, cut off from reality."[21]

Given this complex framework, it is difficult to decide whether postmodernist literature was a *post*-humanist phenomenon.[22] In a sense, with postmodernism literature seemed to have accomplished the same kind of destiny ex-

---

[17] W. J. Harvey, *Character and the Novel* (London: Chatto & Windus, 1965), 132.

[18] Cfr. M. Bradbury, *No, Not Bloomsbury* (London: Arena, 1989), 19–45. The expression "methods of composition," used by Nabokov, is cited on p. 22; the expression "wordmasses," used by E.M. Forster, is cited on p. 43.

[19] Jameson totally disagrees with this idea. According to the American thinker, "distance in general (including "critical distance" in particular) has very precisely been abolished in the new space of postmodernism", *Postmodernism, or the Cutlural Logic of Late Capitalism*, 48.

[20] D. Lodge, *The Modes of Modern Writing* (London: Edward Arnold, 1977), 239ff.

[21] R. Wellek, "Some Principles of Criticism" in *Times Literary Supplement* (July 26, 1963), cited in Hutcheon, *Narcissistic Narrative*, 17.

[22] I. Hassan was among the first to define postmodernism as the advent of a post-humanist era. Cfr. Hassan, *The Dismemberment of Orpheus*.

perienced by other forms of art in the course of the twentieth century: namely, *dehumanisation*. Dehumanisation, as Ortega y Gasset had already announced in his famous 1925 libel, was a *sine-qua-non* condition for a post-human handling of subject. I would like to report some of Ortega y Gasset's statements because, though referred to modernism, they seem to describe what would happen in literature in the following decades: the passage from humanism, as a category already deeply questioned by modernist literature – although in the mood of a vitalistic reassertion of human experience and of the artistic form – to the allegedly post-humanist ethos of postmodernist literature.

> Not only is grieving and rejoicing at such human destinies as a work of art presents or narrates a very different thing from true artistic pleasure, but preoccupation with the human content of the work is in principle incompatible with aesthetic enjoyment proper.[23]

According to Ortega y Gasset, the novel is the narrative form in which the danger of referentiality is most acutely felt: the more realistic the genre, the more necessary it seems to draw a line between fiction and life. Hence Ortega y Gasset's paradox:

> To enjoy a novel we must feel surrounded by it on all sides; it cannot exist as a more or less conspicuous thing among the rest of things. Precisely because it is a pre-eminently realistic genre, it is incompatible with outer reality. In order to establish its own inner world it must dislodge and abolish the surrounding one.[24]

His conclusions are audaciously prophetic: great periods do not "revolve around human contents"; rather, "aesthetic pleasure derives from the triumph of art over human matter."

At the same time, though, it cannot go unnoticed that Ortega y Gasset's conclusions are still founded on humanistic assumptions: a) the power of art over human matters is (still) explained by resorting to the category of the sublime; b) albeit in the form of the "strangled victim,"[25] art cannot avoid representing human figures and passions; c) feeling any sort of sympathy towards this scapegoat sacrificed on the altar of art is a form of generous weakness;[26] d) finally, this dehumanized art is but a shadow which refers to a higher form, as in Plato's allegory of the cave.[27]

---

[23] Ortega y Gasset, *The Dehumanization of Art; And Other Essays on Art, Culture, and Literature* (Princeton: Princeton University Press, 1972), 9–10.

[24] Ortega y Gasset, *The Dehumanization of Art*, 96.

[25] The expression "strangled victim" with reference to Ortega y Gasset's idea of character is used by Malcolm Bradbury in *No, Not Bloomsbury*, 42.

[26] People "are looking for the human drama" (Ortega y Gasset, *The Dehumanization of Art*, 39).

[27] Cfr. D. Carpi ed., *Why Plato? The Influence of Plato on Twentieth Century English Literature* (Heidelberg: Winter, 2005).

In the next part of my study I shall attempt to see how these four points showing an early vision of dehumanisation were further elaborated by post-modernist authors. Starting from these considerations, I shall also attempt to see to what extent postmodernist novelists can be considered mere cynical manipulators of their characters or if, on the contrary, they were sensitive to bioethical problems, since, as Linda Hutcheon remarks, "self-informing narrative does not signal a lack of sensitivity or of humanitarian (or human) concern on the part of the novelist."[28]

The first point, which concerns the power of art to transform painful human matters into something sublime, had never actually abandoned the epistemological background of literary criticism. Not even the early manifestations of a disrupting modernist sensibility had changed the general faith in the transfiguring mission of art. In 1896, George Santayana discusses this issue in his *The Sense of Beauty*: "the arts" wrote Santayana, "must stand modestly aside until they can slip fitly into the interstices of life."[29] It is only by virtue of the aesthetic quality of the artistic experience that the choice of subject is justified and morally redeemed. Significantly, Santayana resorts to a medical metaphor: "any violent passion, any overwhelming pain, if it is not to make us think of a demonstration in pathology, and bring back the smell of ether, must be rendered in the most exalted style." Only so "grief itself becomes [...] not wholly pain" and "the saddest scenes may lose their bitterness in their beauty. [...]." By contrast,

> if ever the charm of the beautiful presentation sinks so low, or the vividness of the represented evil rises so high, that the balance is in favour of pain, at that very moment the whole object becomes horrible, passes out of the domain of art, and can be justified only by its scientific or moral uses.[30]

The clinical atmosphere, the surgical mood, in other words the treatment of painful human issues as such, without any aesthetic sublimation and subsequent pathetic effect, is considered as intrinsically unfit for art. The author who plays with these subjects for their own sake is compared to a child who, watching a shipwreck from the shore, may have only two things in mind: either "without understanding of the calamity, [he] would have a simple emotion of pleasure as from a jumping jack" or "if he understood the event, but was entirely without sympathy, he would have the aesthetic emotion of the

---

28 Hutcheon, *Narcissistic Narrative*, 18.
29 G. Santayana, *The Sense of Beauty. Being the Outline of Aesthetic Theory* (New York: Collier Books, 1955), 153.
30 Santayana, *The Sense of Beauty*, 157.

careless tyrant, to whom the notion of suffering is no hindrance to the enjoyment of the lyre."[31]

The latter statement seems to echo the paradox contained in the title of
this study. Indeed, postmodernist authors have often been compared to tor-
turers and tyrants, to murderous creators.[32] Martin Amis, for instance, who
defines himself as "a comic writer interested in painful matters,"[33] and who
deeply penetrates the interstices of life (he claims that in the world of fiction
"there aren't any 'Do Not Enter' signs")[34] extensively displays his will to in-
flict damage on his creatures, both physically, especially sexually, and psychologically. Not believing that his characters may be related to real people, the
author can torture them "into life" and mercilessly manipulate them, even
kill them "for his and our sport."[35]

In so doing, he flouts one of the maxims of bioethics: the one which describes the relationship between a physician and his/her patients as a covenant based on the principles of autonomy, well-doing (or non ill-doing) and
justice. The parallelism between the figure of the author and that of the
scientist is so common that it is not necessary to explain it here. Also the as-
similation of certain kinds of characters to patients is frequent in literature:
the identification of passive, malleable fictional characters with "patients"
(being their more active and resilient counterparts "agents," those who "act"
rather than those who "are acted upon") was overtly made by Anthony
Powell in his novel *Agents and Patients* (1936). The title of the novel para-
phrases John Wesley's statement in Sermon 62 (published in 1872): "He that
is not free is not an Agent but a Patient." More generally, the name of "patients" has been attributed by psychoanalytic criticism to all characters discussed as real people with consistent psyches and backgrounds, traumas and
desires.

I am not going to propose here either a Powellian or a psychoanalytic
reading of postmodernist novels. The comparisons simply allows me to
argue that the bioethical or biolegal "contractual model,"[36] which claims the

---

[31] Santayana, *The Sense of Beauty*, 155–156.
[32] See, for instance, B. Finney, "Narrative and narrated homicide in Martin Amis's
       Other People and *London Fields*" (1995), available at: http://www.csulb.edu/
       ~bhfinney/MartinAmis.html.
[33] A. Smith, "The Two Amises", *The Listener* 92 [1974], 219–20.
[34] Martin Amis interviewed by Christopher Bigsby, in *New Writing* (Minerva-British
       Council, 1992), 174.
[35] Cfr. Finney, "Narrative and narrated homicide".
[36] R. M. Veatch, *The Patient-Physician Relation. The Patient as Partner* (Bloomington: In-
       diana University Press, 1991).

necessity of an alliance, of a *covenant* between doctor and patient, could also be applied to the relationship between the implied author, or the narrator, and his or her characters. For most postmodernist authors this ideal covenant seems to be deliberately neglected on both sides: not only do authors "kill" their characters, but certain characters themselves ask to be killed.[37] This is particularly true for Martin Amis. According to Kurt Leutgeb, Amis's characters, especially the female ones, "have a bent for self-extinction."[38] Martin Amis suggests that characters may, so to say, acquire a life of their own and impose their will by asking to have things done to them (including violence and murder). In the Prologue to *Other People*, the author tries to justify himself in these terms: "I didn't want to have to do it to her. I would have infinitely preferred some other solution. Still, there we are. It makes sense, really, given the rules of life on earth; and she *asked* for it."[39] In fact, characters are not just passive victims: they also react to the treatment they undergo by "doing things to the author-narrator" ("I've done things to her, I know, I admit it. But look what she's done to me").[40] From this point of view a sort of mutual agreement does exist in Amis's novels, although with a negative sign:

> In both *Other People* and *London Fields* Amis explores the ambiguous position that the narrator of postmodern fiction such as his must occupy. Simultaneously he must be both an instrument for the author and as much a victim of the author's capricious will as any of the other characters. Similarly readers are made to see their active implication in the sadistic treatment met out to characters by narrator and author, and yet to be themselves subject to the wayward will of the author.[41]

The plot and its denouement become a sort of biotestament in which the character *outwrites*[42] the author. In *London Fields* the narrator claims that

---

[37] We could discuss the author-character relationship resorting to the bioethical and biolegal distinction between "killing" and "letting die" theorised by Veatch et al. See R. M. Veatch, "Death and Dying: Euthanasia and Sustaining Life. Professional and Public Policies," in *Encyclopedia of Bioethics*, ed. W. Th. Reich (New York: The Free Press, 1978), 261–86.

[38] K. Leutgeb, "Cynicism as an Ethic and Aesthetic Principle. A Study of Martin Amis's Fiction. With Special Emphasis on *Dead Babies*, *Time's Arrow*, and *Career Move*", available at: www.martinamisweb.com/documents/Leutgeb_Cynicism. doc. See also M. Logaldo, "Il realismo grottesco nella narrativa di Martin Amis", *Merope*, VII. 14 (January 1995): 67–89.

[39] M. Amis, *Other People* (New York: Vintage International, 1994), 9.

[40] Amis, *Other People*, 106.

[41] Finney, "Narrative and narrated homicide".

[42] "When the writer sits down to his desk in the morning, there is nothing so safe and deadly as knowing exactly what comes next. If he is a real novelist by temper of imagination, but weakly hopes to have the security of a controlled symmetry, he

"character is destiny,"[43] and in *Dead Babies* he declares that he could not have behaved otherwise:

> Well, we're sorry about that, Keith, of course, but we're afraid that you simply *had* to be that way. Nothing personal, please understand – merely in order to serve this particular fiction. In fact, things get much, much worse for you later on, so appealingly bad that you'll yearn to be back at the Institute, or even in Parky Street, Wimbledon, with that family you so loathe. It's all far too advanced for us to intercede on your behalf. Tolerate it. You'll turn out all right in the end. Now go and lie on your bunk.[44]

The decisions of the author are ultimately justified in the name of the narrative logic of the text, with the necessities of the plot. We can discuss this point by making reference to other two theoretical models in bioethics: the *engineering model*[45] and the *utilitarian model*.[46] The former explains medical decision only according to facts, leaving aside any kind of emotional involvement or religious implication. Of the two models mentioned, however, this is the least applicable to postmodern writers who, as in the case of Martin Amis, still claim to be moralists[47] and still point out the horror and disgust of the subjects they tackle. The latter, the *utilitarian model*[48] is probably more apt, because it allows us to compare the novelist's commitment to form and argumentation to bioethical theories which emphasise effects rather than a deontology inherent to actions themselves. If this were the case, *consequentialism*, which is at the core of the utilitarian model, could easily be applied to literature, especially hedonistic and aesthetic consequentialism, which judges actions according to the amount of pleasure and beauty achieved. But it would leave important questions open (the same it arouses in the bioethical

will find that his characters shake their fists into the crooks of their elbows, set his schemes on their tails, ruthlessly rewrite him. H. Gold, "The Mystery of Personality in the Novel", in *The Living Novel: A Symposium* (New York: Collier Book, 1962), 107.

[43] M. Amis, *London Fields* (New York: Vintage International, 1991), 7.

[44] M. Amis, *Dead Babies* (Harmondsworth: Penguin, 1984), 162.

[45] See Caplan's definition and criticism of this model. Cfr. A. Caplan, "The Ethics of Evil: The Challenge and the Lessons of Nazi Medical Experiments", in *Dark Medicine: Rationalizing Unethical Medical Research*, ed. W. LaFleur (Bloomington: Indiana University Press, 2007), 50–64.

[46] P. Singer, *The Language of Morals* (London: Clarendon Press, 1952).

[47] According to Christopher Bigsby, Amis has "a moralist's fascination with the corruption of values, taste, style, form" (C. Bigsby, "Martin Amis interviewed by Christopher Bigsby", in *New Writing*, eds. M. Bradbury and J. Cooke (London: Minerva, 1992).

[48] Cfr. P. Singer, *Practical Ethics* (Cambridge: Cambridge University Press, 1979).

field): who's going to judge which consequences are good? Who will bene-ficiate from these actions? John Rawls speaks of an "ideal observer:"[49] could this universal figure be assimilated to that of the implied or ideal reader?[50] Moreover, since characters are incapable of feeling pleasure or pain, would there be no end to the author's manipulations? "It would be nonsense," writes Peter Singer

> to say that it would not be in the interests of a stone to be kicked along the road by a schoolboy. A stone does not have interests because it cannot suffer. Nothing that we can do to it could possibly make any difference to its welfare.[51]

Are characters like stones to be "kicked" along the narrative path, insensible things to which nothing matters? Or should we, paradoxically, respect char-acters and stones as much as humans and animals?[52]

More strikingly, the consequentialist thesis is difficult to hold within the deconstructive framework of postmodern literature. Indeed, against all reas-suring affirmations of cohesion and logical coherence, this literature often denies or inverts the chain of cause and effect deliberately. In Amis's *Time's Arrow*, for instance, the order by which events are presented is chronologi-cally reverted, and so is the cause-effect relationship. Healing processes become destructive. Hospitals – where people go in healthy and come out horribly wounded or sick – are "atrocity producing" situations.[53] Conversely, destructive processes turn into creative ones: for instance, abortion is de-scribed as the implantation of a foetus in a woman's womb. The description is objectively crude, doubly defamiliarizing:

> These prospective ladyfriends arrive quietly. John, who is ready, receives them quietly. They feel cold, and rest and cry for a while, and then mount the cleared table. They assume their half of the missionary position, though John, of course, is busy elsewhere, with the full steel bowl. A rectangular placenta and a baby about half an inch long with a heart but no face are implanted with the aid of forceps and

---

49 See J. Rawls, *A Theory of Justice* (Cambridge, Mass.: Harvard University Press, 1971).
50 Cfr. W. Iser, *The Implied Reader* (Baltimore and London: The John Hopkins Press, 1974); U. Eco, *The role of the Reader: Explorations in the Semiotics of Texts* (Blooming-ton: Indiana University Press, 1979).
51 P. Singer, *Animal Liberation, A New Ethics for our Treatment of Animals* (New York: Random House, 1975).
52 This is the idea held by Joanna Zylinska, who advocates a non-systemic, non-hier-archical bioethics that should include not only human beings and animals, but also objects, especially machines. Cfr. J. Zilinska, *On Spiders, Cyborgs, and Being Scared: the Feminine and the Sublime* (Manchester: Manchester University Press, 2001); *The Ethics of Cultural Studies* (London: Continuum, 2005); *Bioethics in the Age of New Media* (Cambridge, Mass.: MIT Press, 2009).
53 M. Amis, *Time's Arrow* (London: Penguin, 1991), 102.

speculum. He is always telling the women to be quiet. They *must* be quiet. The full bowl bleeds. Next, digital examination and the swab. They can get down now, and drink something, and talk in whispers. They say goodbye. He'll be seeing them. In about eight weeks, on average.[54]

Indeed, in *Time's Arrow* the Protean Nazi doctor Odilo Unverdorben, who appears as a healer rather than as a murderer ("Unverdorben" means "un-contaminated"), is an emblematic figure. The comparison with Nazi medical experimentation has been made both in the field of bioethics and in the field of literary criticism, with reference to utilitarian bioethics (particularly Peter Singer's)[55] and with reference to postmodernism.[56] Although I do not agree with a direct affiliation, the intellectual fascination with nihilism does lurk in the background. And it is difficult to draw a line between a moralist and a cynical attitude. Even when, as in Amis's case, the author claims a moralist intent, this can be seen, as Kurt Leutgeb has pointed out,[57] as an ethicized version of cynicism rather than as an obvious form of morality.

Whatever the ideological interpretation, the writer's power to create be-comes inseparable from a destructive power, giving rise to what novelist Chuck Palahniuk has recently defined as "constructive destruction."[58] Pa-lahniuk definitely takes this paradox to its furthest extent. The following de-scription, in one of the final chapters of *Lullaby*, is, at once, immensely dis-turbing and immensely moving. Helen, mysteriously injured and hospitalized in Intensive Care, reaches the cryogenic unit of the hospital. Once there, she looks for the glass case in which the body of Patrick, her baby son who had been killed by a deadly "culling song" twenty years before, had been pre-served. When she finds it, she starts contemplating it along with her lover, the first-person narrator:

> "Look at him," she says, and touches the grey glass with her pink fingernails. "He's so perfect."

---

54 M. Amis, *Time's Arrow*, 101. Kurt Leutgeb comments that "*Time's Arrow* is much more serious a book than *Dead Babies*. The authorial pose is not one of self-con-scious superiority, but shaped by the narrator's endeavours to understand the world" (K. Leutgeb, "Cynicism as an Ethic and Aesthetic Principle. A Study of Martin Amis's Fiction"). On the subject of Nazi medical experimentations and bioethics, see Arthur Caplan, "The ethics of evil: The challenge and the lessons of Nazi medical experiments", in ed. W. LaFleur, *Dark Medicine*, 50–64.

55 See, for instance, Don Felder, "Professor, Death Will Fit Right in at Princeton", *Jewish World Review* (October 28, 1998).

56 R. Wolin, *The Seduction of Unreason. The Intellectual Romance with Fascism from Nietzsche to Postmodernism* (Princeton: Princeton University Press, 2004).

57 K. Leutgeb, "Cynicism as an Ethic and Aesthetic Principle".

58 C. Palahniuk, *Lullaby* (London: Vintage-Random House, 2003), 148–150.

She swallows, blood and shattered diamonds and teeth, and makes a terrible wrinkled face. Her hands clutch her stomach, and she leans on the steel cabinet, the gray window. Blood and condensation run down from the little window.

She says she's unplugged the cryogenic unit. Disconnected the alarm and backup batteries. She wants to die with Patrick.

Inside the gray window, the perfect infant is curled on its side in a pillow of white plastic. One thumb is in his mouth. Perfect and pale as blue ice.

[...]

With the gray rock in my fist, I punch through the cold gray window. My hands bleeding, I lift out Patrick, cold and pale. My blood on Patrick, I put him in Helen's arms. I put my arms around Helen.

[...]

Helen raises Patrick in her hands. Her child, cold and blue as porcelain. Frozen fragile as glass.

And she tosses the dead child across the room where it clatters against the steel cabinet and falls to the floor, spinning on the linoleum. Patrick. A frozen arm breaks off. Patrick. The spinning body hits a steel cabinet corner and the legs snap off. Patrick. The armless, legless body, a broken doll, it spins against the wall and the head breaks off.[59]

Decidedly, in spite of the beauty of the style, we "smell the ether" here. Or is this, maybe, just the author going as far as he can, exploring the subtle boundary between pity and terror, touching the chords of a new form of sublime? According to Lyotard the sublime, which in modernist literature was an aesthetic event, becomes in postmodernist literature an ethical event. This ethical event stresses "the pain that imagination and sensibility" are not equal to express a missing content, while modernism felt it could express it through the beauty of form.[60] But there is also another interpretation: postmodern sublime may lie precisely in the true/false ambivalence which underlies its poetics. Jameson formulates it as a crucial question: "Can we in fact identify some 'moment of truth' within the more evident 'moments of falsehood' of postmodern culture?"[61] Emily Lutzker tries to articulate a possible answer to Jameson's question:

---

[59] Palahniuk, *Lullaby*, 252.

[60] Lyotard distinguishes between a modernist aesthetics of the sublime, which is nostalgic and based on the substitution of the missing content by the consistency of form, and the real sublime "which is an intrinsic combination of pleasure an pain: the pleasure that reason should exceed all presentation, the pain that imagination or sensibility should not be equal to the concept." (Cfr., "Ethics of the Sublime in Postmodern Culture. A Talk From the International Conference Aesthetics and Ethics (March 18, 1997)" The text of the paper is available at: http://www.egs.edu/mediaphi/Vol2/Sublime.html.

[61] Jameson, *Postmodernism*, 49.

In a postmodern world of language games [...] the sublime experience presents us with an inescapable confrontation with the inner substance which makes us human. This is no game, no fashion. For the split second of the sublime experience, regardless of the source, we humans are faced with an incontestable truth. [...] The essence of natality and mortality, the habitation of a being who lives in the interstices of enormity and microscopic insignificance. Solid space, emptiness, pleasure and pain.[62]

As Lutzker writes in another of her essays, "The proliferation of violence in the media is a symptom of the cultural fascination with dismemberment and fragmentation of the body and blood."[63] From this point of view, the sublime becomes "a feeling dependent on, and a reminder of, the physicality of the body."[64] This vision is shared also by other thinkers: due to a sort of transparent film which wraps us up in the illusory perspective of a painless civilization, the sublime becomes the expression of a "desire of the body,"[65] of the search for a border experience of darkness beyond the horizon of form as mere image.

We have thus reached the second point detected in Ortega y Gasset's argumentation: the acknowledgement that form – particularly narrative form – cannot avoid representing "human figures and passions."

This point is extensively discussed in David Lodge's fiction. And, interestingly, the metafictional discourse is often twinned with a bioethical problem. *The British Museum is Falling Down* (1965), for instance, is both an exploration of the character's narrative status and a story centred on how Catholic couples tried to cope with the interdiction to use artificial contraceptives formulated in Pope Paul VI's encyclical letter *Humanae Vitae,* divulgated in 1962. While Adam Appleby, a post-graduate student who lives on a scholarship, is leaving the British Library to go home, burdened by the thought that his wife might be pregnant for the fourth time, he realizes he is a fictional character just as he surprises himself in the act of thinking. "As he descended the ladder he was conscious of re-enacting one of the oldest roles in literature".[66] The description seems to echo Gilbert Meilaender's extended metaphor (once again, the metaphor of the stone) by which the ontological status of human beings is distinguished from that of inanimate objects:

---

[62] Lutzker, "Ethics of the Sublime in Postmodern Culture", available at: http://www.egsvedu/mediaphi/Vol2/Sublime.html.

[63] Lutzker, "Violence and the Sublime in Mass Media", available at: http://www.egs.edu/mediaphi/pdfs/emily_violence_sublime.pdf.

[64] Lutzker, "Violence and the Sublime in Mass Media".

[65] Cfr. M. Morioka, *Painless Civilization. A Philosophical Critique of Desire* (Tokyo: Transview Publications, 2003). The English translation is available at: www.life-studies.org/painless00.html:21

[66] D. Lodge, *The British Museum is Falling Down* (London: Penguin, 1983), 145.

Drop me from the top of a fifty-story building and the law of gravity takes over, just as it does if we drop a stone. We are finite beings, located in space and time, subject to natural necessity. But we are also free, able sometimes to transcend the seeming limits of nature and history. As I fall from that fifty-story building, there are truths about my experience that cannot be captured by an explanation in terms of mass and velocity. Something different happens in my fall than in the rock's fall, for this falling object is also a subject characterized by self-awareness. I can know myself as a falling object, which means that I can to some degree "distance" myself from that falling object. I cannot simply be equated with it. I am that falling object, yet I am also free from it. Likewise, I am the person constituted by the story of my life. I cannot simply be someone else with a different history. Yet I can also, at least to some degree, step into another's story, see the world as it looks to him – and thus be free from the limits of my history. The crucial question, of course, is whether there is any limit to such free self-transcendence – whether we are, in fact, wise enough and good enough to be free self-creators or whether we must acknowledge destructive possibilities in a freedom that refuses any limit.[67]

From this point of view, as I mentioned in the introduction, postmodernism shares with bioethics the emphasis given to self-awareness and to the notion of person as a narrative construction. However, unlike bioethical self-perception, which entails norms and limitations concerning our actions, the auto-referentiality of postmodern literary characters can also be interpreted as an abdication to representativeness of the characters in favour of the author. The possibility that this might allow boundless creative/destructive freedom to the author is metafictionally thematized. The problem is crucial and variously perceived. While Martin Amis, as we have seen, almost "schizophrenically" shares this authorial responsibility with his characters, David Lodge still lays bare the fact that the handling of narrative techniques is ultimately only the novelist's choice, even when he uses the characters as mouthpieces of his own views. No matter how self-willed and self-aware fictional people are or are made to appear, the great "puppeteer" is still the author. As Linda Hutcheon points out:

> Some have argued that postmodernist art does not aim, as did modernist, at exploring the difficulty, so much as the impossibility, of imposing that single determinate meaning on a text. Yet it is also true that it does so, not so much by means of textual difficulties alone, but – paradoxically – by overt self-conscious control by an inscribed narrator/author figure that appears to demand, by its manipulations, the imposition of a single, closed perspective.[68]

The novel in which Lodge overtly thematizes the relationship between the author and his characters is *How Far Can You Go?* (1980). The narrative pat-

---

[67] G. Meilaender, "Bioethics and the Character of Human Life," *The New Atlantis* 1 (Spring 2003): 67–78.
[68] Hutcheon, *Narcissistic Narrative*, XIII.

tern enacts a sort of Barthesian strip-tease theory in a Catholic key, interspersed with metafictional comments about character. The question of the title is actually a double question: how far can you go with your girlfriend or boyfriend without committing a sin according to the Catholic teachings in sexual matters? And how far can authors go with their characters and narrative situations without infringing an unwritten but binding deontology which prevents authors from treating their characters as *actants*[69] deprived of any relation to moral – as well as aesthetic – categories? This is probably the most explicit comment in the novel:

> Two years after Nicole was born, Dennis and Angela's next youngest child, Anne, was knocked down by a van outside their house and died in hospital a few hours later. I have avoided a direct presentation of this incident because frankly I find it too painful to contemplate. Of course, Dennis and Angela and Anne are fictional characters, they cannot bleed or weep, but they stand here for all the real people to whom such disasters happen with no apparent reason or justice. One does not kill off characters lightly, I assure you, even ones like Anne, evoked solely for that purpose.[70]

This passage reveals the impossibility of entirely escaping the referential dimension of literary characters. In another part of the book, the narrator explains that "we maintain a double consciousness of the characters as both, as it were, real and fictitious, free and determined."[71] As Linda Hutcheon remarks in her analysis of the relationship between the narrator and his character in *The French Lieutenant's Woman*, "possibility" in narrative, is not "permissibility:" John Fowles' character "refuses to allow him [the author] his creator's liberty" and "he is morally as well as aesthetically bound to obey her."[72]

Finally, the responsibility is *also* towards the reader. For, in spite of the blurring of the boundaries between the figure of the author and that of the reader which characterizes postmodernism, no one can deny that the reader's identification with the characters is still at work: not only fictional modelisation is one of the principal patterns of cognition, mimetic immersion is essential to the very existence of fiction.[73]

This is true not only for the novel but for media discourse in general. David Lodge was among the first novelists to immerse his narrative in a complex world perceived as multi-modally mediated. The relationship is no longer between the "real" world and the fictional one: it is between likewise

---

[69] Cfr. A.J. Greimas, *Les actants, les acteurs et les figures,* in *Sémiotique narrative et textuelle,* ed. C. Chabrol (Paris: Larousse, 1974).

[70] D. Lodge, *How Far Can You Go?* (London: Penguin, 1981), 240.

[71] Lodge, *How far can you go?*, 240.

[72] Hutcheon, *Narcissistic Narrative*, 58.

[73] J. M. Shaeffer, *Pourquoi la fiction?* (Paris: Seuil, 1999), 198.

mediated worlds. In Chuck Palahniuk we find a further exploration of the problem. In *Non-Fiction*, a sort of re-writing of Barthes' *Mythologies* which shows only the Dyonisian side of myth and the physical factuality of pain,[74] Palahniuk outlines his theory on character and on the function of story-telling. This theory, explained with a legal simile, coincides with the postmodern idea of subject as discursive formation:

> We live our lives according to stories. About being Irish or being black. About working hard or shooting heroine. [...] Each time you create a character, you look at the world as that character, looking for the details that make that reality the one true reality. Like a lawyer arguing a case in a courtroom, you become an advocate who wants the reader to accept the truth of your character's worldview. [...] This is how I create a character."[75]

In *Lullaby*, the ethical problems related to media communication as to any form of storytelling comes out powerfully. The narrator talks about the Ethics test at his final exam to become a reporter. The examiner poses him the following question: on a Christmas Eve, would he call the parents of a baby who has just choked to death on a Christmas decoration, only to ask what colour the decoration was?

> My answer was to call the paramedics. Items like this have to be catalogued. The ornament had to be bagged and photographed in some file of evidence. No way would I call the parents after midnight on Christmas Eve. The school gave my ethics a D. Instead of ethics, I learned only to tell people what they want to hear. [...] There are so many people with infants, my editor said. It's the type of story that every parent is too afraid to read and too afraid not to read. (...) We'd show how this could happen to anyone"[76]

The co-existence of industrial journalism with ethics is a paradox, writes Andrew Belsey.[77] Reporters want to give readers what they want. From this standpoint, we deduce that the tyrant is the reader, who indulges in a voyeuristic pleasure, either with regard to the mediated world represented by the news or the fictional world represented by the novel. Because if, on the one hand, characters are "other people", the other people we hear about in the news are like characters in a novel. But even without indulging in such a negative attitude, we can conclude that the novel too, like any kind of media

---

[74] In the chapter on the wrestling, entitled "{Demolition}," we read: "Tonight is about breaking things and then fixing them. About having the power of life and death." C. Palahniuk, *Non-Fiction* (London: Vintage-Random House, 2005), 40.

[75] Palahniuk, *Non-Fiction*, XX–XXI.

[76] Palahniuk, *Lullaby*, 11–12

[77] A. Belsey, "Journalim and Ethics: Can They Co-exist?", in *Media Ethics*, ed. M. Kieran (London and New York: Routledge, 1998), 13.

discourse, is founded on the ambiguous response triggered by the text: one of sadism or *Schadenfreude*, one of identification and sympathy.

> The novel's structure is a structure of suggnômê, of the penetration of the life of another into one's own imagination and heart. It is a form of imaginative and emotional receptivity, in which the reader, following the author's lead, comes to be inhabited by the tangled complexities and struggles of other concrete lives.[78]

As also George Lavis suggests, it is precisely by acquiring awareness of the fictional referentiality of literary discourse that the novel appeals "d'une manière plus large, à ma connaissance du monde, à mon experience d'être humaine."[79] The sympathy that we feel towards characters, even considered as "strangled victims" or scapegoats sacrificed on the altar of art, can also be seen as a form of "generous weakness" (and we have thus come to the third point of Ortega y Gasset's argumentation). As reported in Plato's *Protagoras*, Socrates claimed that we show a weakness of will whenever we behave in a way that runs contrary to what our judgement tells us we should or should not do. This state is defined as *akrasia*. Socrates treats it as a theoretical impossibility, because he believes that knowing virtue means behaving accordingly. Plato seems to share Socrates' intellectualistic view. At the same time, in Book 4 of *The Republic*,[80] he seems to foreshadow Saint Paul's account of *akrasia* as something which "is not only a theoretical possibility" but "in some sense inevitable for post-lapsarian humanity, although its effects can be mitigated."[81] Now this dilemma between what we know to be rationally right and morally good and the actions we actually take is at the core of bioethics. For instance, the way we practise abortion or euthanasia does not correspond to what we judge to be uncompromisingly correct.

A literary analogue for this ambiguity could be found in what Coleridge defined as "willing suspension of disbelief."[82] Our rationality tells us that it is only fiction, but we temporarily believe in what we read. No matter how con-

---

[78] M. Nussbaum, "Equity and Mercy", cited in H. Antor, "The Ethics of Story-Telling and Reading: Literature, the Law, and the Principle of Equity" in *The Concept of Equity*, ed. D. Carpi (Heidelberg: Winter, 2007), 157.

[79] G. Lavis, "Le Texte littéraire, le référent, le réel, le vrai", *Cahiers d'analyse textuelle*, 13 (1971), 7–22, 19; cited in Hutcheon, *Narcissistic Narrative*, 96.

[80] In *The Republic*, Socrates argues that we may know what is good, but fail to be moved by it because we are not able to master our temptations or fears. See K. Dorted "Weakness and Will in Plato's Republic", in *Weakness of Will from Plato to the Present*, ed. Tobias Hoffmann (Washington D.C.: The Catholic University of America Press, 2008). Available at http://ndpr.nd.edu/review.cfm?id=14067

[81] Dorted "Weakness and Will".

[82] S. T. Coleridge, *Biographia Literaria* [1817].

scious we are of the fictionality of the characters, we still react to their actions and situations as if they were *somehow* human. Even Ihab Hassan, who, as I mentioned before, defined postmodernism as "posthuman" or "transhuman", nevertheless wonders whether "the dismemberment of Orpheus prove(s) no more than the mind's need to make but one more construction of life's mutabilities and human mortality."[83] No matter how removed from humanity characters may be, they still retain the power of catalysing man's fear of pain and death, they still demand the reader's "generous weakness."

Finally, we have to see whether, beyond evidence, this form of art is but a shadow which refers to a higher form, as in Plato's allegory of the cave. As we have seen, in *The Dehumanization of Art* Ortega y Gasset wrote that the price to be paid for the triumph of art over everything is that it may turn into a never-ending farce, into "a system of mirrors which indefinitely reflect one another" so that "no shape is ultimate, all are eventually ridiculed and revealed as pure images."[84] In another essay, "On Science," he also pointed out the limitations of scientific truth, which is exact but turns "a deaf ear to the last dramatic question [...]: where does the world come from, and whither is it going? Which is the supreme power of the cosmos, what the essential meaning of life?"[85]

In *Yellow Dog*, by Martin Amis, the narrator tells the story of a character who loses his memory after having being assaulted by thugs in Camden Town, London. Most of the novel is set in the hospital where he is being cured both physically and psychologically. The world is for him a space without references, peopled with shadows:

> Intensive Care felt like a submarine or an elderly spaceship: dark compartments where important devices whirred and ticked – electrocardiograms, panting ventilators; the churning of life and death in shapes and shadows.[86]

In another part of the book we read:

> It was like an investigation into the very early universe, that infinitesimal fragment of time which was obscured by the violence of the initial conditions. You couldn't quite reach the Big Bang – no matter what you did.[87]

In a world deprived of a god, of an essence, of a sense of identity, of a form, of an origin to refer to, and in which ideas are only inter-referential copies of

---

[83] Hassan, *The Dismemberment of Orpheus*, 271.
[84] Hassan, *The Dismemberment of Orpheus*, 48–49.
[85] Ortega y Gasset, "On Science", in *History as a System* (New York: Norton, 1962), 13–15.
[86] M. Amis, *Yellow Dog* (New York: Vintage-Random House, 2005), 36.
[87] Amis, *Yellow Dog*, 97.

each other, Plato's metaphor of the Cave is bound to remain an unanswered question. What should shadows stand for?

These shadows can be associated with the postmodernist version of *mise-en-abyme*: "as repeated reduplication 'in infinitum' in which the [...] mirroring fragment bears within itself another mirroring fragment."[88] They can also be identified with an unfulfilled desire of meaning, like Lutzker's 'reminder' or 'desire' of the body;" or with Lacan's "shadows [...] departing from our-selves".[89] Likewise Derrida's idea of "supplement", "which supplements the failing origin,"[90] can express the sense of void.

Finally, we can add to these possible interpretations Hassan's metaphor of the "Holy Ghost," a postmodernist figure that the scholar, in his famous list of binary oppositions[91], opposes to the modernist "God the Father." I have long considered this opposition and found it difficult to arrive at an interpre-tation. The only one I could find is that, being the Third Person of the Trin-ity, the Holy Ghost could be assimilated to the postmodernist idea of the Other.[92] At the same time, though, since it connects man to the mystery of God's death and Resurrection, the Holy Ghost could also be assimilated to a medium which provides a bridge between the known and the unknown, the visible and the invisible, the living and the dead. In McLuhan's *The Medium is the Massage. An Inventory of Effects* (1967), the interpretation of media as both means of communication and ghostly presences is very strong, at the visual as well as at the verbal level, as Quentin Fiore's graphics suggests. The fa-mous phrase "The medium is the message,"[93] is illustrated by a black-and-white out-of-focus photograph closing-up on a ghostly appearance (only after a while do we recognize in the eerie figure the face of someone whose eyes are blotted out by dark sunglasses and who is talking through a mega-phone). In *Lullaby* Palahniuk's thesis is precisely this: language (information) may kill, it can lull us to death. Silence is impossible, "You can't think. There's always some noise worming in. Singers shouting. Dead people laugh-

---

88 L. Hutcheon, *Narcissistic Narrative*, 55–56.
89 "Me and my shadow getting farther and farther apart" (Lutzker, *Lullaby*, 221).
90 Cfr. J. Derrida, *De la grammatologie* (Paris: Les Éditions de Minuit, 1967), 207–218.
91 Hassan, *The Dismemberment of Orpheus*, 268.
92 Cfr. E. Levinas, *La traccia dell'Altro* (Milano: Franco Angeli, 2008).
93 "The medium is the massage" is a parody of the phrase "The medium is the mes-sage," coined by M. McLuhan in *Understanding Media: The Extension of Man* [1964] (Cambridge: The MIT Press, 1994). The change of the word "message" into "massage" was either the publisher's mistake or a suggestion from Quentin Fiore. In either case, McLuhan decided to adopt the term "massage" instead of "mes-sage" in the book because it denoted very effectively both the "manipulative" and "pampering" effect media have on our senses.

ing. Actors crying. All these little doses of emotion."[94] The media are like parasites that use *us* as media: "We're all of us haunting and haunted. Something foreign is always living itself through you. Your whole life is the vehicle for something to come to earth. An evil spirit. A theory."[95]

Maybe this Holy, or un-Holy, Ghost can be identified with postmodernism itself. In the end, as Hassan writes: "What was postmodernism, and what is it still? I believe it is a revenant, the return of the irrepressible; every time we are rid of it, its ghost rises back. And like a ghost, it eludes definition."[96]

---

[94] Lutzker, *Lullaby*, 19.
[95] Lutzker, *Lullaby*, 258–9.
[96] I. Hassan, "Postmodernism to Postmodernity: the Local/Global Context". The text is available at: http://www.ihabhassan.com/postmodernism_to_postmodernity.htm.

Chiara Battisti
University of Verona

Fay Weldon's *The Lives and Loves of a She Devil*: Cosmetic Surgery as a Social Mask of Personhood

> "She looked at the body that had so little to do with her nature, and knew she'd be glad to be rid of it."[1] "I do not put my trust in fate, nor my faith in God. I will be what I want, not what He ordained. I will mould a new image for myself out of the earth of my creation. I will defy my Maker and remake myself." [LLSD, 186].

Body, nature, will, mould, creation, remake: these are words that bring us into the heart of our discussion, i.e. the possibility of creating, enhancing and transforming our body according to our will.

Throughout the second half of the XX century, feminist theorists belonging to various currents of thought have analysed Western representations of the body in an analytical perspective. Thanks to their efforts we have learned to read all the texts Western culture has to offer – literary works, philosophical texts, works of art, medical literature, films, and fashion – in a much less ingenuous and more aware manner.

In analysing the body's representation, cosmetic surgery has been dealt with as a problematic issue for feminist theory on femininity and women's relationship with their bodies. Whereas critics provide convincing explanation as to why feminine beauty norms and practices are so oppressive, they have some drawbacks when it comes to understanding cosmetic surgery as a practice that is harmful and degrading and yet fervently desired and actively pursued by thousands of women and girls each year. Feminists themselves are deeply divided on the issue. Some of them describe cosmetic surgery as an act of "taking control". They acknowledge that our choices are made within social and cultural contexts but they respect the individual's choices as a locus of personal power, creativity, and self definition. In contrasting themselves to those feminists who view the industries involved in feminine self-improvement as totalizing agents of women's oppression, they insist that

---

[1] F. Weldon, *The Lives and Loves of a She-Devil* (London: Hodder and Stoughton, 1983), 242.

these industries have in fact an important role in empowering women. The position of feminism that views cosmetic surgery as an agent of women's oppression, underlines, on the contrary, the power of cultural construction and argues that cosmetic surgery is more than an individual choice. It is, in fact, a normative cultural practice that encourages individuals to see themselves as defective.

The result of such a critical indecision is that cosmetic surgery and the normative and political topics related to it seem to be silent, and silenced, which are issues both in mainstream bioethics and in feminist medical ethics. In the attempt to break this silence, I believe it important to quote Hans Jonas' words in order to provide a brief conceptual preamble to my considerations. In his article "Toward a Philosophy of Technology", Jonas introduces a distinction between premodern and modern technology. A peculiar trait of modern technology is that the relationship between means and end, far from being unilinear as in the past, is now circular, so that "new technologies may suggest, create, even impose new *ends*, never before conceived, simply by offering their *feasibility*"[2].

If for centuries Western scientific medicine has considered the human body a machine that, as such, can be treated in an atomistic and mechanical fashion, the *feasibility* offered by the spectacular rise of new technology in cosmetic surgery has introduced some implications, among which the idea and purpose of transforming the human body into an ever more perfect object. A form of biotechnology that fits the dialectical picture of modern technology, in which we regard ourselves as both technological subjects and objects that are transformable and literally creatable through biological engineering, is thus introduced.

From this point of view, Fay Weldon's novel *The Lives and Loves of a She-Devil* (1983) can be considered a source of inspiration for an alternative point of view that shifts the debate on cosmetic surgery away from the perhaps unsolvable dilemma to its positivity or negativity. This allows us to focus it, first and foremost, on the ethical and biological borderlines of intervention on the body and especially on the connected implications for new possibilities of enhancing human traits.

Before facing the ethical challenges introduced by Fay Weldon, I would like to briefly summarise the novel for those who are not familiar with it.

When Ruth's unfaithful husband Bobbo calls her a "she-devil", she decides to appropriate that identity with a vengeance and take up a different

---

2 H. Jonas, "Toward a Philosophy of Tecnology", in *Hastings Center Report 9*, (1 February 1979): 34–43, 35.

position in the world's power relations. She wants revenge, power, money, and "to be loved and not love in return" [LLSD, 49]. In particular Ruth wants to bring about the downfall of her husband's lover Mary Fisher, a pretty blonde romance novelist who lives in a tower by the sea and lacks neither love, nor money and power. Ruth begins her revenge by burning down her own house and dumping her children on Mary and Bobbo. She goes on a literal shape-shifting quest during which she changes identity, gains skills, power, and money and explores and critiques key sites of power and powerlessness in contemporary society, including the church, the law, the geriatric institution and the domestic home. By the end of the novel, Ruth achieves all her goals. Her success, however, raises complex ethical questions, not only because she uses the same strategies of manipulation and cruelty of which she was a victim, but also because of the painful physical reconstruction of her body that is the tool of her victory. Ruth's willingness to mortify and mutilate her own body to achieve power has induced critics to raise serious questions about the value of the power she achieves, and who she oppresses most through her success. In fact, critical readings of the novel have often highlighted that Weldon's work leaves the reader with several enigmas. It is a feminist novel about sexual politics, replete with shocking examples of female oppression and male treachery. It is the story of a woman who suffers so much due to cultural conceptions of feminine beauty that she is willing to undergo the excruciating pain and staggering expense of cosmetic surgery to alter every part of her body. Nevertheless, the heroine also uses cosmetic surgery as a source of empowerment, a way to regain control over her life. Ruth is both a victim of the feminine beauty system and one of its most devastating critics. Her decision to undergo cosmetic surgery both supports the status quo of feminine inferiority and shifts the balance of power – temporarily, at least – in her own relation. Weldons' tale becomes, in this sense, a bitter commentary on the constraints of normal femininity and on heterosexual relationship. It offers an intricate, hilarious, but also increasingly disturbing remark on women's compliance with the concept of beauty and their willingness to undergo terrible suffering for a man's love. I think, however, that the section of the novel in which Ruth uses her wealth to turn herself into the clone of her enemy Mary Fisher thanks to cosmetic surgery opens several and interesting avenues of inquiry on cosmetic surgery, going far beyond the standard plot of the plastic surgeon like Pygmalion.

The protagonist of the novel is perfectly conscious of the fact that female access to forms of power and empowerment appears to be, and very often is, so limited that cosmetic surgery is the primary domain in which a woman can experience some semblance of self-determination. In this sense, on reflecting upon women's increasing recourse to cosmetic surgery Robin Lakoff and Rachel Scherr suggest:

> The only way to understand the situation is to agree that those conditions are, in fact, perceived as life threatening, so dangerous that seriously damaging interventions are justified, any risk worth taking, to alleviate them[3].

Cosmetic surgery becomes a way an individual woman can choose as a kind of subjective transcendence against a background of constraint and limitation.

While discussing her plans to have cosmetic surgery with the judge – parenthetically, a man she has seduced in return for assistance in causing her husband's financial downfall – Ruth compares herself to Christian Andersen's little mermaid:

> In this particular case I am paying with physical pain. Hans Andersen's little mermaid wanted legs instead of a tail, so that she could be properly loved by the Prince. She was given legs, and by inference the gap where they join at the top, and after that, every step she took was like stepping on knives. [LLSD, 173]

Fay Weldon introduces a fairy tale, *The Little Mermaid*, in her novel that links, as no other can, women's subordination in heterosexual relationship with beauty, as if to strengthen the idea that the success of all women as women is defined in terms of interlocking patterns of compulsion that determine the legitimate limits of attraction[4]. Ruth knows, as does the little mermaid, that such compulsory attractiveness implies suffering since it is not based on self-determined ideals of integrity and health; it is, on the contrary, defined as attractive to men.

This fairy tale has a subtext, however. It is also a story about feminine wiles and subterfuge behaviour – a woman who purposely deceives in order to get her way. The little mermaid knows the rules of the game and plays by them. So does Ruth.

> "Of course it hurts," she said. "It's meant to hurt. Anything that's worth achieving has its price. And, by corollary, if you are prepared to pay that price you can achieve almost anything" [LLSD, 172].

Both the mermaid and Ruth are well aware of the pain involved in magic/ cosmetic surgery, but they know that women have traditionally regarded their bodies, especially if they are beautiful and young, as a locus of power:

---

3 R. T. Lakoff and R. Scherr, *Face Value: The Politics of Beauty* (Boston: Routledge and Kegan Paul, 1984), 165–166.

4 K. P. Morgan "Women and the Knife: Cosmetic Surgery and the Colonization of Women's Bodies" *Hypatia*, 6: 3 (Autumn 1991), 25–53, 32.

the affirmation of a woman's beauty brings with it privileged heterosexual affiliation, and privileged access to forms of power that are unavailable to the plain, the ugly and the aged.

The piercing pain, endured in silence by the little mermaid at each step, becomes a telling metaphor of the discomfort and risks that accompany every cosmetic surgery operation. Cosmetic surgery is always painful, sometimes requiring hospitalization for several days, not counting the possible side-effects that might even include death. In addition to the pain involved in any major operation, cosmetic surgery patients may be considered complainers by the hospital staff. Their suffering is viewed as self induced and therefore not taken as seriously as that of patients with cancer or heart disease[5].

Kathryn Pauly Morgan attempts to explain women's involvement in cosmetic surgery and the natural acceptance of the "self induced" suffering of pain connected with it by observing how a woman's pursuit of beauty through transformation is often associated with experiences of self creation and self-fulfilment in our society:

> Electing to undergo the surgery necessary to create youth and beauty artificially not only appears to but often actually does give a woman a sense of identity that, to some extent, she has chosen herself. Second, it offers her the potential to raise her status both socially and economically by increasing her opportunities for heterosexual affiliation. [...] third, by committing herself to the pursuit of beauty, a woman integrates her life with a consistent set of values and choices that bring her wide-spread approval and a resulting sense of increased self-esteem.[6]

According to Sara Goering, those who undergo cosmetic surgery gain – or believe to gain – much more than just an aesthetic advantage. How one looks affects not only one's self-esteem and confidence, but also how others regard

---

[5] The distinction between self-induced suffering and the suffering of patients with cancer or heart disease is retraceable to the distinction between enhancement and treatment. If we define treatment "interventions responding to genuine medical needs", the concept of enhancement defines a situation where "the desire for self-improvement, but no medical illness or impairment, motivates these interventions." (D. DeGrazia, "Enhancement Technologies and Human Identity" *Journal of Medicine and Philosophy*, 30: 3 (New York and London; Routledge and Kegan Paul, 2005), 261–283, 263). However, such a definition, which clarifies the concept of enhancement by way of contrast with treatment is compelling only if the treatment/enhancement distinction is meaningful. Even if treatment has to do with real physical problems, one might question the meaningfulness of the treatment/enhancement distinction in a range of cases because the very distinction between normal and abnormal health is often arbitrary.

[6] Morgan "Women and the Knife: Cosmetic Surgery and the Colonization of Women's Bodies", 34.

one's competence, personality and likelihood for success.[7] It is, therefore, essential not to underestimate, as Sandra Lee Bartky, Kathryn Pauly Morgan and others have argued, the identity-confirming role that femininity plays in bringing women as subjects into existence while simultaneously creating them as patriarchally defined objects[8]. They stress how, under those circumstances, "refusal may be akin to a kind of death, to a kind of renunciation of the only kind of life-conferring choices and competencies to which a woman may have access"[9]. Considering the circumstances we have highlighted, in the novel Ruth seems to choose the lesser of two evils: her recourse to cosmetic surgery becomes a way to conform to the beauty norm rather than fall victim to it, a voluntary act to control her identity.

By combining the contradictory and disturbing dimension of cosmetic surgery and a feminist critique of the power relations between the sexes, Fay Weldon demonstrates how ambivalences (women as victims) of cosmetic surgery versus cosmetic surgery as a resource of sorts (in the power struggle between the sexes) can be embraced rather than dismissed or avoided.

However there is an aspect of Weldon's open quotation of the fairy tale which offers a completely innovative and extremely interesting reading perspective by suggesting an interaction between cosmetic surgery, biolaw and the concept of personhood. And it is precisely the reference to the voice that discloses another level of reading the novel. As a matter of fact, in the *The Little Mermaid*, in addition to physical pain the price the little mermaid must pay in order to have legs is to lose her voice. The Sea-witch demands it in exchange for her spell:

> "But I must be paid also," said the witch, "and it is not a trifle that I ask. You have the sweetest voice of any who dwell here in the depths of the sea, and you believe that you will be able to charm the prince with it also, but this voice you must give to me; the best thing you possess will I have for the price of my draught. My own blood must be mixed with it, that it may be as sharp as a two-edged sword."
> *"But if you take away my voice," said the little mermaid, "what is left for me?"*
> "Your beautiful form, your graceful walk, and your expressive eyes; surely with these you can enchain a man's heart. Well, have you lost your courage? Put out

---

7 S. Goering, "The Ethics of Making the Body Beautiful: What Cosmetic Genetics Can Learn from Cosmetic Surgery" *Philosophy and Public Policy Quarterly*, 21: 1, (Winter 2001), 21–27, 23.

8 See: S. L. Bartky "Foucault, Femininity, and the Modernization of Patriarchal Power" in eds. I. Diamond and L. Quinby, *Femininity and Foucault: Reflections of Resistance* (Boston: Northeastern University Press, 1988); Morgan "Women and the Knife: Cosmetic Surgery and the Colonization of Women's Bodies".

9 Morgan "Women and the Knife: Cosmetic Surgery and the Colonization of Women's Bodies", 43.

your little tongue that I may cut it off as my payment; then you shall have the powerful draught." "It shall be," said the little mermaid.[10]

Although, according to certain feminist perspectives, "the tale becomes a metaphor for emotional dead ends and repetition, the mermaid's voiceless-ness a sign of inability to speak in a way that might change things"[11] I believe that in the fairy tale the voice is an emblematic indicator of personhood, thus becoming the subject's very essence:

> The voice corresponds to that which is most hidden and true in the single person. However, it does not consist in an intangible treasure, an indescribable essence, nor much less in the self's secret core, but rather in the profound vitality of the unique being who enjoys his/her revelation through the voice's emission.[12]

In a parallel manner, and not by chance, in *The Lives and Loves of a She-Devil* the surgeon who operates on Ruth's body significantly decides not to change the tone of her voice, almost as if to reiterate the wish not to operate on the person's most profound essence:

> He thought he might leave her vocal chords untautened: a voice that sounded harsh out of a massive frame might sound husky, and by inference sexy, from a slighter one. It was the balance of male and female in the body that attracted, he observed. A male desire in a fragile body, a deep voice in conjunction with a delicate gesture, and so forth: duplicity and artifice, and not simplicity at all. [LLSD, 248]

Nevertheless, after all the cosmetic surgery that transformed Ruth into Mary Fisher's physical clone even her voice "sounded as Mary Fisher's had, long ago" [LLSD, 261].

It is precisely the significant change in Ruth's tone of voice – which, as previously, asserted says much about its owner and enables us to recognise it – that invites us to turn our considerations to cosmetic surgery from an ex-tremely subjective level of judgement to a consideration on the relationship between cosmetic enhancement and identity and allows us to expand the im-plications of the concept of *persona*.

The reference to a change in voice and Ruth's own claim, to which I shall return further on, of having an exceptionally adaptable *personality* and of having tried to fit herself to her original body in many ways [LLSD, 236–237] induces us to ponder the idea of the body as an arduous mask that protects

---

[10] H. Ch. Andersen, *The Little Mermaid*, [1836].

[11] C. Kay Steedman, *Landscape for a good Woman: A Story of Two Lives* (New York: Rutgers UP and New Brunswick, 1997).

[12] A. Cavarero, *A più voci. Filosofia dell'espressione orale* (Milano: Feltrinelli, 2003), 10. My translation.

and enables us to function better as a public person by endowing us with a voice and visibility. What I therefore wish to suggest is that the body itself may become a mask, which is moulded like a voluntary act or perceived as the obligation to ostentatiously represent one's self. Clothes, cosmetics, perfume and cosmetic surgery further shape such a mask.

Etymology and semantics allow us to properly grasp this connection between voice, visibility, persona and mask. According to *The Oxford English Dictionary* the word *person* "traces back at least to the Latin *persona*: a mask, especially as worn by an actor, or a character or social role".[13]

This concept is closely associated with the idea of voice; in fact, in Greek 'pro'sopon indicates both 'that which one looks at in front' and 'that which consents one to see through'. It therefore represents a sort of optical instrument which is privileged in establishing a particular sort of relation. In Latin it is possible to perceive an analogous expressive function: 'person' in this case means 'to play through'. Theatrical masks attained the effect of amplifying the voice's emission by altering it and, in doing so, bestowed force and power, thus making it audible.

The expansion of the term *persona* from theatrical lexicon to legal terminology is due to its fundamental intermediate role between person and role, intimate experiences and public representation, individual identity and collective identity as performed by the *persona*/mask. Lawsuits and litigations therefore become performances in which everyone acts within a representative function, both of others and of oneself. "In general, the word [*persona*] is preferably used compared to other analogous words when referring to that which the individual is in the eyes of others or in his or her function amongst others; role, office, dignity, rank";[14] these are elements that enable us to understand how a word meaning "mask" can also be used, in certain relationships, to show what man is in front of other men.

Even if in the modern age the mask has essentially become an artistic object, the metaphor of the mask still helps us understand the necessary mechanisms to construct identity, the dialectic between hiding and revealing and, with it, the ambivalences of modern subjectivity. In today's society, the subject's representing his/herself is in function of the expression of a profound identity, for he/she must create his/her own masks and make them recognizable, he/she must give life to a social performance that is situated in determined contexts. This way, the metaphor of the mask enables us to conceive identity not as "a subjective feeling", but as "a commission by others,

---

13 *The Oxford English Dictionary*, (Oxford: Clarendon, 1989) 596–7.
14 A. Pizzorno, *Sulla Maschera*, (Bologna: Il Mulino, 2008) 76.

an act of recognition".[15] Identity therefore is considered a process of learning and negotiating reality. "And if the subjects do not act according to the masks they are entrusted with it will be necessary to mould new masks, [...] working on a repertoire of enduring references that will gradually be adjusted and reinterpreted to re-identify the actor and re-establish social order".[16]

Highlighting this function of the mask and recalling that in their essay, "The Feminine Body and Feminist Politics"[17] Beverly Brown and Parveen Adams argue that the body should be considered not as a site of political refusal, but as a site for feminist action through transformation, appropriation, parody and protest, we could consider Ruth's protest an anticipatory version of what Judith Butler calls "Gender Performatives". Ruth, by undergoing cosmetic surgery, wears a mask that endows her with a voice on a social level, and enacts a performance of gender which shows that cosmetic surgery is simply revealing what is true for all embodied subjects living in cultures, that is, that all human bodies are cultural artefacts. What is more, the practices of cosmetic surgery exemplified by the novel provide "extremely public and quantified reckoning of the cost of beauty, thereby demonstrating how both the processes and the final product are part of a larger nexus of women's commodification"[18]. Rather than agreeing that participation in cosmetic surgery will result in the further victimization of women, Weldon's feminist strategy advocates appropriating such technology for feminist ends, that is, to highlight the arbitrariness of cultural norms that currently lead women to choose cosmetic surgery.

Ruth however, in putting her identity to the test through an indispensable game based on an alternative representation of herself, gives shape to an experimental and extreme way of existing in which the metaphor that sees cosmetic surgery as a mask also becomes a significant expression of the construction of biographical identity. Ruth's choice to be represented by her body which – thanks to cosmetic surgery – had been transformed into and disguised as Mary Fisher's body becomes the possibility for Ruth to recognize herself as a human being by participating in her identity and taking possession of her social presence/voice. In this sense, the very expression *"to be represented"* by the mask is determining when indicating such a creation

---

[15] A. Pizzorno, *Le radici della politica assoluta e altri saggi*, (Milano: Feltrinelli, 1993) 92.

[16] R. Sassatelli, "Postfazione" in Pizzorno, *Sulla Maschera*, 110.

[17] B. Brown and P. Adams, "The Feminine Body and Feminist Politics" *M/F* 3 (1979) 35–50.

[18] K. P. Morgan "Women and the Knife: Cosmetic Surgery and the Colonization of Women's Bodies", 44.

of the notions of person and social person, in that it contains the words "*to be*", which suggests the idea of an identified and concluded presence, and "*represented*", in other words made present to others.

Through the "cosmetic mask", the representation – Ruth's "being represented" – in her presumption of identity submits to the interference enacted by the interpretation of appearance. That which appears and lives in fact are the actions of a new identity which is immediately inserted in the presence of others and made present by other people's gazes.

These observations underline how human beings tend towards the visible constitution of his/herself by constructing and narrating an identity and biographical continuity. In this sense, Ruth's enhancement project, which is explicated through the taking on of a "cosmetic mask", appears to be connected with her narrative identity by affecting her self-conception and reveals the importance of exploring enhancement technologies in relation to identity and self-creation. The enhancement process, which is desired and realised by Ruth, draws the veil of incognito over her personal identity and introduces a different sense of identity, thereby transforming her narrative identity, that is, "the way she organizes the story she tells herself about herself"[19].

For the procedure that concerns our considerations, I think it is important to clarify the concept of enhancement and provide a framework to understand human identity. In biomedicine, enhancements are commonly understood as "interventions designed to improve human form or functioning beyond what is necessary to sustain or restore good health"[20]. Because it is often difficult to understand where this beyond can be located, the shade DeGrazia attributes to the concept of enhancement is essential, that is: "Enhancements are interventions to improve human form or function that do not respond to genuine medical needs, where the latter are defined: 1. in terms of disease, impairment, illness, or the like 2. as departures from normal (perhaps species-typical) functioning or 3. by reference to prevailing medical ideology"[21].

This conception categorises enhancements by the goal of improvement in absence of medical need. It is consequently immediately comprehensible that many scenarios involving cosmetic surgery are paradigm instances of enhancement.

---

[19] D. DeGrazia, *Human Identity and Bioethics*, (Cambridge: Cambridge University Press, 2005) 8.

[20] E. Juengst, "What does enhancement mean?" in ed. E. Parens, *Enhancing Human Traits*, (Washington, D. C.: Georgetown University Press, 1998) 29.

[21] DeGrazia, "Enhancement Technologies and Human Identity" 263.

"I am here to improve my body" [LLSD, 249] says Ruth to the plastic surgeon. Furthermore:

> I have an exceptionally adaptable **personality** [...], I have tried many ways of
> fitting myself to my original body, and the world into which I was born, and have
> failed. I am no revolutionary. Since I cannot change the world, I will change
> myself. I am quite sure I will settle happily enough into my new body. [LLSD,
> 236–237]

This passage introduces the relationship between enhancement and identity,
which is powerfully called into play due to the cosmetic surgery, and the
above analysed idea of person, in its accepted meaning of mask which be-
stows a social voice.

However, as David DeGrazia points out in his essay "Enhancement
Technologies and Human Identity", "enhancement via biotechnology is
inherently problematic for reason pertaining to our identity"[22]. If biotechnologies' rapid growth increasingly creates many possibilities for enhancing human traits, it has also caused a sense of urgency to attend to ethical
issues.

Fay Weldon's novel, as already mentioned, specifically and accurately analyses and questions some of these issues. As a matter of fact, one of the most
problematic features of enhancement technologies is precisely their connection with central aspects of a person's identity and their possibilities to alter
one's identity. In this sense Carl Elliott in *A Philosophical Disease: Bioethics, Culture and Identity* asserts:

> What is worrying about so-called "enhancement technologies" may not be the
> prospect of improvement but the more basic fact of altering oneself, of changing
> capacities and characteristics fundamental to one's identity [...] Deep questions
> seem to be at issue when we talk about changing a person's identity, the very core
> of what the person is. Making him smarter, giving him a different personality or
> even giving him a new face – these things cut much closer to the bone ... They
> mean, in some sense, transforming him into a new person[23].

The concern is that enhancement technologies, such as cosmetic surgery,
threaten to alter the self fundamentally, thereby changing someone's identity,
transforming him/her into a new person- and that such a change is objec-
tionable/questionable. At this point, referring to the novel, the provocative
question is: after a long ordeal of operations who is the person who leaves
the operating room? Ruth or Mary?

---

[22] DeGrazia, "Enhancement Technologies and Human Identity" 261.
[23] C. Elliott, *A Philosophical Disease: Bioethics, Culture and Identity* (New York: Routledge, 1999) 28–29.

Has Ruth's objectivising need to substitute her face and appearance with Mary's therefore hidden the person named Ruth, thus also interrupting her personal identity?

In facing such questions the novel weaves together with issues concerning the philosophical literature on human identity – or personal identity, as it is usually called. What sense of identity, however, is at hand? According to most critical positions there are two distinct senses of human identity which prove to be important to certain ethical issues: numerical identity and narrative identity.

Numerical identity points to the relationship an entity has with itself over time in being one and the same identity. Such concept of identity requires the possibility of a continuity of knowledge about our own identities and identifies itself in memory and anticipation, major resources which make self-knowledge possible over time, and are two constitutive elements of our identity. David DeGrazia exemplifies how numerical identity consists in the continuity of a mental history over time, where present and past transient moments of awareness are connected by memory "A human person undergoes enormous changes over the years, yet a single individual may reflect on all of these changes as having occurred to him or her"[24].

On the one hand, in analysing the concept of numerical identity, the psychological approach traces the persistent conditions of our numerical identity in some part of the psychological continuity, where this relevant type of continuity is one of the experiential contents, or the maintaining of psychological connections. On the other hand, over time the biological approach, instead, sustains that our continuing existence over time requires no psychological continuity: rather, we continue to exist so long as we remain alive.

Besides numerical identity there is also what we have already defined as the narrative identity, which involves an individual's self-conception, that is to say his/her implicit autobiography, self-image, and sense of oneself as the protagonist of one's life story. According to David DeGrazia it is important to reflect on the role of enhancement projects in the narrative identities and self-creation of individuals, whereas the metaphysical relationship that an entity has to itself over time, that is to say the numerical identity, is not at issue because the person who exists after the enhancement will remember life before the intervention.[25]

---

[24] Elliott, *A Philosophical Disease: Bioethics, Culture and Identity* (New York: Routledge, 1999) 264.

[25] See DeGrazia, *Human Identity and Bioethics*, 229.

The novel demonstrates that some of Ruth's traits were so basic to who she was that they represented a kind of core of her narrative identity; changing parts of that core changed her self-narrative so profoundly that in a sense (that of narrative identity) the result is a different person (Mary). If realistic scenarios of cosmetic surgery do not imply changes as drastic as the ones suggested by the novel, and their logical possibility is "only" an alteration of a person's narrative identity, what the novel seems to further suggest is the possibility of a transformation of the person's numerical identity. In my opinion, the enormous changes Ruth undergoes prevent her from corresponding to the above mentioned definition of numerical identity: Ruth, in fact, cannot reflect on all of these changes as having occurred to *her*, for at the end of the novel Ruth's gestures, tastes, movements, way of thinking, personality, interests and even voice have all naturally merged into Mary's.

If wearing a mask suggests identifying oneself with the other who enables us to develop a "process of experimental and virtual simulation that allows an extension of subjectivity"; imitation should however only constitute the evolutionary requirement for such manifestations. Far from increasing the self's subjectivizing, in Ruth's case the use of masks/cosmetics realises her total identification with Mary, the being that the mask represents.

Moreover, with the gradual transformation that brings Ruth to become a copy of Mary comes Mary's progressive physical decline, almost as if to reaffirm the impossibility of two people with coinciding numerical identities to exist. It is precisely at Mary Fisher's funeral that Ruth witnesses the consecration of her new identity: old Mrs. Fisher, in fact mistakes her for her daughter and declares: "I thought it was her for a moment. [...] Just like her to send her own ghost to her funeral!" [LLSD, 272].

In conclusion, though recognising that the case considered in the novel is referred to as an alteration of numerical identity that is currently impossible to execute, to face cosmetic surgery from the point of view of its enhancement/identity relation implies recognising the necessity to reflect on the possibility of extreme changes that can create a new person with a new way of socially interacting. Ruth's case shows that we have arrived at a stage where the possibilities offered by the spectacular rise of new technology in cosmetic surgery, and in operating radical changes in the body represent a shift not only in one's experience of one's own body, but also in one's experience of one's self per se. In fact, the theme of identity is closely intertwined with that of the body and the development of bodily identity represents the plot upon which to construct the self as persona.

Appendix

Leif Dahlberg
Kungliga Tekniska Högskolan, Stockholm

Mapping the Law – reading old maps of Strasbourg as representing and constituting legal spaces and places

> Die Griechen haben kein Wort für "Raum". Das ist kein Zufall; denn sie erfahren das Räumliche nicht von der extensio her, sondern aus dem Ort (τόπος) als χώρα, was weder Ort noch Raum bedeutet, was aber durch das Dastehende eingenommen, besetzt wird.
>
> (Martin Heidegger)[1]

> Wir sind so gewöhnt, Gesetz und Recht im Sinne der Zehn Gebote als Gebote und Verbote zu verstehen, deren einziger Sinn darin besteht, daß sie Gehorsam fordern, daß wir den ursprünglich räumlichen Charakter des Gesetzes leicht in Vergessenheit geraten lassen.
>
> (Hanna Arendt)[2]

Introduction

Conrad Morant's 1548 woodcut map of the city of Strasbourg (Strateburg, Strassburg) is only the third surviving map portraying an entire city North of the Alps in the Holy Roman Empire.[3] The other two German cities portrayed are Augsburg (cut by Hans Weiditz after a drawing by Jörg Sedel, 1521) and Cologne (by Anton Woensam von Worms, published by the editor and printer Peter Quentel, 1531).[4] The first known representation of an en-

---

[1] M. Heidegger, *Einführung in die Metaphysik* [1953] (Tübingen: Max Niemeyer Verlag, 1998), 50.

[2] H. Arendt, *Was ist Politik? Fragmente nach dem Nachlaß*, hrsg. Ursula Ludz (München: Piper, 1993), 122.

[3] Map of Strassburg (87,5 × 68,2 cm), woodcut by Conrad Morant, 1548. Now in Germanisches Nationalmuseum, Nürnberg. The historical discussion of Morant's plan over Strasbourg is based on L. Châtelet-Lange, *Strasbourg en 1548. Le Plan de Conrad Morant* (Strasbourg: Presses Universitaires de Strasbourg, 2001).

[4] Map of Augsburg (59,2 × 352,6 cm), woodcut by Hans Weiditz after a drawing by Jörg Sedel, 1521; map of Cologne, Anton Woensam von Worms, published by the editor and printer Peter Quentel, 1531. See H. Appuhn & C. von Heusinger, *Riesenholzschnitte und Papiertapeten der Renaissance* (Unterschneidheim: Verl. Alfons Uhl, 1976) 45–50; woodcut of Köln reproduced in fig. 33 & 34. See also F. Bach-

tire city in modern times is a woodcut view of Florence made by Francesco Roselli between 1478 and 1482 (surviving in a single print from 1505).[5] The second surviving and perhaps most well known representation of a modern city is Jacopo de' Barbari's woodcut view of Venice (1500).[6] Other important early representations of modern cities include woodcut views and plans of Antwerp (1515),[7] Amsterdam (1544),[8] Lyons (1545–1453),[9] Paris (1550),[10] and Norwich (1558/1559).[11]

Equally significant are the printed books with profile views, elevated bird's-eye views and plans of cities, including Sebastian Münster's *Cosmographia* (1544), which contains a profile view of Strasbourg; Johannes Stumpf's *Schweizer Chronik* (1548); and in particular Georg Braun and Franz Hogen-

---

mann, *Die alten Städtebilder. Eine Verzeichnis der Graphischen Ortsansichten von Schedel bis Merian* (Leipzig: Verlag Karl W. Hiersemann, 1939).

[5] Map of Florence (57,8 × 131,6 cm), woodcut by Francesco Roselli, 1478–1882, surviving in a single print from 1505. See Appuhn & Heusinger, *Riesenholzschnitte und Papiertapeten der Renaissance*, 44, fig. 32. There exists earlier bird's-eye views of cities, such as Cristoforo Buondelmonte's picture of Constantinople (1420) and Bernard von Breydenbach's of Jerusalem (1480), but these are considerably less detailed and accurate.

[6] Map of Venice (135 × 282 cm), woodcut by Jacopo de' Barbari, 1500. Reproduced in Appuhn & Heusinger, *Riesenholzschnitte und Papiertapeten der Renaissance*, 39–44. See J. Schulz, "Jacopo de Barbari's View of Venice: Map Making, City Views, and Moralized Geography before the Year 1500", *Art Bulletin* 60 (1978), 425–474.

[7] "Antverpia mercatorum emporium" (53 × 216 cm), woodcut view by unknown artist, 1515. Now in Prentenkabinet, Museum Plantin-Moretus, Antwerpen. Reproduction in Appuhn & Heusinger, *Riesenholzschnitte und Papiertapeten der Renaissance*, fig. 31. See also Châtelet-Lange, *Strasbourg en 1548*, 27.

[8] "De vermaerde Koopstadt van Amstelredam", woodcut plan by Cornelis Anthoniszoon, 1544. British Museum (S.T.A (4)).

[9] Map of Lyons, (170 × 220 cm), woodcut plan, 1545–1453. In Les Archives municipales de Lyon.

[10] The woodcut plan of Paris exists in two different versions, 133 × 96 cm, and 82 × 68 cm, both from 1550. The two woodcut maps of Paris are based on a mural map (442 × 514 cm) from 1524–1529 (now destroyed). The larger woodcut map of Paris ("Plan de Bâle") is in the Universitätsbibliothek Basel (Kartenslg. AA 124); the smaller woodcut map ("Plan de Saint-Victor") is in the Bibliothèque nationale, Paris. See J. de Rens, "Notes sur les plans de Paris au XVIe siècle", *Bulletin de la Société de l'histoire de Paris et de l'Ile-de-France*, 107 (1980): 71–86; P. Pinon & B. le Boudec, *Les Plans de Paris. Histoire d'une capitale* (Paris: Atelier parisien d'urbanisme, 2004), 30–33.

[11] The woodcut view of Norwich is found in W. Cunningham, *The Cosmographical Glass* (London, 1559). See J. Elliot, *The City in Maps: urban mapping before 1900* (London: British Library, 1987), 39–41.

berg's *Civitates orbis terrarum* (1572), which contains a bird's-eye view of Strasbourg based on Morant's map.[12] However, it was not only chorographical representations of cities that became increasingly popular during the sixteenth century: the cartographic hype is manifest in the production of maps of regions, countries, and the world, as well as in the printing of atlases containing maps of every known part of the planet. In fact, during the one hundred year period 1475–1575, both through geographical explorations (including the "discovery" of the New World) and the development of cartography as an art and a science, the European perception of the world changed more radically than it did in any comparable period.[13]

This cartographic development in Early Modern Europe was accompanied by a displacement and radical transformation of our practical understanding of social space or lifeworld, as has been argued by philosophers from Martin Heidegger to Henri Lefebvre and Edward Casey. Whereas social space in earlier periods primarily had been defined by unique, concrete place (τόπος) or locality (χώρα), it increasingly was conceived as an abstract, measurable geometrical space (from Latin *spatium* (via French *espace*), meaning *room, distance, interval*).[14] According to Heidegger, this abstract notion of space does not exist in Greek philosophy, but appears much later, in Descartes' understanding of space as three-dimensional geometric and measurable space (where every location can be defined by co-ordinates).[15] However, Lefebvre maintains that this abstract, scientific representation of space began earlier and became dominant already with the Tuscan painters, architects and theorists in the Renaissance.[16] Heidegger further argues that in the Modern period – in the seventeenth century – things in the world began to be presented as objects in space and the world itself as a "world picture" (*Weltbild*).[17]

---

12 J. Stumpf, *Die gemeiner loblicher Eydgnoschafft Stetten, Landen und Volckeren Chronik wirdiger thaaten bescheybung* (known as the *Schweizer Chronik*), Christoph Froschauer, Zürich, 1548; Georg Braun & Franz Hogenberg, *Civitates orbis terrarum* (Köln, 1572); see also Elliot, *The City in Maps*, 26–37.

13 N. Thrower, *Maps & Civilization. Cartography in Culture and Society*, 3rd ed. (Chicago: University of Chicago Press, 2007), 64; J.-M. Besse, *Les Grandeurs de la Terre. Aspects du savoir géographique à la Renaissance* (Lyon: ENS Editions, 2003); J. Rennie Short, *Making Space. Revisioning the World, 1475–1600* (Syracuse: Syracuse University Press, 2004).

14 E. S. Casey, *The Fate of Place. A Philosophical History* [1997] (Berkeley: University of California Press, 1998).

15 M. Heidegger, *Sein und Zeit* [1927] (Tübingen: Niemeyer, 1986), 89–101 *et passim*.

16 H. Lefebvre, *La Production de l'espace* [1974] (Paris: Anthropos, 2000), 51.

17 M. Heidegger, "Die Zeit des Weltbildes" (1938), in *Holzwege* [1950] (Frankfurt am Main: Vittorio Klostermann, 2003), 75–113.

In Kant, still according to Heidegger, one finds this Cartesian understanding of space as three-dimensional applied to human perception, in the transcendental categories, with the claim that man perceives the physical world in three spatial dimensions and one temporal dimension.[18] Against this view, Heidegger and other phenomenological philosophers argue that man's sense of place and of his/her living body is primary (and that geometric space is secondary).[19] Similarly, Lefebvre describes the scientific representation of space as a code that silences a lived experience of space, which nevertheless remains intact.[20]

In contrast to Heidegger, for whom this new understanding of space began with Descartes, Lefebvre argues that this re-presentation of space had been in the making for centuries and goes back to the Greeks and Romans. If one takes into account the cartographic discoveries in Antiquity – such as the invention by Hipparchus of Nicaea in the second century BCE of a grid of parallels and meridians (longitude and latitude) crossing each other and enabling the exact location of places, a system further developed by Ptolemy in the second century CE – the historical and intellectual development and transformation of place and space indeed appears more complex and gradual than in Heidegger. On the other hand, even though the geometric paradigm of space in some modern scientific discourses may appear hegemonic and the chorographical paradigm subordinate or even absent, it would be incorrect to propose that the geometrical paradigm has replaced or erased the chorographical conception of space as place. As Lefebvre argues, the cultural history of space should not be conceived in terms of a discrete (or dialectical) movement from one spatial regime to another, but should be understood in terms of an ongoing interaction between the spaces produced in and through everyday life (*la pratique spatiale*), the spaces lived and experienced through images and symbols of space (artistic and non-artistic) (*les espaces de représentations*), and the abstract, geometric and scientific representations of space (*les représentations de l'espace*).[21]

---

[18] M. Heidegger, *Kant und das Problem der Metaphysik* [1951] (Frankfurt am Main: Vittorio Klostermann, 1998), 44–50; M. Heidegger, *Beiträge zur Philosophie (Vom Ereignis), Gesamtausgabe, Bd. 65* (Frankfurt am Main: Vittorio Klostermann, [1989] 2003), 370–387. See also F. Dastur, "Heidegger: Espace, lieu, habitation", in *Les Temps Modernes* 650 (juillet-octobre 2008): 140–157.

[19] M. Merleau-Ponty, *Phénoménologie de la perception* (Paris: Gallimard, 1944); M. Heidegger, "Bauen Wohnen Denken", in *Vorträge und Aufsätze* [1954] (Stuttgart: Klett-Cotta, 2004), 136–156.

[20] Lefebvre, *La Production de l'espace*, 51.

[21] Lefebvre, *La Production de l'espace*, 48–50.

The historical and intellectual displacements and transformations of our understanding of the social world in Early Modern and Modern Europe have far-reaching cultural, social, and political effects. During this period there emerges a perception of the state as a territorial entity and is delineated on maps with borders; cities and provinces are represented as belonging to countries (rather than as independent political entities) connected by networks of communication and transportation; and the national territory is represented increasingly as a geometrical and exactly measured space. In this essay I will argue not only that we can see a parallel development and transformation in legal space – from law being primarily local and regional to law becoming increasingly territorial and even universal – but also that these developments are connected with each other. This connection between law and cartography enables a new and different understanding of the social and legal development of constitutional nation states (*Rechtsstaaten*) in Europe and America in the Modern period. Although the relation between the developments of cartography and the nation state in Early Modern and Modern periods has been been well explored in recent research, to my knowledge the connections between cartography and law have not previously been studied. In this essay I have chosen to focus on Strasbourg, a city well situated to illustrate the development of legal places and spaces.

In order to understand what a map says about a city – and also what it does not say and keeps silent about[22] – one needs to learn more about the city itself. But one may also choose to focus on the map and put social reality within parenthesis. In this way the map is studied both as a "way of seeing" (an optics) and as a historical document. This does not mean that one should ignore social reality altogether, but it opens the study of old maps to discourse analysis in which maps appear as historical objects in their own right, and maps and map-making are viewed as an integral part of the social, political and geographical discourses that constitute the (hegemonic) social power structure.[23]

---

[22] As has been emphasized by J. B. Harley (among others), the meaning of maps is constituted as much by what they depict and describe (in image and word) as by what they leave out, what he calls cartographic 'silences'. See J. B. Harley, "Silence and Secrecy. The Hidden Agenda of Cartography in Early Modern Europe", in *The New Nature of Maps: Essays in the History of Cartography*, ed. Paul Laxton (Baltimore: Johns Hopkins University Press, 2001), 83–107. See also Jai Sen, "Other Worlds, Other Maps. Mapping the Unintended City", in *An Atlas of Radical Cartography*, ed. Lize Mogel & Alexis Bhagat (Los Angeles: Journal of Aesthetics & Protest Press, (2007) 2008) 13–26.

[23] For the notion of discourse analysis, see e.g. N. Fairclough, *Critical Discourse Analysis* (Boston: Addison Wesley, 1995). For discursive studies of maps and map-mak-

In pursuing this approach, this essay looks at old maps and asks what they can tell us about the law and its historical and institutional development. The fact that prior to the nineteenth century law courts were often not explicitly marked on maps (or by architectural traits) is revealing of the decentralized and subordinate role of law in these societies, but also revealing of the fact that law (and legal space) was not seen as distinguishable from political power. Furthermore, one can explore how maps depict legal facts (like property), administrative and political borders, as well as other traces of law in terms of organization, shape and categorization of spaces, both public and private. For instance, on Early Modern maps the term *forest* (or *forêt* in French, from Latin *foris*, outside) does not denote the presence of woods, but denotes land (of any kind) *outside* the common law, traditionally for the use of the crown or the nobility.[24]

In investigating how old maps represent social or anthropological space, in relation to the representation of law and legal institutions, this essay also tries to understand how maps construct, and thus constitute, the reality they depict. As already mentioned there existed in Early Modern and Modern Europe a parallel development between the science and industry of cartography and the emergence of the nation state, and there also existed a close connection between these developments in that the emergent state apparatus needed more and better maps to govern more efficiently.[25] In other words, maps and map-making were instrumental for the constitution of the modern nation state. It has also been suggested, by James Akerman and Michael Biggs among others, that as maps during the seventeenth and eighteenth centuries increasingly began to represent political borders, the graphic representation of the state may have contributed to a changing understanding of the nature of the state itself, which changed from being defined as govern-

---

ing, see Harley, *The New Nature of Maps*; D. Wood, *The Power of Maps* (New York: The Guilford Press, 1992).

[24] C. Delano-Smith, "Signs on Printed Topographical Maps, ca 1470-ca 1640", in *The History of Cartography, Vol. 3. Cartography in the Renaissance, Part 1*, ed. David Woodward (Chicago: University of Chicago Press, 2007) 552.

[25] See e.g. the essays in D. Buisseret (ed.), *Monarchs, Ministers and Maps. The Emergence of Cartography as a Tool of Government in Early Modern Europe* (Chicago: University of Chicago Press, 1992); Josef Konvitz, *Cartography in France 1660–1848. Science, Engineering and Statecraft* (Chicago: University of Chicago Press, 1987); and R. L. Kagan and B. Schmidt, "Maps and the Early Modern State: Official Cartography", in *The History of Cartography, Vol. 3. Cartography in the Renaissance, Part 1*, ed. David Woodward (Chicago: University of Chicago Press, 2007) 661–679.

ment over people to a territorially bounded entity.[26] Similarly Morant's map of Strasbourg underlines and even exaggerates the boundary line between city and countryside and thereby accentuates the difference. In this way it can be argued that certain features of maps (directly or indirectly) affect both the perception and the constitution of political and legal reality, and hence that maps have a performative aspect, creating reality rather than merely describing it.[27]

Although maps were made for a number of different reasons (architectural, celebratory, military, legal, proprietary, scientific), an important use of maps in the period of emergent nation states is connected to what Michel Foucault has called "biopower" and "biopolitics": the practice in modern states of developing and deploying diverse techniques to achieve the subjugations of bodies and the control of populations.[28] The discussion of the relation between maps and law will therefore also include the concept of biopower – biopolitics, biolaw and bioethics, and in particular discuss the affinity between how law constitutes social space and how it conceptualizes the human body. This essay therefore also investigates to what extent affinities and analogies can be traced between how law constitutes legal space and legal conceptions of persons and the human body. In this respect the essay will employ key notions from bioethics and biolaw – autonomy, dignity, integrity, and vulnerability[29] – which serve to define and distinguish the legal subject (*persona sui iuris*) from the non-legal subject (*persona alieni iuris*), the

---

26 See J. Akerman, "The Structuring of Political Territory in Early Printed Atlases", *Imago Mundi*, 47 (1995), 138–154; M. Biggs, "Putting the State on the Map: Cartography, Territory, and the European State Formation", *Comparative Studies in Society and History*, 41, 2 (1999), 374–405. The same argument is made in Kagan and Schmidt, "Maps and the Early Modern State". See also D. Buisseret, "The Cartographic Definition of France's Eastern Border in the Early Seventeenth Century", *Imago Mundi*, 36 (1984), 72–80.

27 J. L. Austin, "Performative utterances", in *Philosophical Papers*, 3rd ed. (Oxford: Oxford University Press, 1979) 233–252.

28 M. Foucault, *Surveiller et punir* (Paris: Gallimard, 1975); M. Foucault, "Bio-histoire et bio-politique" [1976], in *Dits et écrits II* (Paris: Gallimard, 2001) 95–97; M. Foucault, *Naissance de la biopolitique. Cours au Collège de France, 1978–1979* (Paris: Gallimard/Seuil, 2004). See also A. Giddens, *The Nation State and Violence. Volume Two of A Contemporary Critique of Historical Materialism* (Berkeley: University of California Press, 1987).

29 Cf J. Dahl Rendtorff & P. Kemp, *Basic Ethical Principles in European Bioethics and Biolaw, Vol. 1, Autonomy, Dignity, Integrity and Vulnerability*, Report to the European Commission of the BIOMED-II Project Basic Ethical Principles in Bioethics and Biolaw, Centre for Ethics and Law, Copenhagen, and Borja de Bioetica, Barcelona, 2000, 17–62.

human from the non-human (other animals), the animal from the non-animal (things), as well as cross-over categories (slaves).

The argument in this essay is pursued along several paths simultaneously: it follows the cartographic development from antiquity to modern times; it traces the legal history and development in Western Europe under the same periods; it identifies and thematizes a series of intersections between these two paths, i.e. how legal places and spaces appear (and disappear) on maps and how the intersection of cartography and law produces jurisdictions (first local and regional, later national and universal) as well as legal subjects with human bodies. The essay is divided into three parts. The section "Strasbourg Seen From the Top of the Cathedral Tower", presents a reading of Morant's map. The next section, "Maps, Law, and Society", sketches the historical background for the cartographic and legal development in the Early Modern and Modern periods. The third section, "Maps, Technology and Law", resumes the discussion of city maps and analyzes the cartographic representation and construction of social and legal places and spaces in old maps of Strasbourg from a technological and phenomenological point view. The concluding section of the essay attempts to sum up the discussion.

The main argument in this essay is that the displacements and transformations of the understanding of space and place that can be seen in maps in the Early Modern and Modern periods have both direct and indirect effects on the understanding of legal space. From law in earlier periods primarily having been defined in relation to the people living in a certain place (often called the personality principle), in the Modern period law becomes increasingly territorial and nationally unified. During the Modern period there also occurs a separation of political and juridical institutions that is clearly legible on maps. However, when reading maps, old as well as new maps, one should keep in mind, in the words of the cultural geographer Karl Schlögel, that

> maps speak the language of their authors, and they keep silent about [*verschweigen*] that which the cartographer does not or cannot speak. Maps say more than a thousand words, but they also keep silent about [*verschweigen*] more than a thousand words can say."[30]

Since maps in general neither are made by lawyers nor for lawyers, legal place and space often appear obliquely and through empty silences.

---

[30] K. Schlögel, *Im Raume lesen wir die Zeit. Über Zivilisationsgeschichte und Geopolitik* (Frankfurt: Fischer, 2006) 95 [Karten sprechen die Sprache ihrer Verfasser, und sie verschweigen das, wovon der Kartograph nicht spricht oder nicht sprechen kann. Karten sagen mehr als tausend Worte. Aber sie verschweigen auch mehr, als man in tausend Worten sagen könnte.]

Strasbourg Seen From the Top of the Cathedral Tower

Morant's 1548 map of Strasbourg captures the city as seen from the 142 meter-high cathedral tower, completed in 1439 (and between 1647 and 1874 the tallest building in the world). This fact alone makes the map different from most other early examples, which tended to capture objects from a point in space not yet reachable by the human eye. Thus although Barbari's view of Venice, Sedel's plan of Augsburg, Worms' view of Cologne, Cornelis Anthoniszoon's plan of Amsterdam, and the anonymous view of Norwich may appear more realistic to the modern eye, these cities are seen from a point of view that did not yet exist for human beings.[31] And whereas the plans of Augsburg and Amsterdam were made to show the cities from directly above (often called an ichnographic plan), the views of Florence, Venice, Cologne and Norwich represented these cities from an angle closer to 45 degrees (often called a bird's-eye view).[32] The perspective in Morant's map of Strasbourg ranges from zenithal, when looking straight down on the cathedral square and the cathedral roof, to an oblique bird's-eye view of the

---

[31] H. Lavedan, *Les Représentations des villes dans l'art du Moyen Age* (Paris: Van Oest, 1942); R. Wittkower, *Architectural Principles in the Age of Humanism* (New York: Norton, 1962); L. Marin, *Utopiques: jeux d'espace* (Paris: Editions de Minuit, 1973); all three quoted in M. de Certeau, [1990] *L'Invention du quotidien, 1. Arts de faire* (Paris: Folio, 2005) 320 (note 2). It is in a sense ironic that the buildings from which Michel de Certeau watched and described the "everyday life" of Manhattan, the Twin Towers, no longer exist, and that they have become as inaccessible (and imaginary) to us as the point from which Jacopo de' Barbari depicted Venice in 1500. Likewise, most of the houses and ordinary buildings seen in Morant's depiction of Strasbourg, apart from some palaces and official buildings, have long since disappeared. What today constitutes the old parts of Strasbourg was mainly built during the last decades of the sixteenth century.

[32] Although the terminology in cartographic literature is not always consistent, there seems to exist a certain consensus regarding the following three chorographic perspectives: (1) plan (inchnographic or vertical plan) denotes a representation of a place as seen from straight above; (2) bird's-eye view denotes either an *elevated oblique* perspective (as in Barbari's view of Venice) or a *low oblique* perspective (as in Roselli's view of Florence) of a place; and (3) profile view denotes the representation of a place as seen from the side, as in a silhouette. The chorographic representation can furthermore be (more or less) realistic or schematic. See D. Buisseret, "Introduction", in D. Buisseret ed., *Envisioning the City. Six Studies in Urban Cartography. Six Studies in Urban Cartography* (Chicago: University of Chicago Press, 1998) ix–xiii; Delano-Smith, "Signs on Printed Topographical Maps, ca 1470–ca 1640", 541; H. Ballon and D. Friedman, "Portraying the City in Early Modern Europe: Measurement, Representation, and Planning", in *The History of Cartography, Vol. 3. Cartography in the Renaissance, Part 1*, 680–704.

houses close to the city walls. The cartographic historian Liliane Châtelet-Lange has therefore argued that although Morant's plan is more pictorial than the maps of Augsburg and Amsterdam, the zenithal point of view from the cathedral tower represents a more topographical and cartographic interest when compared to the more pictorial images of Florence, Venice, Cologne and Norwich.[33] At the same time it should be noted that the main purpose of large graphic sixteenth-century representations of cities is celebratory rather than instrumental or taxonomic.[34] They are made not to help visitors find their way in an unknown city or to provide property information for taxation, but to show the greatness of the city.

Although Morant's map does not apply the central perspective (found in Barbari's bird's-eye view of Venice), and although the artist has not attempted to give a geometrically exact representation of the city, he does show Strasbourg in great detail. The only extant print of Morant's map (now in Nuremberg) has been hand coloured.[35] The landscape surrounding the city is painted in brown, the water is a pale blue. Most of the important buildings in the city have red painted roofs, with the exception of the cathedral whose roof is coloured dark blue. The roofs of the houses are either red or grey pink, with facades either white or grey pink. The function of this difference in colouring is mainly to permit the distinction of one building from another. The trees in the gardens are green. It should also be mentioned that despite this abundance of detail, which gives the impression more of a cityscape than a modern city map, there are no people or animals shown and the streets are just as empty as in Eugène Atget's daguerreotypes of Paris from the end of the nineteenth century. In other words, Morant's map is descriptive rather than dramatic: there is nothing happening in the picture, the city is not presented as a scene or place for a narrative. This very absence of action and people in Morant's map also has the effect of presenting Strasbourg not as a *civitas*, a community of men joined by social bonds, but as an *urbs*, a walled architectural structure, with buildings, streets, and squares.[36] In fact, during the Early Modern period these two terms, *civitas* and *urbs*, represent two different and often opposing views of what constitutes the greatness of the city: the number and importance of its inhabitants and the quality of its

---

[33] Châtelet-Lange, *Strasbourg en 1548*, 28.

[34] Châtelet-Lange, *Strasbourg en 1548*, 37–48.

[35] See note 3.

[36] On the distinction between *urbs* and *civitas* in the representation of Early Modern cities, see R. L. Kagan, "*Urbs* and *Civitas* in Sixteenth- and Seventeenth Century Spain", in Buisseret ed., *Envisioning the City*, 75–108.

government (*civitas*); or the magnificence of its buildings as well as the layout of its streets and squares (*urbs*). As regards the graphical representation of cities, it seems that during the sixteenth century there was a shift away from the moralized geography of the Middle Ages toward chorographic views offering a more or less accurate description of a particular place.[37]

The city of Strasbourg had originally been founded by the Romans around 12 BCE as a military camp, called Argentorate, on a former Celtic settlement.[38] During Morant's lifetime Strasbourg belonged to the Holy Roman Empire and had constituted (since it was emancipated from episcopal authority in 1263) a *Reichstadt*. Already in 1358 the emperor Charles IV had given the city status of a free city (*freie Stadt*).[39] In judiciary matters this status gave the city sovereignty in criminal cases, whereas in civil law cases the city courts were only the first instances and the emperor's court constituted the court of appeal.[40] In Strasbourg, as in general in continental Europe during the Middle Ages and the Early Modern period, there existed a multiplicity of jurisdictions with loosely defined and variable competences.[41] As a consequence, a person was simultaneously dependent on several tribunals, according to whether it was a question, for instance, of possession of land, of debts, or of crimes. These tribunals would furthermore differ among themselves in their composition and in the law they administered. Although this has often been viewed as an inherent weakness of the legal system dur-

---

[37] Kagan, "*Urbs* and *Civitas* in Sixteenth- and Seventeenth Century Spain", 100. See also Schulz, "Jacopo de Barbari's View of Venice".

[38] J.-J. Hatt, "Strasbourg Romain", in G. Livet & F. Rapp (eds.), *Histoire de Strasbourg des origines à nos jours 1* (Strasbourg: Ed. des Dernières nouvelles d'Alsace, 1980) 73–284.

[39] P. Dollinger, "La Ville libre à la fin du Moyen Age (1350–1482)", in G. Livet & F. Rapp eds., *Histoire de Strasbourg des origines à nos jours 2* (Strasbourg: Ed. des Dernières nouvelles d'Alsace, 1981) 108.

[40] According to Philippe Dollinger the city of Strasbourg did not appreciate this situation and made the parties promise not to "prolong" their cases to higher court. (Dollinger, "La Ville libre à la fin du Moyen Age (1350–1482)", 108.)

[41] G. Wunder, *Das Straßburger Gebiet. Ein Beitrag zur rechtlichen und politischen Geschichte des gesamten städischen Territoriums vom 10. Bis zum 20. Jahrhundert* (Berlin: Duncker & Humblot, 1965) 19–27 *et passim*; J. Bart, *Histoire du droit privé de la chute de l'Empire romain au XIXe siècle* (Paris: Editions Montchrestien, 1998), 103–233. For detailed studies of individual regions during the period, see B. Lemesle, *Conflits et justice au Moyen Âge. Normes, loi et résolution des conflits en Anjou aux XIe et XIIe siècles* (Paris: Presses Universitaires de France, 2008); M. P. Breen, *Law, City, and King. Legal Culture, Municipal Politics, and State Formation in Early Modern Dijon* (Rochester: University of Rochester Press, 2007).

ing the period, it also had the positive value of flexibility, allowing the individual a certain freedom to choose in which court to try a case.[42]

It should also be remembered that although the burghers of Strasbourg were considered free men, this was not necessarily the case with the country people coming to the city to find work and other opportunities. In legal terms these immigrants belonged to the village where they had been born. This situation created an incompatibility between their new social status and their traditional legal status.[43] Although the city did not constitute an entirely separate jurisdiction in relation to the surrounding rural regions and was not entirely independent from the Empire, its laws and legal institutions were different and it even had a constitution (*Verfassung*). This together with other circumstances made the urban space of the city stand out as a distinct and, in most aspects, sovereign legal space, even if this varied over time. As a substantially sovereign city, Strasbourg also executed a certain influence and attraction on the surrounding countryside and villages.

It is therefore understandable that the leading burghers of Strasbourg may have wanted to emphasize the boundaries that separate the city from the countryside. The city map offered this possibility. As has been noted by the cartographic historians Hillary Ballon and David Friedman, the boundaries of Early Modern cities were rarely demarcated on the ground with the clarity they possessed on maps:

> In reality, there was usually a transitional zone between city and countryside, with buildings clustered outside city gates and flanking major roads. Such inconvenient facts were altered in bird's-eye views, and the outlying land was depicted in abbreviated form, so the density of detail in the urban fabric sharply contrasted with the relative blankness outside the city. This boundary line, which helped to establish the visual coherence of the city, was essential to the rhetoric of the image.[44]

It suffices to throw a glance at Morant's map of Strasbourg to realize the veracity of this comment. The map may give a true description of the city itself,

---

[42] It has been suggested that the movement since 1958 towards a unified European legal space is currently fashioning a similar multiplicity of jurisdictions in the EU. See C. Harding, "The Identity of European Law: Mapping Out the European Legal Space", *European Law Journal*, 6, 2 (2000), 128–147. See also H. Lindahl, "A-Legality: Postnationalism and the Question of Legal Boundaries", *Modern Law Review* 783 (2009), 30–56.

[43] However, in many regions the king or local sovereign would actively support the founding and development of cities, and would also grant freedom to people who settled there. See R. E. Dickinson [1961] *The West European City. A Geographical Interpretation* (London: Routledge & Kegan Paul, 1962) 282–283.

[44] Ballon and Friedman, "Portraying the City in Early Modern Europe", 691–692.

but not of the space surrounding it. When reading maps one needs to be aware of the rhetorical intentions of the author(s).

As noted above, it is the point of view of the cathedral tower that commands Morant's view of Strasbourg: all other buildings on the map (with the exception of the city walls) are facing the cathedral. Hence, in order to properly see or take in the city, you have to walk around the map.[45] The cathedral tower itself is made to stand out vertically from the city by a separate print attached to the map surface. Yet the function of the cathedral tower as point of view – the point from which you perceive the city – has the paradoxical effect of diminishing the visual importance of the cathedral itself. This is easily seen when you compare Morant's plan over Strasbourg with the one in *Civitates orbis terrarum* (1572). The latter, which is based on Morant's map, has changed the perspective to a bird's-eye view similar to Barbari's representation of Venice. The second most imposing building on Morant's map is the Prediger Closter Collegium (the former Dominican church and convent, in 1548 the place of higher education). In comparison to the cathedral and collegium, the other ecclesiastical buildings (including the Bischoffs Hoff), though recognizable, do not stand out. It is also noticeable that the civic buildings – the City hall (*Pfalz* or *Pfaltz*)[46], the Mint (*Münz* or *Müntz*), the Custom house (*Das Kauffhaus*), and the Arsenal (*Aerckellerei haus*) – although clearly recognizable do not stand out either.[47] The only civic building that stands out is the Treasury (*Pfennigthorn*).

---

[45] In this respect Morant's map resembles Hans Sebald Beham's woodcut sketch of the siege of Vienna 1529 (printed 1530 on six woodblocks, dimensions 81,2 × 85,6 cm), a 360 degree drawing centered on the church of Saint Stephen. (Beham's woodcut sketch is reproduced in J. Hale, "Warfare and Cartography, ca. 1450 to ca. 1640", in *The History of Cartography, Vol. 3. Cartography in the Renaissance, Part 1*, 732.) This panoramic feature prefigures the maps from the so called Dieppe school during the seventeenth century (see D. Buisseret, "Monarchs, Ministers, and Maps in France before the accession of Louis XIV", in Buisseret ed., *Monarchs, Ministers and Maps*, 103), as well as the panoramas of the nineteenth century in which the spectator is presented with 360 degrees view of a scenery (a landscape or cityscape) from a single elevated vantage point and in which the observer becomes integral to the object and inspection if made from the inside rather than from a distance (see M. Kemp, *The Science of Art: Optical Themes in Western Art from Brunelleschi to Seurat* (New Haven: Yale University Press, 1989) 213–215).

[46] The German word *Pfalz* (or *Pfaltz*) usually denoted a place where the king (or the emperor) would reside while visiting his domains, but this was not its function in Strasbourg. Instead it was the place of both the city council (*Rat*) and the higher law courts. The *Pfalz* in Strasbourg was built in 1321, modified in 1556 and pulled down in 1780. See Châtelet-Lange, *Strasbourg en 1548*, 109.

[47] Châtelet-Lange, *Strasbourg en 1548*, 33.

The highest court in criminal cases in Strasbourg in the sixteenth century was the council (*Rat*), but it would delegate lesser cases to the minor council (*Kleiner Rat*) or the so-called chamber of XIII.[48] Whereas these courts of appeal were located in the City hall (*Pfalz*), the location (and function) of the numerous inferior city courts (*Stadtgerichten*) is not well known. The bailiff's court (*Schultheissengericht*), which only had jurisdiction over lesser burghers (*Kleinburger* and *Schultheissenburger*), would meet in the open in a garden on Lang Strosse (today Grand'Rue).[49] There also existed an ecclesiastical court that not only concerned clerics but also matters regarding succession, civil status, marriage contracts, and other private law cases.[50] This law court probably met at the Bischoffs Hoff. Apart from these courts, each professional organization (guild, company) would have its own tribunal; and there also existed temporary courts, such as "the three honorable men" of the market.[51] Although these various courts were primarily for burghers in Strasbourg, it was possible for foreigners to acquire the "rights" of a burgher (they were called either *Usburger* or *Pfalburger*).[52] This judicial order would remain substantially intact until the end of the eighteenth century, and hence was not much affected by the French conquest of Alsace in 1680 and the siege and capture of Strasbourg in 1681.[53] Instead of being a free city in the

[48] Dollinger, "La Ville libre à la fin du Moyen Age (1350–1482)", 113.

[49] Dollinger, "La Ville libre à la fin du Moyen Age (1350–1482)", 113.

[50] The possibility of the ecclesiastical court (*audientia episcopalis* or *episcopale iudicium*) to try private law cases goes back at least to the Dominate period of the Roman Empire. Although the process had the advantage of being free, it was not possible to appeal the decision to a higher legal instance. See J. C. Lamoreaux, "Episcopal Courts in Late Antiquity", *Journal of Early Christian Studies*, 3, 2 (1995), 143–167; J. D. Harries, "Resolving Disputes: The Frontiers of Law in Late Antiquity", in R. W. Mathisen ed., *Law, Society, and Authority in Late Antiquity* (Oxford: Oxford University Press, 2001) 68–81; N. Lenski, "Evidence for the *Audientia episcopalis* in the New Letters of Augustine", in Mathisen ed., *Law, Society, and Authority in Late Antiquity*, 82–97.

[51] Dollinger, "La Ville libre à la fin du Moyen Age (1350–1482)", 113.

[52] Dollinger, "La Ville libre à la fin du Moyen Age (1350–1482)", 106.

[53] However, there existed at times considerable tensions between the city council and the representatives of the French king. See G. Livet, "La Guerre de trente ans et les traités de Westphalie. La Formation de la province d'Alsace (1618–1715)" and "Le XVIIIe siècle et l'esprit des Lumières", in P. Dollinger ed., *Histoire de l'Alsace* (Toulouse: Edouart Privat, 1970) 259–303, 305–356; I. Streitberger, *Der Königliche Prätor von Strassburg 1685–1798. Freie Stadt im absoluten Staat* (Wiesbaden: Franz Steiner Verlag, 1961); H. Ritter von Srbik, *Wien und Versailles 1692–1697. Zur Geschichte von Strassburg, Elsass und Lothringen*, (München: Verlag F. Bruckmann, 1944); and W. Forstmann et al., *Der Fall der Reichsstadt Straßburg und seine Folgen* (Bad Neustadt: Verlag Dietrich Pfaehler, 1981).

Holy Empire it now became the capital of the French province of Alsace (although the *conseil souverain* was located in Colmar).

As the reading of Morant's map of Strasbourg shows, it is possible to describe and construct a series of more or less loosely defined legal spaces in which legal places are located, some of which may be identified on the map and others not. At the same time, these legal spaces are usually not differentiated from other social spaces (political, professional, religious) and often coincide with these social structures. Before continuing the discussion of the mapping of legal spaces of Strasbourg it is necessary to place them within a wider historical context, both cartographical and juridical, showing the diversity both of cartographic practices and of law. The description of Roman law serves both as a backdrop and as a conceptual road map for the development of law in the Early Modern and Modern periods.

Maps, Law, and Society

There is evidence that human societies have made maps since very early times. The earliest known city plan is a wall painting from Çatal Hüyük (Turkey) datable to c. 6200 BCE.[54] Several maps on clay tablets from Mesopotamia have survived, for instance an Akkadian clay tablet map found in Nuzi dated c. 2300 BCE.[55] There is also a town plan petroglyph (rock carving) found in Bedolina in Northern Italy believed to be from 2000–1500 BCE.[56] It seems that maps usually appear prior to the invention of writing, and/or perhaps can be seen as an early form of writing.

Maps made on less durable material and not protected by oblivion have, for obvious reasons, not survived the erosion of natural and human history. For this reason we have only map fragments from ancient Greece and Rome. For instance, a massive marble map of the city of Rome (18 × 13 meters) was created under the Roman emperor Septimius Severus between 203 and 208 CE, carved into 150 marble slabs mounted on an interior wall of Vespasian's Temple of Peace (Templum Pacis).[57] This map, called simply *Forma*

---

54 J. Mellaart, *Çatal Hüyük: a Neolithic Town in Anatolia* (London: Thames and Hudson, 1967).

55 Thrower, *Maps & Civilization*, 14–16.

56 Thrower, *Maps & Civilization*, 3–4.

57 O. A. W. Dilke, *Greek and Roman Maps* (London: Thames & Hudson, 1985) 103–106; *La Pianta marmorea di Roma antica: forma urbis Romae*, ed. Gianfilippo Carettoni (Roma: Comune di Roma, 1960).

urbis romae (or the Severan marble plan), was detailed enough to show the floor plans of nearly every temple, bath, and block in the central Roman city. The map was gradually destroyed during the Middle Ages, when the marble stones were used for building or for making lime.

An important form of mapping in Roman society was property or cadastral surveys.[58] These maps had the status of legal documents and it is perhaps for this reason that they were often incised in bronze or carved into marble as official records. An example of this are fragments of a first century Common Era cadastral map found in Orange (Arausio) in Southern France.[59] A compilation of Roman surveyor manuals (*Corpus Agrimensorum,* 500 CE) has also survived. The only Roman map that has survived (almost) intact is not a city map but a road map, and it has come down to us in the form of a twelfth or early thirteenth century copy. This is the so-called Peutinger table (*Tabula Peutingeriana*), a fourth century road map showing the main routes from the Northern borders of England, through France, Italy and the Mediterrean, all the way to India.[60]

Although not many maps have come down to us from Antiquity, verbal descriptions of geography have survived, most importantly Strabo's *Geographica* (Γεωγραφικά, c. 24 CE) and Ptolemy's *Geographia* (Γεωγραφιχὴ 'Υφήγησις, second century CE).[61] It has been suggested by the cartographic historian Evelyn Edson that the oral presentation of some of these geographical texts were made in front of quite large world maps (*mappa mundi*), although not as large as the *Forma urbis romae.*[62] Ptolemy's geography manual has survived in several thirteenth century manuscript copies, some of which also contain maps.[63] During the fifteenth century both manuscript and print copies of Ptolemy's *Geographia* were augmented with new maps, including city plans. Of particular interest in this context are the maps by Pietro del Massaio illustrating three manuscript copies made in Florence in the sec-

---

58 Dilke, *Greek and Roman Maps,* 87–101, pp. 108–110; R. J. P. Kain & E. Baigent, *The Cadastral Map in the Service of the State: A history of property mapping* (Chicago: University of Chicago Press, 1992) 2–3.

59 Dilke, *Greek and Roman Maps,* 108–110.

60 Dilke, *Greek and Roman Maps,* 113–120; Thrower, *Maps & Civilization,* 39–41.

61 Dilke, *Greek and Roman Maps,* 62–65 *et passim* (on Strabo); 75–86, 155–166 *et passim* (on Ptolemy).

62 E. Edson, *Mapping Time and Space: How Medieval Mapmaker Viewed Their World* (London: British Library, 1997).

63 Dilke, *Greek and Roman Maps,* p. 80, pp. 198–200.

ond half of the fifteenth century.[64] The artist presumably worked from old models and, with one exception, the ten city maps included in the so-called *Urbino codex* (1472 CE) belong to the medieval tradition of bringing together isolated phenomena, in this case landmark buildings, surrounded by circular city walls.[65] That is, although the buildings represented often (but not always) are recognizable and in the correct location, there is little ambition to represent the internal structure of the city (squares, streets) or its topography. However, in one city map in the *Urbino codex*, a map of Volterra in Italy, we have, in the words of art historian Naomi Miller, "moved from the medieval map [...] to a unified view of the city characteristic of Renaissance art."[66]

Apart from the geographical and chorographical works mentioned above, there are also a number of explicit descriptions of maps and the use of maps in ancient dramatic and epic literature. For instance, in Aristophanes' fifth century BCE comedy *Clouds*, cast in contemporary Athens, we find the old man Strepsiades looking at a map of the entire world (γῆς περίοδος πάσης). Next to Strepsiades stands a pupil from Socrates' "school" (pejoratively named the "φροντιστήριον" (Thinkery), pointing out to him the location of Athens. Strepsiades, who obviously never has seen a map before, says incredulously: "What do you mean? I don't believe it; I don't see any juries in session."[67] Although in Antiquity it was not unusual to associate Athens with litigation and law courts, and Athenians were known for being quarrelsome and tricksters, it would be closer at hand to identify the city by the Parthenon or other large public buildings, rather than by the law courts. As a matter of fact, from modern excavations of the Athenian agora, it seems that there was

---

[64] Two of the manuscripts are in the Vatican Library (Vat. lat. 5699 (1469 CE); Vat. Urb. lat. 277 (1472 CE)) and one in the Bibliothèque national, Paris (B.N. lat 4802). See Naomi Miller, "Mapping the City: Ptolemy's Geography in the Renaissance", in D. Buisseret ed., *Envisioning the City*, 34–74.

[65] According to Naomi Miller, "the main distinctions seem to be between the Italian cities and those of the Near East. But even among the former, variations are noted. In the maps of Milan and Venice, reliance on older models is evident, whereas in the maps of Florence and, to a lesser degree, that of Rome, it appears as if the drawing of each building is plotted, perhaps on a coordinate system. And, despite the absence of a street network, we can almost sense a connective tissue as we wander from landmark to landmark. In the Eastern maps, we approach the city from a bird's-eye view and confront a more coherent panorama, albeit a more abstract prospect, seemingly due to lack of data." (Miller, "Mapping the City", 39)

[66] Miller, "Mapping the City", 65.

[67] Aristophanes, *Clouds,* lines 207–208. Quoted from Aristophanes, *Clouds, Wasps, Peace,* trans. J. Henderson (Cambridge: Harvard University Press, 1998).

no single location or building for trials, although they usually took place on or close to the agora.[68]

In contrast to the ancient Greek *poleis*, in Roman cities the law court not only had a specific site adjoining the forum, but also had its own building, the *basilica*.[69] Although the name applied to a wide range of building forms, it was most commonly and characteristically used for the large, multipurpose public halls that regularly accompanied the forum in the Western half of the Roman world, corresponding to the Greek and Hellenistic *stoa*.[70] A typical basilica was an elongated rectangular building, open either along one side or at one end.[71] The Roman basilicas were not only used for legal business, but also for other public affairs as well as for religious purposes.[72] However, both during the Roman Republic and the Principate, just as in Athens, the legal processes could also take place in the open air on the forum.[73]

During its long history, the Roman Republic and the Empire saw a steady development in legal procedure, legislation and jurisprudence, which increased the complexity of the legal system. Initially there were two parallel legal systems, one for Roman citizens and another for interactions between Romans and foreigners (*hostes, peregrini*). As Roman citizenship was gradually extended and finally (through the *constitutio Antoniniana*, 212 CE) came to include all peoples within the Empire, this distinction would in principle only

---

[68] A. L. Boegehold, *The Athenian Agora. Results of Excavations Conducted by the American School of Classical Studies at Athens, Vol. 28, The Law Courts at Athens. Sites, Buildings, Equipment, Procedure, and Testimonia* (Princeton: The American School of Classical Studies at Athens, 1995) 89–150 *et passim*.

[69] This can be seen, for instance, in the remains of the Roman city Pompeii, destroyed by the eruption of Mount Vesuvius in 79 CE. For location and layout of the *basilica* in Pompeii, see A. di Franciscis, *Pompei* (Novara: Istituto Geografico de Agostini, 1968) 14, 30.

[70] In fact, in archaic times the two words were used together to denote the tribunal of a king, *basiliké stoà*.

[71] For descriptions of the design of the basilica, see Vitrivius, *The Ten Books on Architecture*, trans. M. Hicky Morgan [1914] (New York: Courier Dover, 1960) V.i.4–10 (132–136); Leon Battista Alberti, *On the Art of Building in Ten Books*, trans. J. Rykwert et al. [1980] (Cambridge: The MIT Press, 1989), VII.14 (230–236); J. B. Ward-Perkins [1981], *Roman Imperial Architecture* (New Haven: Yale University Press, 1994) 492 *et passim*.

[72] During the Empire the term came to be used of any hall that was basilican in plan, irrespective of its purpose. After the Roman Empire became officially Christian, the term *basilica* came by extension to specifically refer to a large and important church that has been given special ceremonial rites by the Pope. Thus the word retains two senses today, one architectural and the other ecclesiastical.

[73] Quintilian, *Institutio oratoria*, XI.iii.27.

have been relevant in frontier regions.[74] This extension of citizenship emphasizes the central importance of the personality principle in older law. That is, in contrast to modern law, in which the law is tied to the land where one – permanently or temporarily – resides and conducts one's business (called the territoriality principle), in older societies law was tied to the person.[75] Hence, Roman citizens would be tried according to Roman law (*ius civile*) even when in the provinces (prior to 212 CE), and Athenians residing in Rome according to Athenian law.[76] However, since the person was defined by the place that he/she came from, in certain respects it would be more appropriate to talk of the "place principle" and of an opposition between place and space (rather than person and territory). The originary place-bound character of law is also shown by the fact that in several languages the words for law and justice carry connotations of place and region.[77] This connection between place and law can also be found in words for jurisdiction, for instance in the West-Germanic word *berek* (surviving in Middle-Dutch), which was used to identify cities with independent jurisdiction.[78]

---

[74] C. R. Whittaker, *Frontiers of the Roman Empire. A Social and Economic Study* (Baltimore: Johns Hopkins University Press, 1994), 98–131, 243–278, *et passim*.

[75] There are remnants of the personality principle in modern law, for instance in the law of succession. See L. Pålsson, *Svensk rättspraxis i internationell familje- och arvsrätt*, 2nd edition (Stockholm: Norsteds juridik, 2006) 230f; M. Bogdan, *Svensk internationell privat- och och processrätt* (Stockholm: Norstedts juridik, 2008) 148–178.

[76] M. Kaser, *Römisches Privatrecht* 19 Aufl. (München: Verl. C.H. Beck, 2008) 32–33. See also A. Watson, *The Law of Person in the Later Roman Republic* (Oxford: Clarendon, 1967); S. Guterman, *The Principle of the Personality of Law in the Germanic Kingdoms of Western Europe from the Fifth to the Eleventh Century* (Frankfurt am Mein: Peter Lang, 1990); J. Binder, *Das Problem der juristischen Persönlichkeit* (Leipzig: Deichert, 1907); R. Saleilles, *De la personnalité juridique: histoire et théories*, 2. éd. (Paris: Rousseau, 1922).

[77] Hence, the Greek notion νόμος derives from the verbs νέμω ('distribute', 'deal out', 'assign', 'grant', especially in the sense of 'assigning property', 'apportioning pasture or agricultural land') and νέμηειν ('to distribute') (see E. Laroche, *Histoire de la racine nem- en Grec ancient* (Paris: Klincksieck, 1949); J. Svenbro, *Phrasikleia: anthropologie de la lecture en Grèce ancienne* (Paris: La Découverte, 1988), 123–137); the Latin word *equitas* derives from *aequitas* (meaning 'uniformity', 'evenness'); the English word *law* derives from Old-Norse *lag* (meaning 'something laid down or fixed') and the German *Gesetz* has a similar (if more abstract) meaning as *law/lag* (see L. Dahlberg, "Achilles foot and the Law: Legal Space(s), Striated and Smooth", in *Practising Equity, Addressing Law: Equity in Law and Literature*, ed. D. Carpi (Heidelberg: Winter, 2008) 119–120).

[78] E. Wadstein, "Birka och bjärkörätt", *Namn och bygd. Tidskrift för nordisk ortnamnsforskning*, 2 (1914) 92–97.

Intimately connected to the personality principle was the notion of status, which was of central importance not only in Antiquity and the Middle Ages, but throughout the Early Modern and Modern periods until the egalitarianism of the American and French revolutions in the eighteenth century. In Roman society the notion of status was constituted by three elements: freedom (*libertas*), citizenship (*civitas*) and position within the family unit (*familia*).[79] The person of full status possessed all three: he/she had freedom; he/she had Roman citizenship; and he/she belonged to a Roman household. If he/she lost any of these he/she would lose some of his/her status (*capitis deminutio*). This is well captured by the Roman jurist Iulius Paulus (as quoted in the *Digesta*):

> There are three kinds of change in civil status: the greatest [*maxima*], the middle [*media*], and the least [*minima*]. For there are three things which we have: freedom, citizenship and family. Therefore, when we lose all three [...] the change of civil status is the greatest. But when we lose citizenship and retain freedom, the change of status is middle. When both freedom and citizenship are retained and only family is changed, it is plain that the change of civil status is the least.[80]

In Roman law, a slave was at the same time an "object of right" (*res*) and a person (*persona*), meaning that although he/she may be treated well and in fact have a large amount of independence, it had no legal personality and was owned by its master (as *res mancipi*).[81] In civil proceedings a slave could neither be a party nor a representative. However, the master did not have the right to abuse or kill the slave without reason.

Being a Roman citizen implied a number of rights, most importantly the right to participate in legal processes and transactions (*commercium*), the right to enter a civil law marriage (*conubium*), and the right to make a valid will and the capacity to be made beneficiary (*testamenti factio*).[82] As a general rule, these rights were possessed by both men and women, although with important limitations for the latter.[83] There were also other distinctions between citizens, most importantly between persons *sui iuris* and persons *alieni iuris*. Whereas the former were legally independent (i.e. not in the power (*potestas*)

---

[79] *Justinian's Institutes*, trans. P. Birks & G. McLeod (Ithaca: Cornell University Press, 1987) I.ii-v 39–41; Kaser, *Römisches Privatrecht*, 83–85f.

[80] Justinian, *Digest*, ed. Theodor Mommsen, English trans. Alan Watson (Philadelphia: University of Pennsylvania Press, 1985) 4. 5. 11.

[81] Kaser, *Römisches Privatrecht*, 90–93.

[82] Kaser, *Römisches Privatrecht*, 32, 362.

[83] Cfr. Justinian, *Digest*, 1.5.9: "There are many points in our law in which the condition of females is inferior to that of males." (Aemilius Papinianus)

of a *paterfamilias*), the latter were subject to *potestas*, they were "of another's law" (*alieno iuri subiecti*).[84]

The Roman family (*familia*) was a central feature of Roman society and Roman law. The traditional *familia* constituted a monocratic legal unit consisting of the head of the household (*paterfamilias*) and the persons subjected to his extensive powers: his wife, his children, his bondsmen (*clientes*) and slaves.[85] The Roman household also formed a religious unit through the worship of certain deities. Although the *potesta paterfamilias* was gradually circumscribed, first by sacral law and moral sanctions, and then through legislation, especially regarding his power over other persons, he retained exclusive property rights.

The central ideological importance of the *familia* in Roman society is evidenced in the founding myth of the Roman Republic, the rape of Lucretia. In the Roman historian Livy's (Titus Livius) account of this event, Lucretia is raped in her home by the royal prince Sextus Tarquinius. As soon as he has left the house she sends for her father and husband, together with witnesses. When they have arrived, she tells them of the crime: "The impress of another man is in your bed, Collatinus; yet only my body was defiled; my soul is not guilty. Death will be my witness to this. But pledge with your hands and swear that the adulterer will not go unpunished."[86] The assembled men promise to revenge her honour, but plead with her not to blame herself since, they argue, "the mind sins, not the body, and there is no guilt when intent is absent." It is noteworthy that Livy's account is clothed in precise legal language, foreshadowing the modern principle of *mens rea* ("the act does not make a person guilty unless the mind is also guilty").[87] Lucretia replies by saying that she absolves herself of wrong but not of punishment, again inserting a subtle legal distinction, before she plunges a knife into her breast, saying: "Let no unchaste woman hereafter continue to live because of the precedent of Lucretia." Immediately after she has collapsed in death, one of the witnesses, Lucius Iunius Brutus, pulls out the knife dripping with blood and gives an oath that signals the end of the Regal period:

---

[84] *Justinian's Institutes*, I.viii–ix 41–43; Kaser, *Römisches Privatrecht*, p. 81.

[85] Kaser, *Römisches Privatrecht*, 80; Watson, *The Law of Person in the Later Roman Republic*, 98–100.

[86] Livy (Titus Livius), *The Rise of Rome [Ab urbe condita]*, Books 1–5, trans. T. J. Luce (Oxford: Oxford University Press, 1998) I.58.

[87] E. Coke, *Institutes of the Lawes of England* (1628–1644): "actus non facit reum nisi mens sit rea".

By this blood, so pure before defilement by prince Tarquinius, I hereby swear –
and you, O deities, I make my witnesses – that I will drive out Lucius Tarquinius
Superbus together with his criminal wife and all his progeny with sword, fire, and
whatever force I can muster, nor will I allow them or anyone else to be king at
Rome.[88]

The assembled men repeat his oath and follow Brutus' lead in bringing the
monarchy to an end and the founding of the Roman Republic.

Livy's account reveals not only that the family domain was perceived as
beyond the powers of the ruler and therefore outside the public domain, but
also the intimate association between the household as private space and the
inviolability of the individual body. Prince Tarquinius entered the house as a
guest, but since he did not follow the rules of guestfriendship, his presence
marks a violent entry even before he has violated the hostess. As Lucretia
shows herself to be immune to his threats of killing her with his sword if
she does not comply with his desire, he threatens instead to kill her and then
defile her honour. After Tarquinius has left and she has cleared her honour
in front of her father and husband, she herself performs the act that he
threatened, plunging a sharp weapon into her body. Livy's account of the
rape of Lucretia thematizes the autonomy of the household, the dignity and
integrity of the Roman citizen, but also the vulnerability of the family institution and the need and duty to protect it and its members.

It is a commonplace in both ancient and modern thought to view the
family as one of the basic units in human society, and it has often been argued that society is constituted through the bonds between family units.[89]
However, when comparing different family organizations in different societies, past and present, one may inversely argue that it is society that constitutes the family structure. For instance, the family structure in classical
Athens and Sparta were radically different from each other, as were, in turn,
the Roman family unit.[90] The most important difference between these
ancient family models in the present context is that it is the Roman family
that has provided the conceptual model for the modern legal notion of the

---

[88] Livy, *The Rise of Rome*, I.59.
[89] Cf Aristotle, *Politics*; G. W. F. Hegel, *Phänomenologie des Geistes* (1806/1807) and
      Grundlinien der Philosophie des Rechts (1821). However, Plato, in the fifth book of the
      Republic, described the family as an obstacle to the realisation of the ideal society.
[90] G. Sissa, "The Family in Ancient Athens (Fifth-Fourth Century BC)", in eds.
      A. Burguière et al., *A History of the Family. Volume One. Distant Worlds, Ancient Worlds*
      (Cambridge: Harvard University Press, 1996), 194–227; Y. Thomas, "Fathers as
      Citizens of Rome, Rome as a City of Fathers (Second Century BC – Second Cen-
      tury AD)", in *A History of the Family*, 228–269; A. Rousselle, "The Family under the
      Roman Empire: Signs and Gestures", in *A History of the Family*, 270–310.

person.[91] As we have seen, the Roman family defines an autonomous social sphere simultaneously *within* the political community and *outside* its legal jurisdiction.[92] In this way a private space has been created, as a clearing within the public space of the city. However, the boundary of a space is not that at which something ends, but, as Heidegger explains, the boundary from which something "*begins its essential unfolding*": "Space is in essence that for which room has been made, that which is let into its boundaries."[93] Since the extra-legal position of private space is constituted by the public (political) order itself, it can be argued that the private sphere (a territory within a territory) is constituted by a folding of the political body onto itself, creating a pocket simultaneously inside and outside. In this way, the private sphere reversely appears as a projection of the individual body onto the social space of the city: defined by a withdrawing from public life (*privatus*) and a spacing of the common social space, opening for the appearance of the individual (*privus*) in the public domain.[94]

The topical role of this privative figure can also be compared to the Enlightenment philosopher John Locke's famous attempt to legitimize the ownership of the result of one's labour through the connection to a person's body: "every Man has a *Property* in his own *Person*. This no Body has any Right to but himself. The *Labour* of his Body, and the Work of his Hands, we may say, are properly his."[95] Although Locke describes the connection in

---

[91]   Kaser, *Römisches Privatrecht*, 83: "Den Begriff der Rechtsfähigkeit hat die neuzeitliche Lehre aus den römischen Quellen abgeleitet, obwohl die Römer selbst ihn noch nicht formuliert haben."

[92]   In an essay on the relation between legal space and political space as articulated in the Greek notion of νόμος, which in contrast to the Roman notion of *lex* has a spatial connotation, the legal philosopher Hans Lindahl has described a reflexive dimension of νόμος. According to Lindahl, "*nemein* denotes the act by which a community closes itself as an inside over against an outside" and "accordingly, the distribution (*nemein*) of space into public and private places is itself a political act or, to put it another way, the distinction between public and private is itself public." H. Lindahl, "Give and Take. Arendt and the Nomos of Political Community", *Philosophy Social Criticism* 32 (2006) 881–901.

[93]   Heidegger, "Bauen Wohnen Denken", 149. ["Ein Raum ist etwas Eingeräumtes, Freigegebenes, nämlich in eine Grenze, griechisch πέρας. Die Grenze ist nicht das, wobei etwas aufhört, sondern, wie die Griechen es erkannten, die Grenze ist jenes, von woher etwas *sein Wesen beginnt*. […] Raum ist wesenhaft das Eingeräumte, in seine Grenze Eingelassene."]

[94]   Cf H. Arendt, *The Human Condition,* 2nd ed. [1958] (Chicago: University of Chicago Press, 1998), 41, 56, 58.

[95]   J. Locke, "Of Property", in *Two Treatises of Government*, ed. P. Laslett [1960] (Cambridge: Cambridge University Press, 2003) 287–288.

terms of the body having "mixed his *Labour* with" the thing, the basic idea is that a person's body is his property, which metonymically extends (or incorporates) "propriety" through work. However, the metonomy is ruled by metaphor since the result of the work is compared to the person's ownership of his body. It should be noted, of course, that the property metaphor does not quite work since a person may not sell his body (or parts of it) in the same way as he may sell the result of his labour.[96]

As Locke's conception of the person in terms of ownership of one's body shows, the modern notion of a person – who has the right to make decisions about him or herself, about his or her things, about his or her body – is founded on the Roman metaphor that the body is a private legal space/territory within a political space/territory. Although the distinction between private and public space was not unique for Roman society, the legal and political interpretation was different. The Roman legal distinction between private and public space became a fundamental dimension in the construction of legal space in European societies.

As noted above, Roman citizenship was gradually extended to more and more peoples and in 212 CE was extended to all free people in the Empire. However, in practice the application of Roman law fell considerably short of these principles. In Italy and the Western provinces, where there was no culture equal to the Roman, Roman law was applied as of old. Nonetheless, the farther one moved away from the capital the cruder and more hybridised was the law that was actually applied.[97] In the Eastern provinces the position of Roman law was especially difficult to maintain. Here the Roman culture had come into contact with a highly developed Hellenistic culture that had produced a legal order which in several ways was radically different from Roman law. In the words of the legal historian Max Kaser,

> these Hellenistic legal ideas [*Rechtsgedanken*] stubbornly asserted themselves against Roman law which, especially because of its united *familia* under the *paterfamilias* [*geschlossenen Hausverbänden*], but also because of its peculiar formalism, its connection with the Latin language and its close ties with Roman procedure, was looked upon as foreign and cumbersome.[98]

---

[96] This anomaly was noted already by Immanuel Kant in *Die Metaphysik der Sitten* (1797).

[97] Kaser, *Römisches Privatrecht*, 35–36. See also J. G. Wolf, "The Romanization of Spain: The Contribution of City Laws in the Light of the *Lex Irnitana*", in *Mapping the Law* [2006] (Oxford: Oxford University Press, 2008) 439–454.

[98] Kaser, *Römisches Privatrecht*, p. 36. [Diese hellenistischen Rechtsgedanken behaupteten sich zäh gegenüber dem römischen Recht, das vor allem mit seinen geschlossenen Hausverbänden, außerdem mit seinem eigentümlichen Formenstil, seiner

In other words, although Roman law in principle, and by legislation (the *constitutio Antoniniana*), applied to the entire Empire, in reality there existed a plethora of jurisdictions, local legislations and legal procedures – as well as hybrids like the Roman law for foreigners.

This non-systematized, pragmatic quality was also a characteristic of Roman private law itself, which had developed organically since the early Republic. During the later Roman Empire, under the Dominate, it was increasingly felt necessary to bring more order in the domain of private law. Attempts to produce codified law were made by several emperors, but it was only in the sixth century, under Justinian, that this ambition came to fruition in the *Corpus iuris civilis*. This "body" of civil law consists of the *Codex Justinianus, Digesta* (or *Pandectae*), *Institutiones* and *Novellae*. This work is often seen as the greatest accomplishment of Roman law, but primarily concerns us in this context because of its effect on later legal development and as conceptual map for codifications in the eighteenth and nineteenth centuries.

During the gradual decline of imperial rule in the Western parts of the Empire, the administration of justice was put in the hands of the local chiefs or kings.[99] Although they were still formally under Roman rule, local custom would increasingly find its way into the application of Roman law. However, during the periods when Germanic kings ruled parts of the former Empire (Visigoths, Ostrogoths, Francs), they would often take over substantial parts of Roman law (e.g. *Codex Euricianus*, 475 CE; *Lex Romana Visigothorum*, 506 CE; *Lex Romana Burgundionum*, early sixth century), but this "vulgar Roman law" would only apply to Romans (*romani*), the Francs and Goths had their own customary law.[100] In most parts of the former Empire local custom became increasingly important, especially in parts that were either distant from Rome or had only been subject to Roman law for short periods. It is also important to remember that this was a period of migration and resettlement and that hybridization of Roman law and customary law was the rule rather than the exception.

During the early Middle Ages (ninth and tenth centuries) the cities declined and were no longer administrative or judiciary centres. There is no evidence indicating that there were specific and separate places for the administration of law. Instead law tended to co-exist with (or be subordinate to) the

---

Bindung an die lateinische Sprache und seiner engen Beziehung zum römischen Prozeßverfahren als fremdartig und unbequem empfunden wurde.]

[99] Whittaker, *Frontiers of the Roman Empire*, 243–278.

[100] R. C. van Caenegem, *An Historical Introduction to Private Law* (1988), trans D. E. L. Johnston [1992] (Cambridge: Cambridge University Press, 2003) 16–26.

execution of political power. In rural and sparsely populated areas, like in Northern Germany and Scandinavia, people would gather at the assembly (*Ting*) to resolve both political issues and legal conflicts. It is probable that the *Ting*-places often were located either on or close to holy places.[101] In most parts of continental Europe, the administration of justice was, with a few significant exceptions, in the hands of worldly masters (chiefs, lords, bishops, city councils, kings) with greater or lesser respect for the rule of law.[102] There was a system of local jurisdictions (*mallus*) and the assemblies took place either in the open or at the residence of the masters (or in cities, on the market square or in the city hall).[103] The same was true for ecclesiastical law. Beginning in this period, feudal law developed and spread first through the Frankish kingdom and then to other Western lands.[104] This was an original system of law, which although neither connected with Roman law or Germanic national laws nevertheless was Germanic in character, lacking written and formal legislation and giving importance to personal relations and landed property. During the feudal period the king would often spend a considerable time traveling in his kingdom, and hold court while residing with his lords. He would then also conduct legal business. This could take place at a local residence, or in the open, by a spring, a large rock or an imposing tree.[105]

The re-discovery of the Justinian *Corpus iuris civilis* in the eleventh century was an important event in European legal history.[106] The news of the event spread rapidly, but due both to the inefficient and expensive means of reproducing texts and to the inherent complexity of the work itself, the dissemination and reception of the code was slow and gradual. It is also important to stress that reception does not mean implementation. It is interesting to compare the reception histories of the *Corpus iuris civilis* and Ptolemy's *Geographia*.

---

[101] J. E. Almquist, *Svensk rättshistoria I. Processrättens historia*, tredje reviderade upplagan [1971], Oförändrat omtryck 1976 (Stockholm: Juridiska Föreningens Förlag, 1976) 8–9. For an account of law and society in Medieval Iceland, see W. Ian Miller, *Bloodtaking and Peacemaking. Feud, Law, and Society in Saga Iceland* (Chicago: University of Chicago Press, 1990).

[102] J. L. Nelson, "Dispute settlement in Carolingian West Francia", in W. Davies and P. Fourace eds., *The Settlement of Disputes in Early Medieval Europe* (Cambridge: Cambridge University Press, 1986) 45–64.

[103] Caenegem, *An Historical Introduction to Private Law*, 25.

[104] Caenegem, *An Historical Introduction to Private Law*, 20.

[105] Caenegem, *An Historical Introduction to Private Law*, 26.

[106] For a recent critical study of manuscripts, see C. M. Radding & A. Ciaralli, *The Corpus Iuris Civilis in the Middle Ages. Manuscripts and Transmission from the Sixth Century to the Juristic Revival* (Leiden: Brill, 2007).

Whereas the latter had an almost immediate impact on cartography and also that by the end of the seventeenth century European cartography had surpassed the ancient master, at the same point in time Europe was still struggling to digest the *Corpus iuris civilis*, although re-discovered over a century earlier than Ptolemy's *Geographia*. By the sixteenth century the Justinian code was studied at universities throughout Europe, but regional laws (and there was only regional law) remained either primarily or entirely based on customary law, except in Italy and Southern France.[107] In France the dividing line between predominantly customary law and "written law" (i.e. vulgar Roman law) went from La Rochelle on the Atlantic coast to Genève in the Alps.[108] South of this line the law was predominantly Roman in character, while North of the line it was customary and Germanic. However, in higher courts (*parlements, Hofgerichten*), where magistrates often had university formation and knowledge of Roman law, the conceptual influence of the *Corpus iuris civilis* was quite strong. The influence of Roman law was sometimes more oblique, for instance through motivating putting the oral customary law into writing, a process that transformed the dynamic custom into static statutes. This occurred already in the Early Middle Ages, and increasingly became part of a political and administrative effort to control and unify the law (usually called "homologation").[109] There are also many examples of how kings – and also parliaments – referred to Roman institutions to define and legitimize their powers.[110] There also existed "resistance" to both homologation and Roman law that was as much political and cultural as it was juridical.[111]

During the later Middle Ages (eleventh and twelfth centuries), when the economy improved and trade increased, the cities increased in number, size and importance, and many became quasi-independent political and legal entities, just like Strasbourg.[112] New founded cities would normally copy the

---

[107] Caenegem, *An Historical Introduction to Private Law*, 34.

[108] Bart, *Histoire du droit privé de la chute de l'Empire romain au XIXe siècle*, 105–141.

[109] Caenegem, *An Historical Introduction to Private Law*, 36–38.

[110] K. Weidenfeld, "Le modèle romain dans la construction d'un droit public médiéval: 'assimilations et distinctions fondamentales' devant la justice aux XIVe et XVe siècles", *Revue historique de droit français et étranger*, 81, 4 (2003), 479–502.

[111] Caenegem, *An Historical Introduction to Private Law*, 41 *et passim*; Bart, *Histoire du droit privé de la chute de l'Empire romain au XIXe siècle*, 130–135. See also G. Strauss, *Law, Resistance, and the State. The Opposition to Roman Law in Reformation Germany* (Princeton: Princeton University Press, 1986).

[112] Dickinson, *The West European City*, pp. 279–289; D. Friedman, *Florentine New Towns. Urban Design in the Late Middle Ages* (Cambridge: MIT Press, 1988) 81–116.

codes of an older city. Although there was a steady development of private law in the Early Modern period, mainly in the cities and often influenced by Roman law, these developments remained largely insular. Excepting within the domain of canon law there existed no strong movement towards legal integration, and within each region there would exist multiple and poorly defined jurisdictions. Also later, during the period of absolute monarchy in the seventeenth and eighteenth centuries, the states usually did not have enough political power to unify the legal systems.

During the Early Modern period, a new understanding of social space developed, as mentioned in the introduction. According to the philosopher Edward Casey, during this period there occurred a series of material and cultural changes that caused the displacement of place as the primary model for understanding the social lifeworld, and the emergence of geometrical space as the new model.[113] Casey maintains that in this period there was a transition from one medium to another, from text to image, and that this transition also denoted a movement from narrative to description as the primary mode of representation. Casey illustrates this change by the flourishing of the visual arts in the Early Modern period, and in particular by the emergence of landscape painting as a novel genre in the seventeenth century. According to Casey this double shift of media and mode also brought with it an intellectual change of focus, from time to space. In the view of Casey, then, these changes underpinned a transition from concrete place to abstract space as the primary model for understanding the social lifeworld. Although I find that Casey's argument points to real and important historical processes, at the same time I have serious reservations against parts of his reasoning. For instance, I would argue that what happened in the period was not a transition from text to image, but from a primarily oral-visual culture to a textual-visual culture, which increasingly became based on print, which in turn individualizes the act of reading,[114] and I also would argue against an understanding of the distinction between place and space as absolute.

It is plausible that the larger and more accurately depicted world created by the map-makers of the Early Modern period, just as in political and commercial areas, contributed to new ways of conceiving legal space. The cities were represented as belonging to larger political entities, which in this way

---

[113] E. S. Casey, *Representing Place. Landscape Painting and Maps* (Minneapolis: University of Minnesota Press, 2002) 154–170.
[114] See e.g. E. Eisenstein, *The Printing Press as an Agent of Change* (Cambridge: Cambridge University Press, 1979), and W. J. Ong, *Orality and Literacy. The Technologizing of the Word* (London: Routledge, 1982).

were defined geographically rather than chorographically. At the same time, the development of the constitutional nation state was also bound with the emergence of capitalism and bourgeois civil society.[115] These developments – cartographic, commercial, juridical and political – appear to have began in Northern Italy during the fourteenth and fifteenth centuries and were centered in the powerful city-states of Florence, Genoa, Milan, and Venice. In the following centuries, the same developments, to a greater and lesser degree, can be found in England, France, the Low Countries, Spain and in the cities and states constituting the Holy Roman Empire.[116]

It is also important to stress that although there was a great increase in the manufacture of maps as well as a development of the manner and method for representing land and sea during the period 1475–1575, map-making remained for a long period an inexact science, and more often than not maps were based on other (and older) maps rather than on reality. The development of cartography into a more exact science began with graphic representations of places (chorography) during the seventeenth century, and gradually made its way into geography during the eighteenth century. As will be discussed in the next section, the development of more exact and technically sophisticated maps changed both the perception of reality and reality itself, directly and indirectly. These technical advancements in mapping places and spaces can in certain ways be compared with the mapping of the human genome during the 1990's. Whereas the former contributed to transforming our perception of nature and social space, the latter contributed to reconfiguring our understanding of the human body and thereby altered the relations between biology, ethics and anthropology.[117]

Maps, Technology and Law

The maps of Strasbourg from the period 1650–1750 do not identify legal places and spaces to a higher degree or in more detail than Morant's 1548 map of the city. There are two main reasons for this: first, juridical institutions remain distributed and tied to political and professional institutions; second, the maps from this period are often military maps, showing little or

---

[115] J. Habermas, *Between Facts and Norms. Contributions to a Discourse Theory of Law and Democracy*, trans. W. Rehg (Cambridge: The MIT Press, 1996).

[116] See (again) the essays in Buisseret ed., *Monarchs, Ministers and Maps*.

[117] P. Rabinow and G. Bennett, "From Bioethics to Human Practices, or Assembling Contemporary Equipment", in B. da Costa & K. Philip, eds., *Tactical Biopolitics. Art, Activism, and Technoscience* (Cambridge: The MIT Press, 2008) 389–400.

no interest in civic institutions.[118] On the other hand, during the seventeenth
century military maps become increasingly exact and comprehensive. This
cartographic development reflected a radical transformation of the appearance and functions of the early modern European city as an effect of new
military strategies, brought about by cannon warfare.[119] Not only was it
necessary to construct new fortifications and restore old ones, but to do this
in accordance with the latest military technologies and theories required accurate topographical site plans.[120] The same was true for planning a siege of a
town. For these reasons the science of surveying and map-making developed
a higher degree of exactness (through precise measurement and triangu-
lation) and completeness than ever before.[121] The intimate relationship between the developments of map-making and military technology directly affected the reality represented. This happened in two ways: the depiction in a
more exact and comprehensive way of the topographic and urban structure
of the city made possible more geometrically exact individual constructions
and urban complexes; and the development of new architectural and urban
structures, many with direct military functions (like wider streets, angular
city walls, and clearings outside the city walls). During the seventeenth and
eighteenth centuries these new chorographic techniques were increasingly
used to measure and map larger areas, regional and later national territories.
In this way the state acquired exact and comprehensive cartographic knowledge of its domains. This information was in turn used to plan and build
roads, canals and, in some countries, national systems of fortifications. Thus,

---

[118] See e.g. "Grundriß der Stadt und Vestung Straßburg [...]" (1682), Krigsarkivet,
Stockholm (Strasbourg #1), also in Kungliga biblioteket, Stockholm (Kartavd.
M2); "Grundriß Der weiland Vortreflicher Freiÿen Reichs Statt Strasburg [...]"
(1682), Krigsarkivet, Stockholm (Strasbourg #3); "Strasbourg Ville Fameuse
Située sur la petite Riviere d'Ill [...]", no year, but marked "1680?", Krigsarkivet,
Stockholm (Strasbourg #5); "Plan von Strasburg Nebst der Citadelle [...]", hand
drawn, no year, Krigsarkivet, Stockholm (Strasbourg #8); "Strasbourg Ville Forte
Située sur le Rhein [sic] dans l'Alsace [...]", copperplate engraving, no year, Krigsarkivet, Stockholm (Strasbourg #16); "Plan de la Ville de Strasbourg [...] 1726",
signed "I.A. Friedrich fc. a.v.", copperplate engraving, 1726, Krigsarkivet, Stockholm (Strasbourg #17), also in Kungliga biblioteket, Stockholm (Kartavd. 17f
181:1); "Plan et profil de la ville et Citadelle de Strasbourg [...]", copperplate engraving, no year, Krigsarkivet, Stockholm (Strasbourg).
[119] M. Pollak, "Military Architecture and Cartography in the Design of the Early
Modern City", in D. Buisseret ed., *Envisioning the City*, 110
[120] Pollak, "Military Architecture and Cartography in the Design of the Early Modern
City", 119.
[121] Pollak, "Military Architecture and Cartography in the Design of the Early Modern
City", 121. See also Hale, "Warfare and Cartography, ca. 1450 to ca. 1640".

this metrically (and geometrically) exact measuring of national territories had the effect of transforming political territory into measurable space, and to transform cities into points (like co-ordinate points) in geometrical space. This development may be described as a dialectical movement where new chorographic technologies (new ways of measuring and depicting place) affect national geography (metrically exact maps of national territories), and the conception of place is affected by this new geometric space.

These examples show that maps not only served to depict reality (in one way or another), but increasingly functioned as instruments to change reality, local and national. In this way maps themselves became part of the technical production of things, buildings and the social lifeworld. In order to understand the significance of this technical or technological aspect of maps and map-making, it is helpful to use Heidegger's discussion of equipment and technology. In *Sein und Zeit* (1927), Heidegger makes his well-known distinction between the two different relations we have to entities (*Seiende*): "presence-at-hand" (*Vorhandenheit*) and "readiness-to-hand" (*Zuhandenheit*).[122] Whereas the former denotes a factual description of *what* something is (*existentia*), the latter denotes our relation to what he calls equipment (*Zeug*) and tools (*Werkzeug*), stuff one uses in everyday life (like shoes) or for specific purposes (like a hammer). As long as one uses things as instruments to accomplish something, or as extensions of oneself, one does not normally pay much attention to them (except that the shoes are comfortable or that the hammer is well balanced): they are ready-to-hand but not present-at-hand.[123] A good example are the spectacles on my nose, which are less conspicuous to me than the painting hanging on the wall (but without which I would not be able to see the painting).[124] However, at the moment when the equipment or tool stops functioning properly it not only becomes unusable but also presents itself as an "un-readiness-to-hand" (*Unzuhandenheit*), as unusable stuff.[125] To the extent that "it just lies there", it shows itself not as a tool but as an "equipment-thing" (*Zeug-ding*) with a certain appearance that has constantly been factually present-at-hand (*vorhanden*) although one did not pay attention. Apart from malfunctioning there are of course several other ways in which stuff can become present-at-hand, while still being bound up with

---

[122] Heidegger, *Sein und Zeit,* 42–42, 67–88.

[123] In another context, Heidegger indeed describes this as an instrumental attitude to technology (*Technik*). See M. Heidegger, "Die Frage nach dem Technik" (1953), in *Vorträge und Aufsätze,* 10.

[124] Heidegger, *Sein und Zeit,* 107.

[125] Heidegger, *Sein und Zeit,* 73.

the readiness-to-hand of equipment. For instance, things and clothes go out of fashion and old technologies are replaced by new (like when the gramophone was replaced by the CD-player). The point of Heidegger's distinction is to show that our everyday relationship to things is practical rather than theoretical, and that our practical dealings with things are part of a world of activity, constituting a lifeworld.

In the same section in *Sein und Zeit* Heidegger also discusses signs as equipment, among others the "adjustable red arrow" sometimes fitted upon motor cars and "whose position indicates the direction the vehicle will take, for instance at an intersection."[126] As Heidegger points out, the sign is not authentically grasped if one just stares at it and identifies it as an "indicator-thing" (*Zeigding*) occurring for no reason.[127] That is, one will fail to understand the meaning of the arrow if one looks in the direction to which it is pointing, or if one ponders the mechanics of its construction. The same kind of phenomenological description and distinctions can be applied to maps, but with some differences. In order to use a map, for instance a tourist map, one has to fit the graphic and textual information on the map to one's surroundings, and vice versa. As long as one just *looks* at the outward appearance of the map as a collection of shapes and colours, one will not discover anything ready-to-hand.[128] The failure to accomplish this task – due to poor map-design, inexperience with maps, foreign language, etc – makes the map unusable, turns it into "junk".[129] On the other hand, when one successfully uses the map to orient oneself in the city, the map must in a sense withdraw (*zurückzuziehen*) and become transparent. One no longer sees the map itself, but sees the city through it, just like the spectacles on my nose. One can also use a map to visualize a city that one has never visited, but plans to visit, or to recollect experiences from a city already visited; and one can also use maps for imaginary travels, or as a substitute for travelling, as des Esseintes' im-

---

[126] Heidegger, *Sein und Zeit*, 78. [An den Kraftwagen ist neuerdings ein roter, drehbar Pfeil angebracht, dessen Stellung jeweils, zum Beispiel an einer Wegkreuzung, zeigt, welchen Weg der Wagen nehmen wird.]

[127] Heidegger, *Sein und Zeit*, 79.

[128] Cf. Heidegger, *Sein und Zeit*, 69: "Das schärfste Nur-noch-*hinsehen* auf das so und so beschaffene 'Aussehen' von Dingen vermag Zuhandenes nicht zu entdecken. Der nur 'theoretisch' hinsehende Blick auf Dinge entbehrt des Verstehens von Zuhandenheit." Although it is certainly possible to contemplate the beauty of maps and view them as works of art, especially old maps in which the aesthetic and the instrumental are not separated, this is obviously different from having them being ready-to-hand.

[129] Cfr. Heidegger, *Sein und Zeit*, 74: "Das Zeug wird zu 'Zeug' im Sinne dessen, was man abstoßen möchte […]."

aginary travels in Joris-Karl Huysmans' *À Rebours* (1884).[130] The act of using a map is therefore a cognitive performance and in contrast to the spectacles on my nose as soon as one has familiarized oneself with a new place, real or imaginary, one no longer needs the map to find the way.[131] Using a distinction this time from Walter Benjamin, one can say that one absorbs the map rather than is absorbed by it.[132]

Whereas in *Sein und Zeit* the analytic function of equipment and tools serves to let us see how our relationship to them constitutes a world (or lifeworld), and hence is not thematized as such, in Heidegger's essay "Der Ursprung des Kunstwerkes" (1935/1936) the objective is to understand what characterizes the artwork, and in particular the distinction between artifact and artwork.[133] However, and in contrast to the analysis in *Sein und Zeit*, as has been pointed out by Hans Ruin, the mode of equipmentality is here described as lying in the way of a genuine grasping both of nature and of the human lifeworld, which on the other hand can be grasped through the event of the artwork.[134] In "Der Ursprung des Kunstwerkes" it is also possible to discern an ambivalent relation to technology that will become more developed in the essay "Die Frage nach dem Technik" (1953).

Although both the craftsman and the artist produce things, and despite the accomplished artist's high regard for craftsmanship, Heidegger maintains that they are involved in radically different activities. In order to describe the difference between them, Heidegger brings in the Greek notion τέχνη, which he argues means "neither artifact nor art" and is "not at all the technical in a modern sense".[135] That is, Heidegger contrasts the ordinary instrumental and anthropological understanding of technology (*Technik*), which he describes as a practical activity (*praktischer Leistung*), to the Greek

---

[130] J.-K. Huysmans, *À Rebours* [1884] (Paris: Flammarion, 1978) 174: "A quoi bon bouger, quand on peut voyager si magnifiquement sur une chaise?"

[131] G. Deleuze & F. Guattari, *Mille plateaux. Capitalisme et schizophrénie* (Paris: Editions de Minuit, 1980) 20.

[132] W. Benjamin, "Das Kunstwerk im Zeitalter seiner technischen Reproduzierbarkeit" [1936], zweite Fassung, in *Gesammelte Schriften I:2*, ed. R. Tiedemann *et al.* (Frankfurt: Suhrkamp, 1974) 504.

[133] M. Heidegger, "Der Ursprung des Kunstwerkes" (1935/1936), in *Holzwege*, 1–74.

[134] H. Ruin, "Ge-stell: Enframing as the Essence of Technology", in B. W. Davis (ed.), *Martin Heidegger: Key Concepts* (Durham: Acumen Publishing, 2010) 183–194.

[135] Heidegger, "Der Ursprung des Kunstwerkes", 46 [denn τέχνη bedeutet weder Handwerk noch Kunst und vollends nicht das Technische im heutigen Sinne].

notion τέχνη, which he defines as "a form of knowledge".[136] This form of knowledge has the form of "having seen", in the "most wide sense of seeing": "to perceive the present as such".[137] According to Heidegger, the Greek way of thinking knowledge is connected to their use of the word ἀλήθεια (unconcealedness), as "the unconcealing of entities".[138]

> τέχνη, as knowledge experienced in the Greek manner, is a bringing forth of entities [des Seienden] in that it *brings forth* present entities as such entities [das Anwesende als ein solches] *out of* concealedness and *into* unconcealment and specifically *into* the unconcealment of their appearance; τέχνη never signifies the action of making.[139]

In other words, in Heidegger's analysis of the Greek understanding of τέχνη, and in contrast to the instrumental understanding of technology, it does not signify the act of making something. Instead, τέχνη means bringing out what is present "as such an entity" (*als ein solches*) out from concealedness and into the unconcealment of their appearance.[140] In the essay "Die Frage nach dem Technik" Heidegger states this opposition between the usual (everyday) understanding of technology as instrumentality and the philosophical understanding of technology as τέχνη (which also defines the essence or truth of technology) in an even more straightforward way: "Technology is therefore no mere means. Technology is a form of revealing."[141]

---

[136] Heidegger, "Der Ursprung des Kunstwerkes", 46 [eine Weise des Wissens]. See also Heidegger, "Die Frage nach dem Technik", 10–17.

[137] Heidegger, "Der Ursprung des Kunstwerkes", 46 [gesehen haben, in dem weiten Sinne von sehen, der besagt: vernehmen des Anwesenden als eines solchen].

[138] Heidegger, "Der Ursprung des Kunstwerkes", 47 [der Entbergung des Seienden].

[139] Heidegger, "Der Ursprung des Kunstwerkes", 47. I here use the translation of Karsten Harries in *Art Matters. A Critical Commentary on Heidegger's "The Origin of the Work of Art"* (No place: Springer, 2009), 141, although I use *entities* (instead of *beings*) as translation of *des Seienden*. [Die τέχνη ist als griechisch erfahrenes Wissen insofern ein Hervorbringen des Seienden, als es das Anwesende als ein solches *aus* der Verborgenheit *her* eigens *in* die Unverborgenheit seines Aussehens *vor* bringt; τέχνη bedeutet nie die Tätigkeit eines Machens.]

[140] In fact, and as has been emphasized by Andrew Feenberg, in order to more fully understand what Heidegger understands by τέχνη in both these essays it is necessary to put them in the context of his extensive reading of Aristotle in the 1920's and 1930's. See A. Feenberg, "The Question concerning Techné: Heidegger's Aristotle", in his *Heidegger and Marcuse. The Catastrophe and Redemption of History* (New York: Routledge, 2005) 21–45.

[141] Heidegger, "Die Frage nach dem Technik", 16 [Die Technik ist also nicht bloß ein Mittel. Die Technik ist eine Weise der Entbergens.] Heidegger here also maintains that τέχνη defines the essence, and therefore the truth, of technology. See "Die Frage nach dem Technik".

Although one may think that the bringing forth of a silver chalice out of a piece of silver (Heidegger's example) is quite different from the kinds of bringing forth occurring in the artistic-poetic bringing into appearance in verbal and visual art, in Heidegger's reading of τέχνη they belong together.[142] For this reason Heidegger's reading of τέχνη may be applied to the art and science of cartography. Although maps certainly constitute tools (technology in the instrumental sense), map-making is also a knowledge and an art that brings forth something "seen" into the open. Maps reveal "present entities as such entities", they bring forth representations of the earth and of individual places. But maps are different from tools and equipment in yet another way: as architectural constructions they affect the coming into being of spaces and relations between spaces.[143]

However, maps not only reveal to us places and make it possible for us to see them, or at least certain aspects of these places, but maps also transform the reality they represent. As we have seen, seventeenth and eighteenth century military maps made possible exact and comprehensive urban designs, but they also turned cities into targets. In presenting a comprehensive view of a place and giving one a sense of location and orientation, a map offers the possibility to interact with reality in various ways, also in terms of control. Maps can be used to regulate the physical appearance of a city and thereby acquire the status of law – and analogously laws are instruments of the state and the judiciary. But maps are often instruments of change in an even more literal sense, depicting new structures not yet realized. Maps and map-making are therefore, as a form of τέχνη, as much a way of knowing the world as a way of making worlds, both in a phenomenological sense and in a concrete, everyday sense.[144]

The increasingly exact and comprehensive map technologies in the seventeenth and eighteenth centuries transformed the cultural perception of place and space. In this sense, the "revealing" of reality taking place in map-making constituted a projection of the mapping technologies onto the reality represented. This should not be confused with the distortion caused by the problematics of representing a spherical object on a flat sur-

---

[142] Heidegger, "Die Frage nach dem Technik", 15.

[143] Heidegger, "Bauen Wohnen Denken", 155–156.

[144] This also applies to the poetic and aesthetic function of artistic maps, representing imaginary places, or re-imagining existing places, and of seeing artworks as maps. See D. Birnbaum, "We are many", in *Making Worlds. Fare Mondi. Φτιάχνοντας Κόσμους, 53rd International Art Exhibition*, ed. Daniel Birnbaum & Jochen Volz (Venezia: Marsilio, 2009) 185–191.

face, although this can be quite as serious. What happens is instead the transformation of our understanding of place as inherently measured or measurable space that is available to a comprehensive, visual description. The cartographic "revealing" of abstract geometrical space brings about a concealment of place as experienced in an immediate part of the human lifeworld. Lefebvre describes this process as a "silencing" of spatial practice.[145]

The new map technologies were also employed for non-military purposes, for instance cadastral surveys of property relations. A good example is the Plan Blondel (1764–1768), which identifies and gives exact measurements of every property and building lot in Strasbourg.[146] This provided the authorities, both the local government and the central government in Paris, with information regarding ownership of private property and, since each numbered entity denotes one or several households, it is also possible to use the map for a demographic and fiscal survey of the city. This indicates that the function of cadastral survey in the eighteenth century differs from earlier periods. Whereas for the Romans the cadastral map was a legal document (like a title), in the eighteenth century the property survey reveals the human and fiscal resources of the city. The same is true of the extensive rural land surveys taking place in several European countries in the same period, with the double intention to rationalize agriculture and increase fiscal revenue.[147] Not only does this plotting of demographic, economic and other information onto chorographic and geographic maps foreshadow the thematic maps of the nineteenth century, but also both the eighteenth and nineteenth centuries cadastral plan and the nineteenth century thematic map represent explicit uses of cartography as instruments of what Foucault called "biopolitics".[148] These maps constitute crucial techniques for the administrative apparatus (*dispositifs*) to regulate subjects (or citizens) and to efficiently identify and develop human and natural resources.[149] In particular, the maps define a strategic measure for management of populations, under-

---

[145] Lefebvre, *La Production de l'espace*, 51.

[146] Jacques-François Blondel, "Plan Blondel" (1764–1768), Archives de la Ville et de la Communauté Urbaine de Strasbourg.

[147] C-J. Gadd, *Den agrara revolutionen 1700–1870, Det svenska jordbrukets historia, band 3* (Stockholm: Natur och kultur, 2000).

[148] The city of Strasbourg made new and more detailed cadastral surveys in 1837 and in 1897.

[149] For an excellent discussion of Foucault's notion of *dispositif*, see G. Agamben, "What Is an Apparatus?", in *"What Is an Apparatus?" and Other Essays* (Stanford: Stanford University Press, 2009) 1–24.

stood as resource, a technology that Foucault sometimes called "govern-mentality" (*governementalité*).[150]

However, in using maps as part of an apparatus to gather information and set demands on human and natural resources, one can argue that car-tography becomes caught up in what Heidegger characterizes as the essence of "modern technology" (*die modernen Technik*): "enframing" (*Ge-stell*).[151] With this term Heidegger not only wants to name the new demands that modern technology puts on both nature and man, but also the way in which this technology is gathering and interconnecting resources, and in this way turning everything into a resource for something else. According to Hei-degger, there exists a sharp difference not only between the everyday under-standing of technology (as instrumentality) and the Greek, philosophical understanding of technology (as τέχνη), but "modern technology" (al-though instrumental in a certain sense) differs in fundamental ways from both traditional handicraft and τέχνη in its relation to both nature and man.[152] According to Heidegger, modern technology is defined by a new re-lation to nature and to man, which are revealed as resources. Although Hei-degger places this event in history, he maintains that it happens prior to the invention of modern physics, and that in fact modern physics occurs as a re-sult of this way of perceiving nature as resource.[153] The Plan Blondel is a good example of technology as enframing (*Ge-stell*) in Heidegger's sense.

When scrutinizing the detailed picture that the Plan Blondel gives of Strasbourg, one also notices that some areas are more densely covered by buildings than others, mainly the older parts of the city; that some areas have small enclosed parks, probably belonging to palaces; others have large, seem-ingly open parks, either public or belonging to a public institution; but also that some areas appear to contain arable land. Indeed, not all the buildings

---

[150] M. Foucault, *Sécurité, territoire, population. Cours au Collège de France 1977–1978* (Paris: Gallimard/Seuil, 2004). However, maps can also be used as means of resis-tance and of empowerment, see e.g. J.-M. Besse, "Cartographie et pensé visuelle. Reflexions sur la schématisation graphique", in Isabelle Laboulais ed., *Les Usages des Cartes (XVIIe – XIXe siècle). Pour une approche pragmatique des productions cartograp-hiques* (Strasbourg: Presses Universitaires de Strasbourg, 2008) 22.

[151] Heidegger, "Die Frage nach dem Technik", p. 23. On the connection between Foucault and Heidegger, see S. Elden, *Mapping the Present. Heidegger, Foucault and the Project of Spatial History* (London: Continuum, 2001) 93–153; T. Rayner, "Biopower and Technology: Foucault and Heidegger's Way of Thinking", *Contretemps* 2 (2001), 142–156.

[152] Heidegger, "Die Frage nach dem Technik", 10.

[153] Heidegger, "Die Frage nach dem Technik", 18. See also Heidegger, "Die Zeit des Weltbildes".

and lots shown on the Plan Blondel are individual property, many buildings belong to companies, public institutions, social associations, and some entities belong to everybody and nobody. In order to get an idea of the complex (legal) property structures existing in civil society, it is suggestive to invoke the property categories in Roman law, which not only increasingly informed legal practice during the seventeenth and eighteenth centuries but have received renewed actuality today in discussions of immaterial property rights.[154] In Roman law, then, one distinguishes between things that can be owned privately and publicly, there being several categories of property that could *not* be privately owned.[155] Thus things common to all men, like air, running water, the sea, were considered *res communes omnium*; things belonging to the state, like public roads, harbours, ports, certain rivers, enemy property captured in military action, including land, were considered *res publicae* (and under Justinian things intended for public use, public streets, buildings, theatres, parks, were considered *res universitatis*); divine things could be either things protected by the gods, like city wall and gates (*res sanctae*), tombs, sepulchres, mausoleums, cenotaphs and land used for burials (*res religiosae*), or formally consecrated things, dedicated to the gods, like temples, shrines and sacred groves (*res sacrae*). Although these legal property categories belong to Roman society, in the eighteenth century Europe they existed as epistemic and administrative categories both among churchmen and university jurists and as practical social categories. However, with the exception of buildings owned by the city or the king of France, it is only private (including collectively) owned property that is identified on the Plan Blondel.[156] Assuming that everything that is not privately owned is public, it may be noted that the plan does not differentiate between different kinds of public property (as in Roman law), nor does it let us know what public or private institutions are responsible for maintaining the city walls, the streets or the squares. Nevertheless, it is not difficult to map these legal property categories (at least approximately) on the Plan Blondel.

It is not possible here to discuss in any detail the social and cartographic development after the eighteenth century. It should nevertheless be noted that the French revolution had a strong and lasting impact on the legal system (and not only in France). On maps of Strasbourg from the first decades

---

[154] C. M. Rose, "Romans, Roads, and Romantic Creators: Traditions of Public Property in the Information Age", *Law and Contemporary Problems* 66 (2003), 89–110.

[155] *Justinian's Institutes*, II.i 55–61; Kaser, *Römisches Privatrecht*, 104–106.

[156] See "Plan Blondel: Propriétaires", on the website Maisons de Strasbourg. Étude historique sur les maisons de Strasbourg entre le XVII° et le XX° siècles, http://maisons-de-strasbourg.fr.nf/?page_id=57 (accessed December 27, 2009).

of the nineteenth century one suddenly finds a separate courthouse building in the form of a Palais de Justice.[157] Although the post-revolutionary maps of Strasbourg in this way show the physical and chorographical separation of public powers, they do not depict a politico-legal change of equal significance: the introduction of the first French national code in 1804. However, since the creation of a unified national legal space in France was made possible not only by the revolution in 1789, but also by the cartographic revolution in the Early Modern period, one can argue that the Code Napoleon is metonymically present in these maps. Although the civil code was a child of the political revolution, the content was to a large extent a reaction to it. In fact, the Code Napoleon is quite conservative in character, where many of the paternalistic features of Roman private law were introduced. Although Alsace and Strasbourg became part of the national French legal space (in which everybody was equal, although men certainly were more equal than women), during the concordat the region was granted a number of exceptions, some of which persist today.[158]

Conclusion

The city represented on the plans of Strasbourg has changed considerably during the past 500 years. Most of the houses that today constitute "old" Strasbourg were not even built when Morant made his woodcut map in 1548, but would be constructed during the following century. The city has continuously developed, both internally as well as by extending outwards. The present location of the Palais de Justice is for example situated on the grounds of the former city walls. Just as the city has developed and spread, so have the

---

[157] "Plan topographique de la Ville de Strasbourg divisée en quatre Cantons 1819", signed "I. Oberst fer.", Archives de la Ville et de la Communauté Urbaine de Strasbourg; "Plan de Strasbourg […]; Reduit par Ch. Rothé sur le plan général en 1821 par M. N. J. Villot [,] Architecte de la ville. Dessiné et écrit sur pierre par Clément Senefelder, 1823" (F. G. Levrault, Strasbourg & Paris), Archives de la Ville et de la Communauté Urbaine de Strasbourg. The Palais de Justice was located on Rue de la Nuée blue. Less than a century later, during German rule, the Justiz Palast is moved to a new building situated on Quai Finkmatt.

[158] See É. Sander, "Le Droit local alsacien-mosellan: un pluralisme juridique dans un système unitaire", in *Le Code civil français en Alsace, en Allemagne et en Belgique. Réflexions sur la circulation des modèles juridiques. Colloque de Strasbourg et de Colmar, 26 et 27 nov. 2004*, ed. Dominique d'Ambra et al. (Strasbourg: Presses universitaires de Strasbourg, 2006) 117–145; and the website of Institut du droit local Alsacien-Mosellan, http://www.idl-am.org/.

maps of the city proliferated, not only in number but also in kind. Maps have been made for many purposes: celebratory, political, military, pedagogic, historical, proprietary, fiscal, urban planning, and scientific. Since each purpose implies a certain focus and selection of reality, one cannot expect an individual map to be a true or correct representation of the city (whatever that would mean since they never represent the people that live in the city). Rather, the maps should be viewed as instruments that make certain aspects of physical and social reality visible, bringing out certain entities out of concealedness and into unconcealment. In the same way as a telescope lets us see things far away, a map helps us see things as from above, and old maps let us see places far away in time. Both the telescope and the map constitute an optic apparatus that makes reality appear in new ways. However, optic instruments, to a greater or lesser extent, distort the represented object, for instance in the way a telescope makes things look closer to each other than they are in reality. Maps also distort reality in many ways, often intentionally (and sometimes unintentionally), and thus they affect and change our perception and understanding of reality. In doing this, maps also change social reality itself. In other words, the maps of Strasbourg that we have studied are not only (passive) records of physical and social change, they have also (actively) taken part in the process of change.

Although few of the maps examined have been made for legal purposes, it has nevertheless been possible to perform legal readings of them. In the sixteenth century there was no separation of legal space from other social spaces, and the administration of justice typically was co-located with political power, yet there existed separate jurisdictions and Morant's map emphatically demarcated Strasbourg as a sovereign legal space. The most important juridical event in the period from 1500 to the present is the separation of powers and the creation of a unified French legal space in 1804. On the maps of Strasbourg this event is revealed explicitly through the appearance of a separate courthouse, implicitly by the cartographic medium itself. The maps of Strasbourg have also revealed an ongoing history of autonomy, dignity and integrity, as well as vulnerability, both for the city and for its citizens. Situated on the borderland of two powerful kingdoms, Strasbourg has been besieged, conquered, re-conquered, and run-over any number of times. Since the introduction of the Code Napoleon, the city has moved from one national jurisdiction to another four times. Although maps do not present the inhabitants of the city, they bring out other entities out of concealedness and make legible the appearance and disappearance of legal spaces and places.